Standard Poultry For Exhibition
A Complete Manual of the Methods of Expert Poultry Exhibitors and Breeders

by Reliable Poultry Journal

with an introduction by Jackson Chambers

This work contains material that was originally published in 1921.

This publication is within the Public Domain.

This edition is reprinted for educational purposes and in accordance with all applicable Federal Laws.

Introduction Copyright 2018 by Jackson Chambers

The World's Largest Selection of Vintage Poultry Books

www.VintagePoultry.com

Self Reliance Books

Get more historic titles on animal and stock breeding, gardening and old fashioned skills by visiting us at:

http://selfreliancebooks.blogspot.com/

Introduction

I am pleased to present yet another title on Poultry.

The work is in the Public Domain and is re-printed here in accordance with Federal Laws.

As with all reprinted books of this age that are intended to perfectly reproduce the original edition, considerable pains and effort had to be undertaken to correct fading and sometimes outright damage to existing proofs of this title. At times, this task is quite monumental, requiring an almost total "rebuilding" of some pages from digital proofs of multiple copies. Despite this, imperfections still sometimes exist in the final proof and may detract from the visual appearance of the text.

I hope you enjoy reading this book as much as I enjoyed making it available to readers again.

Jackson Chambers

CONTENTS

INTRODUCTION

CHAPTER I
The Service of the Poultry Show 7

CHAPTER II
Fundamental Things About Judging and Judges 17

CHAPTER III
Ethics of Fitting and Fixing Birds for Exhibition 23

CHAPTER IV
The Inalienable Rights of Exhibition Birds 27

CHAPTER V
Early Selection and Care of Exhibition Birds 30

CHAPTER VI
General Care of Exhibition Stock in Summer 41

CHAPTER VII
Preliminary Work in Fitting Exhibition Birds 50

CHAPTER VIII
Definite Selection of Birds for Exhibition 57

CHAPTER IX
Special Fitting of Birds for Exhibition 113

CHAPTER X
Care of Exhibition Birds in Transit and at Shows 139

CHAPTER XI
Analysis of Methods of Judging 153

CHAPTER XII
Philosophy of Judging and Relation of Judging Systems 161

CHAPTER XIII
Logical Development in Judging Practice 168

Index 176

INTRODUCTION

TO realize fully the rewards of ability to produce poultry of improved breeds and varieties, one must have the same mastery of the arts of exhibiting as of the arts of breeding. The truth of this will be readily admitted by anyone who has even a casual acquaintance with conditions affecting the winning of prizes in a competition of any kind.

An equally important fact, not so easily apparent when put in the form of a general statement, is that to acquire surpassing skill in the breeding of any kind one must know by experience many things that observation shows are learned only in the keen competition of the show room.

Theoretically it should be possible for a person with some natural capacity for this kind of work to produce masterpieces witout the stimulus of competition. A few cases might be found where breeders are producing wonderfully fine stock in varieties in which there is in general little interest and no other breeder approaches them. But these are not exceptions to the rule, for in every such case it will be found that the breeder at some stage of his career learned the lessons that are taught only by competition.

So it is necessary for the breeder who would excel in breeding to show frequently against others of about the same experience and skill. If one only cares to have a reputation that will enable him to sell fair-quality stock at a moderate profit he may often protect his commercial interests best by showing where he can easily win, avoiding places where he would have a real contest, but those who do will never become first-class breeders nor will they ever get much honor among breeders, or attain a position where the highest rewards may be theirs.

The rewards of ability to produce Standard poultry of high quality are fame and fortune proportionate to the popularity of the variety shown and the prestige of the show at which it is exhibited. A winning at the smallest local show carries with it some credit. A winning with a popular variety at a show of any importance makes one quite widely known as a breeder and opens the way to the development of quite a business in the sale of eggs for hatching and stock for breeding and exhibiting purposes,—if he is in a position to supply the demand. Consistent winning in any variety at a leading show or shows brings more fame and larger opportunities for doing business. Preeminence in competition in a popular variety at leading shows enables one to develop a business running into six figures annually—provided he has the business ability.

Lacking the business capacity to organize and successfully conduct a big poultry business, the skilled breeder may still realize a very comfortable income from the sale of specimens of extraordinary quality and value. Some idea of the range of top prices for exhibition and breeding poultry is given by the prices mentioned in the legends under the pictures of the high-priced birds in the accompanying illustrations. Such extreme high prices are of course not common, yet the ordinary prices for really good birds in any popular variety are high enough to make the business of breeding them very profitable. Also, contrary to the general opinion, prices for the best stock are remarkably stable. When demand in general slackens, through over-production, or for any other reason, the prices for the best birds are least affected. In a popular variety they will always sell at high prices; not as high in periods of financial stringency as when business in general is good, yet always at a good profit.

The reason for this is that the money value of the birds is determined by their desirability to people who are able to indulge their desire to have the best of everything who take an interest in fine fowls. This class of people may be divided as to their interest in poultry into those who will pay liberally for fine specimens, especially those with a prize record, simply as ornaments for their ground or to satisfy their ambition to have the best; and those who consider either wholly or in part the breeding value of the birds. These two groups bid against each other and against the professional breeder who can often afford to pay a very high price for a bird either to use in his own yards or to keep it out of a competitor's hands.

"OLD CHAMPION"
White Plymouth Rock male for which Wm. Barry Owen paid Harry Graves $1,500—with photographs of the checks given in payment

INTRODUCTION

While the number of birds that command extreme high prices is relatively small, and the demand comparatively limited, the prices established serve to raise somewhat the whole range of prices of Standard-bred stock and eggs for hatching, and to stimulate powerfully the interest in poultry of highest exhibition quality.

The sale of eggs for hatching introduces in the exhibition poultry industry an element not present in any similar line. There is in it an element of chance not elsewhere present but the chances run both ways. A buyer may get less value than he pays for, and he may get very much more. In the writer's experience and observation the buyer of eggs from any reputable breeder gets on the average rather more value for his money than when he buys birds. In buying small lots of eggs he is not as likely to get birds that he can mate to advantage as when he is able to go to a breeder with a large selection and buy mature birds that meet his requirements, but he is very apt to have one or more specimens worth much more than the eggs cost — perhaps a single specimen that he could not buy for many times what it has cost him. And having found the money spent for eggs a profitable investment, he will usually feel justified in buying a breeding bird or birds to mate with what he has at prices he would no doubt have considered prohibitive in the first place.

"SENSATION"
White Wyandotte cockerel for which John S. Martin refused $1,000

Many hundreds of cases occur annually where a very moderate investment in high-priced eggs for hatching furnishes a nucleus of breeding stock from which the owner, by sales of birds he does not need at such prices as he can obtain, and buying with the proceeds a smaller number of better birds, soon has a stock with which he can exhibit successfully in good competition. It is with this possibility in view that men and women have to look at every dollar before they decide to pay $20.00 and $25.00 for thirteen or fifteen eggs, and that in a few cases men have readily paid $5.00 each for eggs secured from special matings.

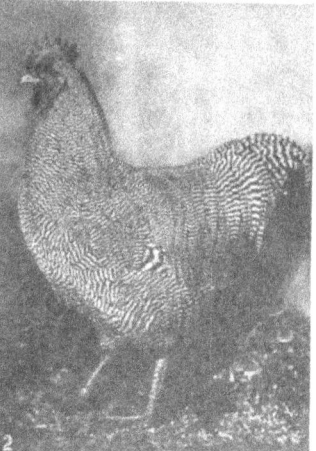

A PRICELESS PAIR OF ROCKS
E. B. Thompson's first Barred Rock cockerel and pullet, Madison Square Garden, 1920
These birds could not be bought at any price

The sale of eggs for hatching gives the people of small means the opportunity to get stock of the best lines. If they are able to breed it successfully when they get it, there is always before them the possibility of getting highly profitable prices for a large part of the birds they raise. If they can produce really first-class specimens they will always get good prices for such of these as they have to sell. There is no reason to anticipate that the readiness of people of abundant means to pay big prices for this class of stock will diminish. On the contrary the prospects are that it will increase,—that more and more people will want the finest creations of the poultry breeder's art and that with an increasing proportion of them higher prices will seem to add to the desirability of the birds.

The two things that can contribute most to this result are breeding to get the highest proportion of birds of finest quality, and then showing the birds in their best possible condition. The class of buyers who will pay $100.00 to $200.00 each for five fowls to put on the lawn of a beautiful country estate can be made many times larger than it is by increasing the supply of the class of birds they want and by showing them in proper form. At the same time the class of well-to-do buyers who become interested in breeding will increase at a similar rate. It is in no way discreditable to them, and it is to the benefit of breeders in general that they buy results for which those who have to count the cost must work for years.

On the other hand, the fact that it is possible with a very small initial outlay for people of moderate means to have Standard exhibition poultry of the highest grade of quality enables them to indulge a taste for fine stock when they could not do it in any other line, and the result has been a community of interest between rich and poor "fanciers" not found elsewhere. Many a small breeder who would consider it the height of folly for him to buy a bird of rare quality at the price he would have to pay another

INTRODUCTION

for it, will see nothing inconsistent in refusing to sell at any price a bird of like character that he has raised. Not having spent money earned in his regular work for his poultry, he feels justified in indulging in a luxury which only the wealthy can afford to buy outright.

Where the small breeder has been most heavily handicapped in his competition with commercial breeders and amateurs of more means has been in the matter of attending shows where the birds were well-fitted and having the opportunity to learn there what constitutes good fitting and the methods employed. In the majority of cases these shows are either inaccessible to him, or when he attends them he does not remain long enough to make the acquaintance of many exhibitors and through them to pick up the professional's knowledge of the arts of exhibiting.

It is the object of this book to give him that knowledge—to thoroughly inform him on everything relating to the subject, and to enable him to show his stock to the best possible advantage. As the reader becomes familiar with the contents of the book he will realize more and more that while it may occasionally profit an exhibitor to use methods that are commonly regarded as unfair, success in exhibiting is, as a rule, dependent upon good work in breeding and growing stock, and upon careful attention to the unquestionably legitimate methods of conditioning. Many more exhibitors are beaten through lack of attention to these than by "faking" on the part of their competitors.

The real service to the exhibitor of the information about practices commonly regarded as unfair is to enable him to detect them, or where they cannot be proved to form a reasonably correct opinion of the probable grounds of suspicion and so be properly on guard in future competitions or dealings with suspected persons. As the reader will observe most of the forms of "wrongful practice" get by when conditions of time or light are against the judge, or when the exhibitor is given the benefit of the doubt. In discussing each practice the author has tried to show fairly the attitude of judges toward it. Reviewing this work as completed and considering the point in the light of the observed tendency of judges, this season to be more severe in doubtful cases, he feels that he should advise the reader that from present indications judges are not going to be as liberal in applying that rule as has been the custom heretofore, and more of them will pass suspected birds by and let the exhibitor invite a most searching examination of his birds by challenging their decisions if he will. Just how far this will go cannot be foreseen, but with that tendency among judges and with the wider knowledge of methods of detecting faking that this book will give, it seems as certain that the hazards of faking will be increased, as that the advantage of legitimate fitting will constantly become greater.

It will also be distinctly to the advantage of the greater number of exhibitors who frequently find themselves in competition with better birds than their own to keep always in mind the fact that in general a well-fitted bird can win over a much better bird that has not been properly conditioned and groomed, and that it is right that it should. The better a bird is, the less excuse for an exhibitor who neglects to do all that the legitimate arts of conditioning can do for it.

A $500 R. I. RED COCKEREL

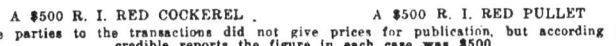
A $500 R. I. RED PULLET

The parties to the transactions did not give prices for publication, but according to credible reports the figure in each case was $500

S. C. WHITE LEGHORN COCKEREL
SOLD FOR $500

POULTRY FOR EXHIBITION

FIFTY YEARS PROGRESS IN STANDARD MAKING

Above—Title page of the American reprint (1867) of the first English Standard; nomenclature as given in it; title pages of the 1871 Standard, known as the Lockwood Standard. **Middle**—Pages of text and illustration from the 1888 edition of the American Standard of Perfection; nomenclature in that edition. **Below**—Nomenclature in 1915 Standard of Perfection; pages of text in that edition. Something of the improvement shown in this series of illustrations is plainly due to better book making. Leaving that out of account, the nomenclature series shows how attention to details of finish steadily increased, and comparison of even these few pages of the 1888 and 1915 Standard of Perfection shows how progress in breeding and progress in Standard making have gone forward together, though the latter has sometimes lagged a little. It should be noted that breeders were not willing to accept the profile outlines of 1888, and the edition containing them was declared obsolete.

CHAPTER I

The Service of the Poultry Show

Early Poultry Shows—Relation to Agricultural Fairs—Reasons for Separate Poultry Exhibitions—Beginnings of Permanent Big Shows—Their Benefits to the Near-by Community—Character of Competition at Different Classes of Shows—The Novice's Chances As an Exhibitor—Conditions of Judging—Attitude of Judges

The Earliest Exhibits of Poultry

THE first recorded instance of a public exhibit of poultry in America that the writer has been able to find takes us back to the year 1826. In that year James Sisson, of Warren, R. I., exhibited at the Rhode Island State Fair at Pawtucket three Emden Geese which he had just imported from Bremen, Germany. If poultry of any kind were exhibited at fairs elsewhere at the time or in the period following, no interest was taken in them by those reporting the shows.

The next record comes twenty years or so later. The date is in doubt, but it may have been in 1845, and it was not later than 1846, that the first competitive exhibit of poultry of which agricultural reports contain mention was held at the New York State Fair at Utica. In the latter year also sixteen exhibitors at the Worcester, Mass., Fair showed "barnyard fowls" which would appear to have been common native stock. The next year at Worcester premiums were awarded on common fowls, ducks and turkeys, and on one exhibit of Chinese fowls. Two years later, in the early fall of 1849, there were large and varied exhibits of poultry at at least four of the important county fairs in Massachusetts.

The interest created by the poultry displays at these fairs was obviously the direct cause of the first exclusive poultry show held in America, at the Public Garden, Boston, November 15-16, 1849. This in turn greatly stimulated further exhibits at fairs in New England, and the publicity given by the press of the country to so unusual an event as a large exhibit exclusively of domestic poultry and pigeons of more interest than the common native stocks awakened interest all over the country, and resulted in nearly every county fair soon having its exhibit of poultry of improved breeds, while special poultry shows were held in many of the large cities.

After the Civil War the interest in good poultry increased more rapidly. Without any diminution of the interest at agricultural fairs, local shows (principally of poultry but also with exhibits of pigeons and pet stock) were held wherever it was possible to bring together a few hundred specimens of improved breeds.

For this early and widespread promotion of special poultry shows there were several reasons. Probably the most forceful was the fact that neither young nor old poultry are usually at their best when the fall fairs are held. In fact, unless the young birds are very early hatched they are likely to be at a most unattractive stage of development in the early fall, and the greater number of the old birds are then in the midst of the annual molt. As two or three months later both old and young are in the prime of condition, it was quite a matter of course

THE CRYSTAL PALACE, LONDON, ENGLAND
The first great poultry show in England was held here about 1845. The "Palace Show" is still England's leading poultry show

that those most interested in poultry and most keen for competition should undertake to promote poultry shows at the season when birds were most fit.

Another almost equally cogent reason was that the interest in improved breeds of poultry early became much more intense among town poultry keepers than among farm poultry keepers. Even the breeders and exhibitors of improved stock who had farms were for the most part city men with whom farming was more or less a hobby. From the first, many of the real farmers took an interest in the improvement and exhibition of poultry, but the early-day fanciers came preponderantly from the ranks of the people to whom poultry keeping was as attractive for the pleasure to be derived from it as for the profit to be obtained. It drew the town people especially because it was the line of live stock breeding best suited to people on small plots of land.

A third important factor in the establishment of small, special poultry shows was the attitude of the management of many fairs toward exhibits and exhibitors of poultry. In the early days of the poultry movement, and long after, poultry was regarded by the great majority of farmers as the least important of farm interests, and among those in charge of agricultural shows there was generally more indifference toward the wishes of poultry exhibitors, and in regard to means of making a poultry exhibit attractive and really useful for the purpose for which fairs are held, than disposition to promote a poultry department. While suitable provision was made for housing exhibits of every other class and displaying them to the best advantage, the poultry usually got some makeshift shelter, that in case of a storm was inadequate. The methods of classification, cooping and judging long followed the crude ways of the first exhibitions, which were so unsatisfactory to poultry keepers acquainted with better methods and having good exhibition stock, that they were unwilling to show under the conditions of competition they usually found at a fair. Improvements in these conditions came very slowly, and usually as the result of the efforts of poultry men who persisted in their determination to get the same consideration for poultry as for other classes of exhibits at fairs in their community, in spite of the indifference of fair managers. In some cases such poultrymen have done gratuitously a large part of the work necessary to put the poultry department at their fair on a satisfactory footing.

Notwithstanding the shortcomings in the management, poultry exhibits at agricultural fairs from the first were on the whole very effective in popularizing improved breeds. The fairs had a stability and regularity that—as we shall see—the special poultry shows of the early days never had. Fair associations were incorporated, had their grounds and buildings, and in most cases received considerable sums from their states to be distributed as premiums or gratuities for the encouragement of all lines of agriculture and household arts. The managers of a fair interested in other classes of live stock or phases of farming might be unwilling to do anything themselves in behalf of poultry interests; but they would not deny the opportunity to exhibit to anyone who brought poultry to a fair. And though they might be negligent or indifferent in various details of some importance to exhibitors, they were ready enough to give the customary small premiums to exhibitors of poultry. In fact, one of the common causes of complaint was that they were much more willing to let premiums go to unworthy exhibits than to pay a small fee to a competent judge and make the award at the fair worth something among those who had some appreciation of the qualities of improved breeds. Still, wherever there was an agricultural fair, a poultry breeder who wished to do so, could make an exhibit with no cost to himself except for transportation and attendance, and with assurance that practically everyone who attended the fair would at least take a passing look at his birds, and that they would have the careful attention of any visitor who took special interest in poultry. Considerable publicity was thus secured usually at only moderate cost to the exhibitor.

OLD HORSE CAR BARN WHERE EARLY NEW YORK SHOWS WERE HELD
This building was on the site of the present Madison Square Garden. It is shown here as used for cars. For show purposes it was remodeled, providing a big amphitheatre in space between wings

INTERIOR VIEW OF THE PORTLAND, ME., SHOW, 1878
From an old engraving

The Risks in Promoting Poultry Shows

The promotion of interest in poultry through special poultry shows was a more serious matter for the poultryman. The expense for a hall, for judges, for premiums, advertising, etc., even for a very small show, would run from one hundred to two or three hundred dollars. The funds for this had to come from exhibitors and visitors, or from the pockets of the promoters, or from the generous business men of the place. The necessary cash outlay includes nothing for the work of the promoters who, according to circumstances and the kind of show they undertake to produce, must give freely of their own services. So it happens that, as a rule, a special poultry show is not projected in a community until a considerable number of persons there are interested enough to give it some support, and a few to give the time necessary to make it a success.

On the face of a statement of the circumstances relating to the promotion of special poultry shows it is obvious that while the enthusiasm of their promoters was fresh and they were somewhat of a novelty in a community, they would attract a good deal of attention and stimulate a great many people to improve their poultry or to engage in the breeding and exhibiting of fine poultry. It is just as plain that in the ordinary course of events most men active in promoting special poultry shows will tire of the work in a few years, or will be obliged to drop it because it interferes with their business. Even where men are willing to continue to carry the burden of a local show, it seldom lasts more than a few years in the average community because after the novelty wears off the general public does not attend.

For some forty years after the first poultry show at Boston the history of all poultry shows, large and small, in great cities and in country villages, was much the same. No show anywhere was held for more than a few consecutive years. But in nearly every community where there was much interest in poultry short series of shows, each under the management of a different group of breeders, came quite regularly with intervals of a few showless years between. This is still the situation with respect to by far the greater number of the shows that are projected, and in the nature of the case it can hardly be otherwise unless a system of management of poultry shows can be devised and put into effect that will give the minor shows greater stability.

The Beginning of Permanent Big Shows

The first show established on a permanent footing was that at Madison Square Garden, New York. The present organization was made in 1889 and the first of the series of shows still held annually at "The Garden" was held in 1890. This show was not a success financially, but its promoters included a number of men of large means who were determined to keep it going until it was established on a sound basis. Mr. Robert Colgate, of New York City, was the first president of the association and Mr. T. Farrer Rackham, of East Orange, N. J., was the secretary and show superintendent. In 1893 Mr. H. V. Crawford, of Montclair, N. J., became secretary and superintendent, and Capt. Chas. M. Griffing, of Shelter Island, N. Y., became treasurer. Both of these men had been directors from the organiation of the association. Mr. Colgate's name appeared in the catalogues of the association for only a few years. His successor in the office was Mr. Thomas H. Terry, of New York. In 1896, Mr. Theodore A. Havemeyer, of Mahwah, N. J., became president, and from that time until Mr. Crawford resigned the secretaryship in 1913 the management of the Garden Shows was in the hands of Messrs. Havemeyer, Crawford and Griffing with Mr. Crawford in actual charge of the shows.

Without disparaging in any way the abilities and service of others it may be said that it was Mr. Crawford more thany any other who, by his own energy, ability and single-minded devotion to the interests of the show, and by the spirit with which he cooperated with his associates in the promotion of its interests, made this show a permanent success and set the standard for great poultry shows in America. He furnished the first example of a successful business man devoting a large part of his time each year, not merely for two or three years but indefinitely, to make a show in the metropolis of America a success in every respect, and a permanent institution. He was a man in the prime of middle life when he took up this work, and he did not lay it down until advancing years obliged him to do so. It is a pleasure to record here that though no more a familiar figure at the Garden, because his winters are passed in milder climates, he retains a keen interest in its welfare and derives much satisfaction from the knowledge of its continuing success.

MADISON SQUARE GARDEN
From a photograph by F. L. Sewell

In point of time the service of Capt. Griffing as treasurer exceeded that of Mr. Crawford, continuing to his death in 1917, at an advanced age. Mr. Havemeyer has held the office of president continuously from his first election. So in the management of this show there has been a permanence and stability in the organization not equaled in

FOUR MEN LONG ASSOCIATED IN THE MANAGEMENT
OF THE GARDEN SHOW

Left to right—W. J. Stanton, T. A. Havemeyer, H. V. Crawford,
Chas. M. Griffing

the history of any other poultry show to the present time, and it is to this as much as to anything else that the poultry interests of America owe what the New York Show has done for them. In this connection too, it is appropriate to refer further to Mr. Rackham, whom Mr. Crawford succeeded as secretary. His retirement from the office was not due to any failure in it, but to the fact that as superintendent for Spratt's Patent his relations to all shows cooped by that company were such that he was not in a position to do the work of the secretary of a show association, nor was it in the interest of the company for him to do so. But he had actual charge of the cooping and other details of management of this and of nearly all the great shows in America to the time when confirmed

invalidism made him the unwilling but cheerful tenant of a wheel chair. In his active days Mr. Rackham was considered remarkably efficient in arranging shows to best advantage, and a past master in the art of smoothing the difficulties that beset show managers. To him more than

MECHANICS' HALL
Where the Boston Poultry Shows have been held since 1890. The entire main floor, a part of the second floor, and some space in the basement are occupied by the show.
Upper inset, Arthur R. Sharp
Lower inset, C. Minot Weld

to anyone else was due the excellence of arrangement, and the general neatness and attractiveness now general, but until very recently characteristic only of important eastern shows, and too often at seen in shows of the same class elsewhere.

Following Mr. Crawford, Mr. Charles D. Cleveland, of Eatontown, N. J., became secretary of the association, holding the position until 1909, when the demands of his own business led him to resign and he was succeeded by Mr. D. Lincoln Orr, of Orr's Mills, N. Y. In the active and responsible management of this show, in the three working offices, there were in a period of thirty years but eight men all told. It was this and the harmony that prevailed in the organization that gave the show its standing and prestige.

Every good poultry show has a stimulating effect on poultry interests, and the larger the show and the greater the importance of the city where, or the fair at which it is held, the greater—for the time being—the influence of the show. But when a show intended to be an annual event is discontinued within a few years there is inevitably some reaction, and many people who watch its course get the impression that the instability of the show reflects instability of the interests it represents. Though the first show of this New York Association had been reported a financial disappointment, the association went on year after year, with each year creating more confidence both in its own permanence and in the standing of those who exhibited at New York year after year. With each year the value of winnings at the Garden increased. The successful exhibitor here could make good sales at the show, and by effective advertising of his winnings in the poultry press could find a good outlet for everything he had to sell at profitable figures. The best possible evidence of the beneficial influence of the standing this show had acquired within those years is found in the announcement made by the present Boston Poultry Associaton, the next to establish a permanent show, upon its organization in 1895. That announcement read in part:

"In order to secure the permanency of an annual exhibition in Boston, it is absolutely necessary that the first show be a financial success, and that the subscribed guarantee fund be returned in full to subscribers; for by a successful annual show alone can we hope to give the prizes won at our exhibition the same importance and value for advertising purposes as those won at New York and other large shows. This is our aim, and we hope to give our home breeders, who cannot afford the expense of sending their birds to Madison Square Garden, the same advantages for disposing of stock enjoyed by those living near New York.

"It is not in any way our purpose to have the Boston Show take the place of any of our local shows. Should the annual show in Boston prove a financial success, it is our intention to encourage our local shows by the offer of special prizes, and by every other means in our power. The local shows can always be run under the score-card method, and will afford ample opportunity for all breeders desiring score cards to secure them. While we appreciate the value of a score card, we realize that in a show of this magnitude the public demands that the prizes be designated as early as prac-

"TATTERSALLS"—WHERE THE EARLY CHICAGO SHOWS WERE HELD

"The COLISEUM"—CHICAGO

ticable. The expenses of running a large show in Boston are so great it is necessary to rely to a great extent on the public to pay a large part of the expenses, and the public will only be satisfied by having the prize ribbons awarded at an early hour, thus very much enhancing the beauty and interest of the show. It must be borne in mind that no show in Boston has ever paid expenses, and we consider that this justifies us in making this departure."

The moving spirits in the Boston Association were Messrs. A. R. Sharp and W. C. Baylies, both residents of Taunton, Mass., but with business offices in Boston. The first president of the association was Mr. Francis H. Appleton, of Lynnfield, Mass., who continued in the office until 1908, when he was succeeded by the late Mr. George B. Inches, of Grafton, Mass., who held the office for five years, being succeeded in 1913 by the present incumbent, Mr. John Lowell, of Boston. From the organization until 1905 the late Mr. C. Minot Weld, of Readville, Mass., was secretary, but the details of the work were handled by Mr. A. R. Sharp as assistant secretary and superintendent, and throughout this period Mr. W. C. Baylies was treasurer, cooperating closely with Mr. Sharp in the management of the show. Messrs. Sharp and Baylies were comparatively young men when the association was organized, and for eight years Mr. Sharp gave almost his entire time for four or five months in each year to promoting the interests of the Boston Show.

While all the shows of this association have been held in the Mechanics Building, the early shows used only one of the two large halls on the main floor, keeping the expense as low as possible. Most of the shows using only part of the building and keeping the rent and other expenses at a minimum were successful financially, and by judicious investment of the surpluses a fund was created which assured the association's ability both to finance larger shows and to weather whatever the future was likely to have in store in the way of adverse conditions for a show. It was then that the association began to rent and use the whole main floor of this big building, with the gallery over one hall and a large exhibition and lecture room under the stage of the other, giving it ample room for the display of a very large show without double tiering the coops as so many large shows are, for lack of space, obliged to do.

F. L. KIMMEY
Who put Chicago Coliseum shows on a sound foundation

In 1904 the association held no show, Messrs. Sharp and Baylies finding it impossible themselves to give it the time necessary, and not being able to find others willing to take over their work. The next year Mr. S. H. Roberts, of Pawtucket, R. I., was elected secretary-treasurer and superintendent of the show. He held the office until 1908, when Mr. W. B. Atherton, of Randolph, Mass., the present incumbent, succeeded him. So we find here again, in a period of twenty-five years, only eight different men in the three or four principal positions affecting the interests of this show. It should be said further that in retiring from active management of the show Messrs. Sharp and Baylies continued to be closely associated with its interests, as also did Mr. Weld.

"TOMLINSON HALL"—INDIANAPOLIS
One of the oldest buildings in the country used for poultry show purposes. Shows were held here earlier than in the "Garden" and "Mechanics Hall"

How Permanent Large Shows Benefit Near-by Exhibitors

I have outlined the history of the two oldest permanent shows in the country because it brings out clearly the essential thing in making poultry shows permanent— harmony in the management and the continuous, united effort of a relatively small group of fanciers more interested in promoting poultry interests than in winning prizes and developing business for themselves. In the evolution of shows it has come about that the large shows pay an appropriate salary to the officers who devote much time to them, but in the early days service was entirely gratuitous, and that is still the case in most of the minor shows. Because that is the case, there is often a feeling that the best way to run a show is to divide the work among as many persons as can be induced to share it. That method has never worked well for a long period. The common result of it is spasmodic interest in the breeding of standard poultry in a community.

A point of the highest importance in regard to poultry shows is brought out in the foregoing quotation from the announcement of the first show of the Boston Poultry Association. The prestige which the New York Show was gaining as a result of its establishment as a permanent annual event gave a great advantage to the breeders in the vicinity of New York. It gave some advantage to all breeders living near enough to that city to be able to move the birds they exhibited from their homes to the show in practically as good condition as they were when they left home; but it gave the greatest benefit to the small breeders of birds of first-class quality, who had not enough birds of this quality to justify the expense of showing at a distance from home. A large show will always get some exhibits from a distance, and the managers of a show are entitled to point to the numbers of such exhibits as indicating the prestige of the show and the character of competition in it. But from the breeder's standpoint it is most important that there should be within quick and convenient shipping distance a permanent first-class show where he can compete on an equal footing with breeders in his own class as producers of standard poultry. And without disparagement to the value of winnings at any large show, it should be said in the interests of the establishment permanently of many more large shows than we now have, that the bulk of the entries at a poultry show always have come, and always will come from breeders in its vicinity, and that—other things being equal— the nearer a breeder is to the show at which he exhibits, the better his chances of winning. In the nature of the case the distribution of prizes at any show is largely among the same group of breeders year after year. The group in any variety changes, but changes slowly as old exhibitors, for one cause or another, drop out and new ones come in. And while a show in a big city has a certain kind of prestige to start with, that the show in a smaller place has not, eventually the prestige of a show is made by the character of the competition and the quality of the exhibits.

Did space permit it would be both interesting and instructive to refer to the history of some of the large

POULTRY BUILDING ILLINOIS STATE FAIR

shows that promised well for a few years but were soon discontinued, and to some of the small shows that were held regularly for quite long terms of years. In their general features, however, these stories would be much the same. It is in nearly every case a question of gratuitous

POULTRY BUILDING AT HAGERSTOWN, MD., FAIR
The first big building erected on a fair ground exclusively for poultry

service from some one until a show has passed through the vicissitudes common to new undertakings of all kinds, and of loyal support by the breeders who benefit by the show. Many instances could be cited of promising shows killed in their infancy by one or more disturbers of the general harmony; of show secretaries who had nothing whatever to gain from the success of a show but such credit as could come to them from having made it a success, and the satisfaction of helping poultry interests, working year after year against the obstacles put in their way by persistent trouble makers, and finally giving up in disgust; of small breeders in moderate circumstances giving freely of their time for many years, and often going into their pockets to pay deficits to keep their local shows going and to maintain interest in their communities. It is not practical to go into such details, but it is pertinent to call the attention of readers to them in a general way, because poultry shows are of fundamental importance in the promotion of poultry interests, and it is equally to the advantage of the breeder of standard poultry to take what they can give and to give as the occasion may require of his service or money to insure the permanent establishment of a show at the most suitable place in his locality.

JOHN L. COST
The man who made a big success of the poultry department at Hagerstown and secured the big building for it

Great Exposition Poultry Shows

The poultry shows held in connection with the World's Fair at Chicago, 1893, The Pan-American Exposition at Buffalo in 1902, the St. Louis Exposition in 1904, Jamestown Exposition in 1907, and the Panama-Pacific Exposition at San Francisco in 1915, were held under different conditions and had a somewhat different influence from other large shows. Such shows are financed from the general funds of the exposition, winnings at them have for the time being extraordinary commercial value, the judges are selected with regard to qualifications, yet at the same time with a view to having every part of the United States and Canada represented in the list. The exhibits, as a rule, come most numerously from comparatively near areas, but there is not the same preponderance of local exhibits as at the great annual shows. Hence, to a much greater extent than any annual show with a fixed habitation, the great exposition brings exhibits and breeders from different parts of the continent together to compare stock and ideas, and helps to harmonize the ideas of different persons and sections.

Great Poultry Shows at Agricultural Fairs

While the agricultural fairs were slow to adjust the management of their poultry departments to meet the

POULTRY BUILDING ON THE MICHIGAN STATE FAIR GROUNDS

wishes of exhibitors of standard poultry, most fairs, large and small, did eventually provide classification as desired, make some effort to secure the services of competent judges and improve housing conditions Some of the large fairs have gone much farther and made their poultry department as good as their other live-stock departments (and in a few cases better), and strong rivals of the leading winter shows. The poultry buildings on some fair grounds, notably at Hagerstown, Md., and at the New York State Fair Grounds at Syracuse, are better adapted to their purpose than any of the buildings in which the large winter shows are held. In number and

W. H. MANNING
Superintendent of Poultry, New York State Fair

for it as for a winter show. Failing to do this he always incurs the risk of being easily beaten by a competitor with more mature birds.

Character of Competition at Different Classes of Shows

As a rule, a novice in showing poultry stands very little chance of winning at a first-class show with birds selected, conditioned, fitted and shown by himself. One who is personally inexperienced in these matters, yet has stock of the breeding and presumably of the quality that is winning at first-class shows, and wishes to exhibit at such, should engage an expert to select and handle his birds until by observation and practice he has learned to do so for himself. But for exhibition at the minor shows a novice, who makes a careful study of what is required, may make his own selection of birds and fit them himself with reasonable assurance that if he is thorough and duly careful in his work he will stand a fair chance in such competition as he is likely to meet, and may possibly win over more experienced breeders and exhibitors who, presuming on the fact that exhibitors generally are likely to be a little careless about putting birds sent to small shows in their best condition, do not give their exhibits the attention they should.

In general, the exhibits in the popular classes at the leading shows are selected with rare skill and judgment and conditioned and fitted very nearly to perfection. But in the classes where competition is not strong, even at the best shows, it will frequently happen that a novice

$100,000 POULTRY BUILDING ON THE NEW YORK STATE FAIR GROUNDS AT SYRACUSE

variety of exhibits few of the great winter shows now surpass the leading poultry shows at fall fairs and, allowing for the fact that they are held at a time when very little stock is in prime condition to show, the quality is as good and the value of winings nearly if not quite as great. In fact, the value of winning at the leading fall fairs is such that the breeders who cultivate all good opportunities are giving more and more attention to hatching early enough to have a good string of young birds at the right stage for showing at one or more of the most important fall fairs. These early-hatched birds seldom have the finish of those hatched at the natural season and growing their adult plumage in cool weather, but they are in the best condition that can be given young birds early in the fall. To the present time only a small proportion of exhibits at fairs are of this class, but there is every reason to anticipate that the practice of hatching early for these shows will steadily increase and the exhibitor who covets winnings at any important fair will find it to his advantage to hatch early enough to have birds as mature

with just ordinary good birds in good condition could come in and make a good winning. This is not said to encourage the novice to break recklessly into competition

OLD POULTRY BUILDING AT SYRACUSE, N. Y.

at a big show at the start, but to show the situation. Conditions in some classes are quite variable from year to year and, as a rule, an exhibitor has no means of knowing what competition he will meet until the birds are cooped.

Considering the question from the point of view of the novice with no experience in showing, who selects and fits his own birds, his best policy is to make his first showing at a small show, and if he gets a fair share of the awards go to a little stronger competition, and pursue this line as far as he finds himself able to meet his competitors on fairly even terms. It is from every consideration much better for a beginner to stay with the smaller shows, provided he has any competition at them, until it is apparent that he has qualified for the stronger competition of larger shows. There are many local shows, in places where breeders of standard poultry are numerous, that can give the majority of exhibitors as strong competition as they are ever able to meet. In general, the small exhibitor, especially in the popular breeds and varieties, is wise to make his exhibits at such shows and to do all that he can to make the minor shows accessible to him prosperous and permanent, for at such shows he is competing on equal terms with others, while at the large show he is usually at a disadvantage in comparison with the veteran exhibitors and the commercial breeders—that is, the breeders who make their living from poultry, whose full strings of first-class birds are put into the show in the best possible condition, and have the care of an expert attendant right up to the hour, and sometimes to the moment, of judging them.

When a breeder has a bird of extraordinary quality, really good enough to win anywhere, but only one—or a very few—such, he may gain a passing satisfaction by winning a high prize at a first-class show, but there is no substantial advantage in it unless he is able to follow it up in succeeding years. Nor, if a breeder is ambitious to attain a standing among the regular exhibitors in the popular classes at leading shows, is there any advantage in selling a single specimen he may possess capable of winning at such a show. A breeder who has such a bird, but no reputation as an exhibitor, is often tempted by the offer of a good price to part with the bird. The man who yields to such a temptation will never get into the front rank of breeders and exhibitors. The wise policy for the breeder who knows he has a superior bird or small line, and is ambitious to get to the front in his variety, is not to show his best birds anywhere, and not to sell them under any circumstances, but to keep them at home and make the most of them as breeders until he can go into a show with at least one or two superior birds in each of the single-bird classes, and still have enough birds close up to these in quality at home to use as breeders in case anything happens to a bird exhibited.

The Novice's Chances As an Exhibitor

The impression prevails very widely among novices that poultry shows are controlled and more or less manipulated by "rings" of old exhibitors, who have some connection with the management, or some sort of pull or influence with the judges. As to the conditions in general, nothing could be farther from the truth. As a rule, judges are both honest and careful to preserve their reputation for integrity as well as for ability. At the same time it must be recognized that various circumstances which are bound to arise in connection with the making of awards at shows often lend some color to suspicions of collusion. There is naturally a good deal of intimacy between the managers of a show, the judges and the regular exhibitors —especially those who come with their birds and remain through the show. In fact, it is this cordiality in the relations of poultrymen that constitutes a large part of the attraction of a poultry show to what might be called the professional elements attending it.

The new or occasional exhibitor who has not made his way into this circle (not at all a hard matter unless he deliberately goes in wrong) is rather prone to take the most unfavorable view of the apparent friendship between a judge in the class in which he exhibits and an exhibitor who wins over him. And the judge who is particularly intimate with an exhibitor is always open to suspicion of having been somewhat biased by his personal feeling when he favors his friend in a close decision. But the regular exhibitors, reporters and show managers, whose estimates of all such matters are based upon all they know of the men involved, rarely misjudge anyone on a single case of what looks like collusion or favoritism, for they know that even though he might be disposed to do so, no judge could follow his profession very long after making a few decisions which competent disinterested persons, or interested exhibitors generally, regarded as plainly biased; nor could the management of any show afford to tolerate —much less connive at—that kind of judging.

Managers do sometimes make poor selections of judges. In that case the sufferer is the exhibitor who does not get what a competent judge would give him, and he is apt to be one of the leading exhibitors. Judges do sometimes show rank favoritism, but such instances are rare, and in the writer's knowledge of shows they have generally reacted promptly upon the judge. A notable case in point is one that happened years ago at an important show. One of the officials of the show was a leading exhibitor in a certain class for years. At a certain show the class was judged by a judge who had officiated at the show, sometimes judging one class, sometimes another, for many years, and who was generally rated competent and fair. In this case he gave the official's birds places which the latter thought should have gone to a competitor. The official, after the awards were made, asked the judge to look over the class with him and, pointing out certain of his competitors' birds as plainly superior to some of his own that were placed above them, asked the judge why he had made the decision as he did. After a little beating around the bush the judge admitted that he had made the awards as he did to favor the official to the extent of giving him a good share of winnings in a class where on merits his exhibit would have stood second all the way through. Having secured this explanation the official addressed him substantially as follows: "Mr. Blank, I personally don't want any award that I am not honestly entitled to, and as an officer of this show I have no use for a judge who will consider anything but the merits of the birds when he makes his awards. You have not done me a favor. You have done me and the show an injury. In the eyes of everyone who sees this class, and who is competent to form an opinion, you have made me appear as in some way influencing you to favor my birds. Both as an exhibitor and as an officer of this association I resent that." The judge never judged at that show again, though the incident was not known outside of the show management until years after, when certain persons who had suggested this judge for a class at the show were told the reason their suggestion could not be accepted.

CHAPTER II

Fundamental Things About Judging and Judges

Standard Descriptions the Basis of Judging—Methods of Judging by the Standard—Merits and Faults of Score-Card Judging—Status of Comparison Judging—Why Different Methods Give Unlike Scores—How the Personality and Ideals of the Judge Influence His Application of the Standards

AN exhibitor of Standard poultry has to consider the subject of judging it at every stage of the production and preparation of his stock. His matings must be made with a view to securing the largest possible percentage of specimens of good standard quality. His chickens must be grown with due consideration of the effects of different environments upon the development of quality in characters which are modified by conditions affecting growth. That he may use his time and equipment to best advantage, it is usually desirable to cull out every chicken that evidently is not of fair standard quality as soon as a fault which eliminates a specimen from further consideration for exhibition or for breeding to standard is noted. To succeed as an exhibitor one must himself be a regularly fixed or graded factors. The whole subject is fully explained in the last three chapters of this book for those who wish a thorough knowledge of the evolution of methods of judging. For our present purpose, and for ordinary use in selecting and preparing birds for exhibition, it is sufficient to describe the common methods of judging without going into detailed explanations of theories of judging Standard poultry.

Standard Descriptions As the Basis of Judging

The description of a breed and its varieties, in the "Standard of Perfection," is designed for use in actual inspection and judging of specimens of the breed and variety which fairly approximate the approved type. The

I. K. Felch and G. O. Brown   H. B. May and Clerk   H. S. Ball
FOUR VETERAN JUDGES AT WORK IN THE SHOWROOM

fair judge of standard quality in his variety of poultry, and must be able to grade his birds with reasonable accuracy according to their merit.

To learn to judge poultry for the purposes specified is not at all difficult when instruction and practice are simplified by considering only the actual essentials in applying the "Standard of Perfection." The confusion of mind into which most people who try to study out the principles of judging for themselves fall, is due to the fact that theoretically the American Poultry Association's system of score-card judging, as developed when the Association was organized, is a complex system with many variable factors, while practically it is a simple system with Standard illustrations, and other illustrations of good models, help to correct understanding and application of Standard requirements, but even with these helps a practical understanding of the Standard specifications is not attained until one compares birds section by section with the Standard descriptions, and with the aid of appropriate explanations and illustrations. The Standard description is not designed to give one who had never seen what is described a correct idea of it, but to enable any intelligent person with a specimen of any variety described in it to determine with reasonable accuracy how closely the specimen conforms to the ideal of perfect form and color upon which the breeders have agreed as the desirable type.

With this object, the "Standard of Perfection" is the guide for both novices and expert judges, but with the difference that the novice uses it in learning to judge values, and every application that he makes of it is tentative and subject to revision as he makes progress in analyzing birds and estimating the proper cuts for the faults he finds; while the expert judge must be able to make a good estimate of the quality of a bird at a glance, and an approximately accurate valuation of all its faults upon a very rapid inspection. The expert is able to do this not by any peculiar faculty of intuition, but as the result of long and thorough training in observing birds and estimating their Standard valuations. Not that professional judges are infallible. All make errors at times and they differ in their ideas of the values of characters but, on the whole, the class of professional judges is made up of men and women whose judgments are considered by those familiar with their work as accurate as estimates of this character can be, and in accord with prevailing opinion as to the best interpretation of the Standard.

If the decision of a judge is questioned he has to justify it by the Standard description, or by an official rule for the interpretation of the Standard, if there are rules applying to the point. Otherwise, it is his privilege to make his own interpretation of the Standard, and also in matters where the Standard recommends certain things but does not specifically require them the judge is at liberty to follow his own opinions—always subject to the consideration that if the application of his own peculiar ideas does not meet the approval of exhibitors collectively he will soon find no demand for his services as a judge. Exhibitors are tolerant of a judge's peculiarities if they are not too numerous, and if they run generally in the line of demanding high standards—of applying the Standard with a view to maintaining the highest standards, rather than by laxity, allowing inferior birds to win. True, an exhibitor, who through the severity of a judge has come out badly in a competition, may in his disappointment exclaim against his severity upon faults, but it is generally recognized that exacting judges do far more for the improvement of Standard poultry than others, and most seasoned exhibitors would much rather show under a judge who is known to be strict in every particular, than under one who has the reputation for leniency with certain faults.

Methods of Judging by the Standard

The first written standards for poultry were made in England, by a poultry club organized for the purpose about 1865-6. Reprints of these standards, modified to suit the ideas of their publishers, were shortly after made in America. The original plan of describing birds was less elaborate and careful in detail, but not materially different from the present "Standard of Perfection."

The practice of "scoring" in judging poultry as presented in the first standards consisted in allotting small numerical credits for excellence in a few principal characters or sections, the sum total of such points being 15. This plan appears to start with the method of computing the score or sum of points representing achievement commonly used in many games and sporting contests, but differs in making a uniform, comparatively low limit. It is simply a crude method of checking up birds competing under comparison judging; but quite naturally, when a maximum limit is placed on the number of points that could be given a specimen, the practice arose of judging by discounting from the maximum number of points allowed a character or section. When this method is used a total of 15 points is too small to make discounts for faults from unless the cuts are all in small fractions. So, when the breeders and judges in America organized the American Poultry Association and revised and amended the standards received from England to suit their ideas, they increased the total number of points to 100, and made definite allotments of points to every section described in the written description of a breed. They considered 100 points the score of a perfect, ideal specimen, and arrived at the value of a specimen compared with this ideal by deducting in each section for the faults found in that section.

THREE WELL-KNOWN JUDGES IN ACTION
Left—T. F. McGrew
Center—M. S. Gardner
Right—C. E. Howell

 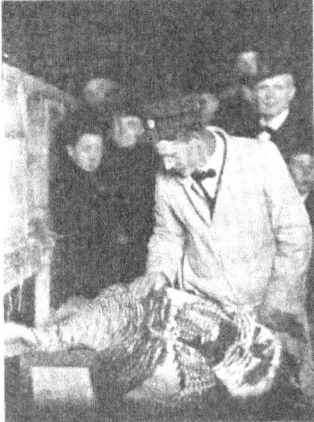 

C. H. Rhodes   S. B. Johnston   Maurice F. Delano
JUDGES HANDLING BIG BIRDS AT ST. LOUIS EXPOSITION

The original theory of cutting for defects was that the number of points assigned a section represented perfection in that section, and that the amount of cut was to be found by computing it as a percentage of the number of points in the section in which it appeared. This process was entirely too intricate for ordinary use. So it came about that in practice the method was to make specific cuts for faults as found—usually ½ for a slight fault, 1 for more pronounced but still not very bad faults, and 1½, 2 or more, for very bad faults. In scoring a specimen then the judge, using the American Association Official Score Card, as reproduced on page 20, marked the cuts he considered appropriate in the proper spaces, added up the totals of cuts in the two columns for shape and color, and deducted the total from 100, the result being the score.

While the unit cut was commonly ½, with heavier cuts in multiples of that fraction, some judges used smaller fractions, and some cut all faults much more severely than others. This and the unavoidable errors in judging gave wide variations in scoring. In an effort to make practice more uniform, the American Poultry Association first inserted in the Standard a list of faults that could be accurately described, as absence of certain feathers, eyes of different colors, etc., and provided definite specific cuts for these. In after revisions of the Standard this list has been increased and made to include faults of varying degrees—the maximum and minimum cuts for such faults as commonly found being indicated. The list of cuts thus mentioned in the 1915 Standard is about fifty. The lowest cut recommended in this list is ½, but it is provided in the general instructions to judges that no cut of less than ¼ shall be made in any section.

The instruction to base cuts on percentages of the value given a section has disappeared from the Standard, though those who use it are still emphatically advised that in cutting for defects careful consideration must be given the scale of points. It is this insistence on the consideration of the scale of points that confuses the novice trying to score fowls by the Standard. If he will simply pay no attention to it, and apply the system of specific cuts which the Standard recommends in the list of special faults in it to all faults, the novice can with a little practice, and testing his results by comparison with scores of professional judges which come under his observation, learn to make very good estimates of the Standard values of his birds.

Practice in scoring is the best training in the systematic observation of the qualities of Standard poultry that an exhibitor can have, whether he is to exhibit at a score-card or at a comparison show. No other way has been devised that so effectively leads to careful consideration of all sections and of every character. It is a great advantage to a beginner to have instruction in scoring from some one more proficient than himself, but he can learn much without such instruction. Many beginners are led to underrate the importance of thorough inspection of specimens they intend to exhibit, by giving ear to advisers who decry the method of searching for defects in judging, and declare that the "true fancier" looks for beauty and merit and is rather above spending time searching for hidden and trivial faults.

Thoroughness of the Poultry Judge

The merits of that view need not be discussed here. The novice in exhibiting may be sure that a competent judge is going to find all the faults in the specimens he exhibits. That is a part of the judge's work. But the novice need not fear that the judge will overlook quality or merit anywhere. A good judge's first glance at a specimen tells him very accurately what its quality is, "if it handles well"; that is, if when he takes it in hand and goes over it, and literally through its plumage, section by section, looking for the faults that are not visible from the aisle, he finds no bad faults.

Practically the only cases where a judge overlooks a really good bird are where the bird is both cooped in a bad light and stands badly as the judge looks at it from the aisle. Such cases are comparatively rare because judges generally realize the possibility of errors due to bad light and carriage, and take precautions accordingly. Many birds naturally take good positions—seem to delight in showing themselves off—whenever anyone stops to look at them. Most birds that come from long-established lines of exhibition stock are either natural posers or can easily be taught to pose when being inspected, and such training is an essential part of the work of fitting birds for exhibition. When a bird under observation by a judge does not take a position that shows good type, the judge, by motions or touches of the hand or a judging stick, tries to make it do so, and makes his decision as to type and carriage according to the best form he can induce it to take. When the light conditions where a bird is cooped are bad, the judge will usually take it from the coop to a good light to examine its color.

An exhibitor watching a judge at a comparison show at work on a class in which he is interested sometimes notes that the judge does not take all specimens from the coops. If some of the birds thus passed happen to be his, he is apt to conclude that the judge has not given them attention enough to know how good they are, and to complain that the judge hardly looked at them. No judge ever passes birds in this way unless as seen from the aisle they plainly show some fault that puts them out of consideration as possible prize winners. Many judges at comparison shows, however, to forestall any criticism of this kind, make it a rule to take every bird in the class in hand and at least "give it the once over."

In score-card shows the judge, even though he may see at a glance that the bird must be disqualified, may handle it and score regularly until the section containing the disqualifiication is reached, then write disqualified on the card and stop. The attitude of a judge in such cases is likely to be determined partly by the time at his command, the grade of the competition, and the advantage to an exhibitor of having the score of the specimen—even though it is disqualified. Judges sometimes complete the scoring of disqualified birds to the extent of marking all the cuts. This is done mostly at small and new shows where exhibitors are mostly novices and whatever helps them to a better appreciation of standard qualities in birds is worth doing.

Merits and Faults of Score-Card Judging

In score-card judging each specimen is mentally compared with the ideally perfect type of its kind as it stands in the mind of the judge, and the placing of numerical valuations upon its faults fixes a numerical valuation of excellence. Theoretically, and also in a rigid application of the method, each specimen judged is considered only in comparison with the ideal, and after the cuts are duly recorded on a score card anyone who can do simple addition and subtraction can find the scores of the birds judged and the order in which they stand. A great deal of judging had been done in this way with fairly satisfactory results in small classes of variable quality. But when applied to large classes of better and more uniform quality such practice often gives results that, when the ribbons are put up, are plainly wrong in the opinion of competent persons comparing the birds in the light of the judge's awards. The scores of all the specimens may be approximately right, yet as nearly all cuts are based on estimates, not on accurate measurements of any kind, and the standard used by the judge is a mental one, it will frequently happen that a nearly equal degree of fault is cut harder in one specimen than in another. The only way to avoid this is for the judge after scoring the birds to compare the birds that get places, and also any others that are close up to them, and by such comparison justify his work, revising it if necessary before finally making his awards.

Exhibitors who did not understand the matter and appreciate the necessity for testing work done under the conditions have been known to imagine and complain that the judge after having made his scores (recorded his unbiased opinions), finding out—supposedly—that the awards were not going where he would like to see them, proceeded to juggle the scores to get the desired result. If a judge is disposed to juggle scores to favor certain birds, he can easily do it when making them in the first place without giving any opportunity for criticism reflecting upon his fairness. All that is necessary is to do systematically and with definite intent what he occasionally does inadvertently—cut a little hard at two or three points on one bird and with leniency on a few points on the other. Even a discrimination of ½ point against one bird may give another the necessary advantage, and the judge can always effectively plead that in his opinion the cut was merited. Judges who want to do unfair things do not do them in ways that are easily visible to novices in the showroom.

Why Different Methods Give Slightly Different Scores

The errors in hasty judging by the regulation score-card method, and the time consumed in careful application of that method in the days when the percentage method of valuing defects was in vogue, led to efforts to improve methods of judging. These took two directions. Some sought to improve the score-card method, others to do away with it entirely and judge birds in competition by comparing them together without making any numerical valuations of defects and excellences. Attempts to improve score-card judging brought out two kinds of cards—the decimal score card which was intended as a substitute for the original style, and a number of more or less elaborate explanatory score cards which were designed to go farther than the old regulation and decimal score cards and show all faults found in each section, and the exact amount of cut for each. The decimal card was the only one that ever came into general use. It differs from the American Poultry Association Official Card in making

A SAMPLE REGULATION SCORE CARD

SOME WELL KNOWN JUDGES AT WORK AT THE ST. LOUIS EXPOSITION, 1904
Upper—Sharp Butterfield, W. S. Russell, F. H. Shellabarger
Middle—Geo. H. Burgott, F. J. Marshall, C. A. Emry
Lower—D. T. Heimlich, Thomas F. Rigg, Chas. T. Cornman

the whole number of sections 10 and allowing 10 points to each section, thus making possible greater facility in computing cuts on a percentage basis, and greater uniformity in some estimates of value. But as it has only 10 sections, where the other card has 13, the tendency is for a judge using the decimal card and making the customary cuts to get a little higher scores than with the other card, though judges who recognize how these higher scores occur easily make their scores the same by either method, by the simple process of making sure that they do not overlook any fault, but cut for each and every one and mark the result in its proper place on the card.

JUDGES CONSULTING
J. H. Drewenstedt and F. L. Platt

Another thing that operated to bring score-card judging into disrepute was the prevalence of the practice of inflating scores by making lighter cuts for defects in poor classes at small shows and in "private scoring" than in ordinary and strong competition. A few judges systematically scored the birds high at small local shows where the exhibitors were mostly novices, because—they said—accurate scoring would discourage the exhibitors. With all the birds at shows scored the practice of selling birds at prices according to what they had scored or would score came into vogue. As there is in principle no valid reason why a judge should not score a bird as accurately in a breeder's yard as in a showroom, it became common for breeders—especially those of little reputation and experience—to engage judges to score all the birds they had to sell and to adjust prices to these scores. Such practice, uncontrolled in any way, led to abuses.

It should be understood that it was the faults of the system in use and the abuses of the use of the score card, and the time required for its application at large shows where competition was close, that led to the adoption of comparison judging, first at New York and Boston, and then at other large shows and many small ones. The often repeated assertion of persons who do not like some features of the score-card system, that the change was based on consideration of the actual merits of the two systems (score card and comparison), is not in accordance with the facts as stated by the management of the first show in this country to return to the comparison method.

Status of Comparison Judging

While that method is either generally acceptable or not unacceptable to exhibitors who are themselves thoroughly good judges of the classes in which they exhibit, and to visitors who know something of the Standard and of the ideas of different judges, it is generally very unsatisfactory to novices and amateurs, and especially so to those who may not be able to attend a show.

The sole objective in comparison judging is to pick the winners of money and ribbon prizes, and the judge makes no records of either these or the other birds in the class except as may be necessary for his own guidance. These are usually in some sort of symbolic system of his own, and have no further value. They are not turned in to the managers, nor do they eventually go to the owners of birds as the score cards do. While a thoroughly systematic and careful judge will make his awards the same by either method, or with any score card, the comparison method undoubtedly encourages slack judging by those at all inclined to slack and slovenly work, and it also gives more latitude to judges disposed to favor particular qualities, either merits or faults.

Under the conditions which existed when the return to the comparison method began, and for some time after, the results appeared to justify the change. There were usually in every large class a few outstanding individuals, and the judge could eliminate most of the others at sight and apply himself without delay to the task of placing these in the right order of merit. But as the number of good breeders increased and average quality in nearly all popular classes rose, comparison judging showed less and less speed. And when some of the large classes at the leading shows reached the stage where they contained many times more birds entitled to consideration than there were ribbons to be awarded, it often took the judge as long to award the prizes by comparison as it would have taken by the score-card method. In fact, it is impossible to judge such a class satisfactorily to the judge himself without some system of scoring preliminary to comparison of the birds entitled to consideration for prizes. The compelling reason for abandoning the use of the score card in large shows was the occurrence of instances where two or three days were consumed in scoring a large class. Everyone familiar with the situation at large shows knows that such instances are becoming more and more frequent now in comparison shows.

The exhibitor at a comparison show need not fear that the judge will overlook a bird that has any chance of winning. The chances of that are too remote to require consideration. What he has to consider is that the personal ideas and preferences of the judge enter into comparison judging more easily than into score-card judging, where he has either to record a cut in every section or by failure to do so record the opinion that the section is in every particular so good that no cut is called for. This may not always completely check a judge's tendency to favor certain things but it is very effective in that direction. In the comparison show there is no direct check upon a judge who is partial to excellence in certain points, or inclined to be unduly severe upon certain faults. The only check on bias in his judging is his own appreciation of his tendency and of the possible criticism and dissatisfaction if it is allowed free play. It has already been explained that the judge has some latitude in the interpretation of the Standard. That being the case, a tendency to partiality has to be tolerated if it is not too pronounced, and the exhibitor's chances of succeeding under a judge are very much improved if he studies the judge's likes and dislikes and selects for exhibition accordingly.

CHAPTER III

Ethics of Fitting and Fixing Birds for Exhibition

Typical Cases of Unquestionably Legitimate, Doubtful and Illegitimate Practices—Analysis of Cases Showing How the Problem of Suppression of Unfair Methods Is Complicated by Peculiar Difficulties of Clearly Determining What Is Unfair—Review of the History of Regulatory Measures—Obstacles to Breeding and Selling Faked Specimens a Powerful Check on Wrongful Practices

THE preceding chapter treats of the attitude of poultry judges toward their work, and tells the novice in exhibiting what to expect for his birds and himself at the hands of the judges. This chapter will tell what he has to meet in the way of competition with expert exhibitors, and will discuss the ethics of the practices which are more or less common among them. As the points and matters which have to be considered in selecting and fitting birds for exhibition are brought up in detail the reader will appreciate (if he has not already learned it by experience) that there are comparatively few birds which upon close expert inspection are found free from faults, and that many of the faults most often found in birds of general attractive quality are of such a nature that they may be easily removed or concealed by artificial means. The practical question in selecting and fitting exhibition birds is: How far is it fair and legitimate to go in the concealment or removal of such faults?

To discuss the question intelligently we must first consider the nature of the faults which can be remedied by artificial means, and the nature of the remedies. It is not necessary at this stage to enumerate and consider all practices. Typical cases will serve the purpose.

Regarding the Removal of Side Sprigs

A small side sprig, that is a round point growing on the side of a single comb, is a disqualification. A side sprig is easily removed. If it is seen and removed at an early stage of the growth of the comb, the resulting scar may be so indistinct as to escape all but eyes of microscopic power. If the removal is made when the comb is full grown, or nearly so, a smooth scar is left, but in many cases the appearance of this is such that it is not possible to say whether the scar was caused by the removal of a side sprig or by an accidental injury to the comb. According to rule, in that case the specimen and the exhibitor have the benefit of the doubt. Thus, in operation, the rule works generally to the advantage of the skilled and clever faker, and punishes the bungler. The exceptions are the cases where judges take the position that scars in the places where side sprigs mostly grow are indubitable evidence that a side sprig has been removed; for while it is not impossible for an injury to cause the scar, it is highly improbable that accidents making similar scars should occur so often. Comparatively few judges are as stiff in the matter as this, but it is noted that stiffness in disqualifying for suspicion of the removal of side sprigs does not hurt the popularity of a judge, unless he is inconsistently lenient toward some other disqualification or fault.

Relations of Washing and Bleaching

In all White varieties the Standard requires that the color shall be a pure white. Pronounced brassiness is cut severely, and creaminess more lightly. Some birds are naturally very white, others slightly creamy, still others decidedly brassy. Practically all white birds need wash-

BEFORE

AFTER

Two photographs of the same bird, taken about an hour apart, show strikingly what an expert "fixer" can do to improve form. The work in this case could have been done in 15 to 20 minutes, but was done very slowly to admit of making photographs of the processes fully described and illustrated in Chapter IX.

ing before exhibiting. Those that are very white, and that have been kept where the plumage gets very little soiled, will look white by comparison with unwashed birds. But a properly washed bird is always much cleaner and whiter looking than an unwashed one. Washing with water and as much soap as is necessary to remove the dirt will reduce creaminess of creamy white plumage. Repeated washings with soap and water will make the creamiest plumage pure white. A less number of washings, or perhaps a single washing with a cleansing compound stronger than soap, will make a creamy plumage pure white.

It is universally conceded that washing with soap and water is legitimate, and that an exhibitor may wash that way as often as he sees fit. It is held by some that if the whitening of plumage by repeated washings is legitimate, it is absurd to take the position that the use of more powerful cleansers, producing the same result with less labor, is illegitimate. Others take the view that it is the nature of the chemicals which should be considered. The issue between them is immaterial: all have the same object and arrive at the same end—the whitening of creamy plumage. Hence the practice of washing white birds is general, and incidentally the creamy birds are made as white as the naturally white ones, and all white birds shown in good competition and in good show condition are practically equal in color, and awards turn upon other points. Washing, with incidental bleaching, is an essential part of the fitting of white birds for exhibition. One who does not do it cannot show with any hope of winning prizes. The removal of side sprigs would in any case apply to only a small proportion of a class.

Removal of "Foul" Feathers and Coloring Feathers

In birds of all colors false-colored feathers—sometimes called foreign colored, or simply foul feathers—appear with more or less frequency, according to the composition of the variety, the length of time it has been bred, and the character of the pattern. When these off-colored feathers are in the soft-feathered sections they may be plucked out, to the great improvement or the appearance of the bird and without detection, unless there are many of them close together. When they are in the hard-feathered sections, as the wings and tail, they cannot be removed or even broken and the fact concealed. It is, however, sometimes possible to stain, dye or paint feathers defective in color so that only close expert inspection will discover indications of the fraud, and a chemical test may be needed to show decisively that the feathers have been artificially colored.

This last form of faking is generally regarded as illegitimate. Its status differs from that of the removal of side sprigs in that the presence of artificial coloring matter cannot possibly be attributed to accident; and it differs from the bleaching of white birds in that the process is in no sense an application of method of treatment sometimes admittedly legitimate. Hence, exhibitors who are careful of their reputations are inclined to let it severely alone, and because that is the general attitude toward it, there is little disposition to tolerate the practice, and there is a strong tendency to make an example of any offender whose guilt can be clearly established.

Removal of Stubs and Down

All the clean-legged breeds that come from crosses or mixtures with feather-legged breeds have a strong tendency to produce specimens with "stubs" or down on the shanks and sometimes also on the outer toes, and with down between the toes, and occasionally on the shanks. The Standard makes the least particle of down on shanks or feet of a clean-legged breed a disqualification, and also makes evidence that stubs or down have been removed a disqualification. The occasion for this severity was the long-continued persistence of stubs as a common fault in all breeds and varieties of the popular American class. In these the fault tended, and still tends, to maintain itself if tolerated at all. But in the smaller clean-legged breeds it is not a serious fault. In all slight cases it is easily removed and the holes plugged so as to escape detection.

Whether such is the real intent of the Standard or not, the provision in regard to the removal of stubs and down encourages their removal in all cases where the fact that they have been removed can be readily concealed, and it is the common practice to do so; while only a bungler, or one willing to take desperate chances on a judge overlooking something, tries to fix up for the show a bird having so many or such large stubs that the holes left by their removal cannot be fully concealed by plugging.

Making Over Bad Combs

The comb of a bird that is defective may be greatly changed in form by a surgical operation. Extra serrations in single combs may be removed, absent rear spikes on rose combs supplied, and even very large and badly formed combs carved to beautiful proportions—preparations being applied to cut surfaces to make them heal with a more natural appearance than they otherwise would. Except as occasionally an operation of little more importance than the cutting off of a side sprig is made to remedy a slight fault in an otherwise uncommonly good bird, exhibitors generally taboo these practices as brutal, and as quite outside the application of the principle which justifies the use of simple measures to remove superficial faults. An expert breeder or judge is rarely deceived by this form of faking.

Splicing Tail and Wing Feathers

A broken or defective tail feather, oftenest a sickle feather, or pair of sickles in cocks of breeds having long sickle feathers, is sometimes cut off, leaving a stub to which a perfect feather can be spliced. The success of this form of faking depends upon the absence of careful examination section by section. It is hardly possible for it to escape the notice of a judge who is looking sharply for the common defects in stiff feathers. Even if the dead spliced feathers are not enough different from the others in appearance to excite suspicion, ordinarily close inspection at the base of the tail shows what has been done. Exhibitors generally consider this practice one of the least tempting forms of illegitimate fitting. There is always more than an even chance of its being detected by the judge, or suspected by a competitor who will demand inspection with certain exposure. There is no explaining it away, and wherever the person known to have practiced it exhibits afterwards his birds will be closely examined for all forms of faking.

The Sum of the Matter

The five illustrations that have been given are representative of the different phases of faking. The statements of the attitude of exhibitors in these cases give a fair idea of the general consensus of opinion of expert breeders and exhibitors as to what is and what is not legitimate in this line. There are a few who take extreme high ground in regard to right and wrong in these practices, but the majority consider them not so much from a rigid abstract moral point of view, as from the point of view of what is fair between exhibitors. With them it is a question of playing a game according to the rules, rather than a question of LAW, and they understand the rules to be the rules as construed and applied by those versed in the game—breeders, judges, show managers, with (of late years) the American Poultry Association as the final interpreter of the rules which it makes.

The common general understanding of the rules is the outgrowth of experience with the practical application of the rules under many different circumstances, as interpreted differently by different judges, as sometimes obviously misapplied, and as variously affecting the interests of individuals and the morale of the fraternity. Novices generally, read the rules and construe them literally in accordance with a logical application of conventional moral precepts. But the more one knows of the conditions of exhibiting, of the character of exhibitors and of the results of what appears to the average novice a low ethical standard among poultry exhibitors, the less ready he is to condemn them and their philosophy.

An important factor in the forming of opinion in matters like this is knowledge of the history of the matter and of the conditions which influenced opinion in the early stages, creating what may be called the traditional sentiment. For a great many years now the debatable and unfair practices of poultry exhibitors have been largely along the line of removal or concealment of superficial faults in specimens of generally superior quality and, as a rule, valuable for breeding purposes in spite of faults which greatly reduce their chances of winning prizes, or perhaps disqualify them for competition. But when the rules against faking were first promulgated the conditions were very different.

Review of Efforts to Prevent Unfair Practices

Just when or where the rule that birds must be shown "in their natural condition" was first made is not known. The earliest statement of it that I have seen in the premium list of the show held in Boston in 1872. In this I find:

"Rule 10—All specimens will be exhibited in their natural condition. Pulling false feathers, coloring, shaping spurs, inserting or splicing a feather, and similar practices are prohibited. Any violation of this rule shall exclude the pen from competition. Games only being the exception so far as to render the ordinary docking and trimming admissible."

Premium lists of some of the small New England shows ten years later contain a similar rule, but others lack it, and it is worthy of note that no such rule appears in the "Rules and Regulations" of the Worcester, Mass., Show of 1883, which was the most important show in the country that year, being held under the joint auspices of the Central Massachusetts Poultry Club, the Middlesex Poultry Association and the American Poultry Association—the latter holding at the same time the convention which considered and adopted the 1883 edition of the "Standard of Perfection."

From the above facts it would appear that the practices called faking were so prevalent in 1872 as to lead the managers of the most important show of that year to make a rigid rule against them, but that as late as 1883 the sentiment of fanciers was so tolerant of it that the promoters of a show bidding for large entries did not deem it advisable to apply the rule. Men who were exhibiting in America at this period and earlier have often stated in recent years that gross forms of faking, now generally rare, were then common, and were quite generally considered excusable, if not absolutely necessary and justifiable. The reasons for this are plainly apparent when we consider how few of the old breeds and varieties were actually bred to such excellence as has since become common, and consider too, that nearly all the new varieties that have since become very popular were in the early stages of making in the period from about 1870 to about 1895.

The faking of those days was not, as now, the removal of small faults, but the removal of large, or numerous, defects by methods, or on a scale, that left evidence of what had been done that was plain to all persons who looked birds over carefully, and that often very materially altered the appearance of a specimen. Those who first made the rule, and those who followed them in adopting it, knew perfectly well that it could not be rigidly enforced. They also knew well just about how far the artificial removal of faults could be carried without detection and with improvement to the appearance of a specimen. They had no expectation that breeders would comply strictly with the rule that all birds must be shown "in their natural condition," with no artificial manipulation at all. Their immediate purpose was to exclude specimens naturally so defective that the removal of their faults left them in a condition which was obviously not natural, and which to a fancier's eye was quite as unsightly as the original fault. Their further object was to emphasize the importance of breeding out defects.

At first the rule appeared very objectionable both to those who held that the shows ought not to make a rule that could not be enforced uniformly against all faking, and to those who wanted the old freedom in fitting for exhibition. But the operation of the rule in practice soon convinced most of those interested in it that so far from leading to an increase of skill in faking and of skilled fakers, it steadily improved the situation, and had the effect of limiting artificial fitting generally to matters requiring little effort and no unusual degree of skill. The rule soon came into universal use as a show rule, and when in 1905 "faking" was for the first time defined in the Standard glossary, the definition—after specifying various forms of faking—concluded with: "In fact, any self-evident attempt on the part of an exhibitor to deceive the judge and thus obtain an unfair advantage in competition."

In this connection it should be observed that the American Poultry Association—representing the views of the body of breeders, exhibitors and judges—does not describe practices as right or wrong. It simply considers what is fair as between exhibitors, judges and other exhibitors, and leaves to each individual to decide for himself as to the right and wrong of matters concerning him. This is not because of any indifference to the right and wrong of such matters, but because the more one knows of the possible applications of unquestionably legitimate methods in fitting fowls for exhibition, the better he appreciates the difficulties of making rigid rules to fit all cases.

Further Observations on the Difficulties of the Question —Prevention of Fading

How the process of bleaching white fowls grew out of the simple practice of washing them has been told. In the fitting of both white and colored birds we have another somewhat similar case — protecting fowls during the growth of the adult plumage, either of young birds or of old birds in molt, from the amount of sunlight that would make some white birds brassy, and would fade most buffs, reds, browns and blues.

Birds of all colors differ greatly in respect to the effect of the sun upon the pigments of their plumage. It is not possible in the case of any bird to say in advance how the plumage will be affected by exposure to the sun. Only exposure will determine that. The birds that sunburn or fade easily soon lose much of the quality of new-grown plumage, and if exposed too much may be spoiled in color for exhibition purposes. Even while the plumage is growing, continuous exposure to strong sunlight may cause sunburn or fading during growth, so that that coat of feathers never has its normal—natural—color. Generally the proportion of specimens that can stand any amount of exposure to the sun without detriment to the color of the plumage is small.

The question that arises here is whether a bird that has its color preserved from fading by keeping it out of the sun—really by keeping it under abnormal and in a sense unnatural conditions—is shown in its natural condition. People can take opposite views of that and argue interminably. The practical phase of the question is that by growing or molting colored birds in the shade a breeder can put fadable specimens into a show on an equality with those that are nonfadable—or practically so; just as by bleaching he can put creamy white specimens in on an equality with pure white. And it does not take a great deal of intelligent consideration to convince that as far as competition at shows is concerned such inequalities have to be tolerated. There is no practical way of distinguishing in the showroom between specimens of pure color, or of sound color, and those that by some simple process can be put into the show looking just like them. This is true also of the removal of small defects generally. The judge must pass upon the birds as he sees them and in suspicious cases he is supposed to give the birds the benefit of any doubt that may exist in his mind.

If the regulation of debatable practices depended, finally upon what the shows and the judges, and even the sentiment of exhibitors, could do to control and check them, it is doubtful whether they would have been reduced in anything like the measure that they have been. The effective regulator is the commercial interest of the exhibitor. In fitting a bird to show an exhibitor may go a little farther in some artificial process than he himself considers absolutely fair—may go beyond the limit he would place if it rested with him to fix the limit—because he feels quite sure that some of his competitors will not be quite so scrupulous. He does not consider that in going as far as they will he is unfair to them or to the judge, who is presumed to have an expert knowledge of all things connected with his work, and who in his sphere is very much of an autocrat and can punish attempts to deceive him severely as long as he does so impartially.

About Selling Faked Specimens

But when it comes to selling birds that have been "faked" in any way for exhibition, or selling eggs from them, breeders generally have a more rigid code of ethics. They are of course, not all equally scrupulous, and occasionally men engage in selling Standard poultry who systematically fake birds sent to customers. But the great majority of those who will strain a point of conscience to win over a competitor they reasonably assume is doing the same thing, are very careful about either selling birds that have faults it was deemed necessary to fix for exhibition, or using them where the results will disappoint customers and damage their own reputations. There are many easy ways of improving birds for exhibition, but comparatively few of these apply to the permanent removal of faults, and in no case can the consequences of selling a customer a bird that has had faults artificially concealed or removed be avoided. The faults are bound to come out either in the bird as faulty plumage grows again, or in its progeny as the natural form of an altered character—as the comb—is inherited.

Whether from the competitive or the commercial point of view, it is to the breeder's interest to make every effort to so breed, and grow, and manage his birds, that occasion for removing or concealing faults will be reduced to the minimum; but where so many characters are concerned, where the standards set are so exacting, where there is an ever present tendency to variation, there will always be many birds needing more or less simple special treatment to put them in perfect show condition, and it is a reasonable presumption that the practices that improve the appearance of good birds, and do not leave plain and unsightly evidence of the process, will always be used as far as it is advantageous and safe to use them.

Realizing this, many interested poultrymen are of the opinion that it would be better to make the rules and regulations applying to the fitting of birds for exhibition to conform to the common interpretations of them in practice—that is, to make rules permitting the things that are commonly done by exhibitors, but prohibiting the practices which the majority of exhibitors do not use. It is much easier to propose a formula like this than to apply it—as the reader will fully understand by the time all the practices in fitting birds to show have been described. The most practical way of dealing with the matters appears to be that which circumstances developed: to make a short and literally rigid general rule, but to apply it as leniently as the sentiment of the fraternity seems to demand— to observe the spirit rather than the letter of the law.

As Lewis Wright points out when discussing the subject of "Faking and Trimming," in the last edition of "The Book of Poultry" (1902): "The evil has always existed; for its springs lie deep in human nature, and not altogether among the baser motives." Farther on, elaborating the idea in the last phrase of that statement he says: "We have already hinted that some trimming was not done altogether from the baser motives. We knew a man who never exhibited, though he sold winners largely to those who did, yet who always removed single foul feathers. The reason he gave was that he 'could not bear to see them.' There are many who have that instinctive fancier's feeling very strongly, whether or not they act on it. We once read in a New England journal: 'It is impossible not to draw a little hard on a feather, when you know that but for that one your bird would be a perfect beauty.' Whoever does not understand that feeling has never been a true fancier; hence it is also that the best fanciers generally feel a little gentle tolerance for that sort of thing. The fancier's passion for perfection in appearance they know is partly at the root of it."

CHAPTER IV

The Inalienable Rights of Exhibition Birds

Modern Exhibition Poultry Are Not Only Bred in Beauty, But by Breeding and Training Have Been Made Acutely Conscious of Admiration and Responsive to Efforts to Make Them "Show Off"—Finest Exhibition Form Is Found Only in Birds of Perfect Development and in the Pink of Condition

WE have considered what the exhibitor has to expect from the judges and from his competitors. Before taking up the multitude of details—often relating to points which to the novice seem insignificant—of the handling and care of exhibition stock at all stages, it is desirable to consider particularly the reciprocal relations of exhibitors and their birds, and the obligations which these impose upon the exhibitor.

I once saw a copy of a little book on the preparation of birds for exhibition, upon the cover of which the owner had written: "Judging from contents the way of the successful exhibitor is a hard one." There is much truth in his observation, but it does not apply exclusively to the exhibitor of poultry, or peculiarly to the exhibition features of live stock production. High excellence and superiority in any phase of animal husbandry calls for similar unremitting attention to countless details day by day. Such work is drudgery or pleasure according to the pleasure given by successful results, but if the results are to be obtained the work has to be done.

EVERYBODY PLEASED
Dr. James B. Paige with a pair of Exhibition Game Bantams

The breeding and feeding of hens for very high egg production, or the growing of extra-fine individual specimens of table poultry, may involve as much detail—in proportion to the number of birds—as the production and fitting of the finest exhibition specimens.

Self-Consciousness in Beautiful Birds

But there is a phase of the handling of exhibition stock that is lacking in the other lines; and that is the apparent capacity of animals and birds to respond to admiration of their appearance. We can go even farther than that and say that many beautiful animals and birds constantly invite human admiration, and are always ready to show off their attractions. This observation still holds good for the seeming exceptions where the characteristic features are not considered beautiful. That is because of changes in the tastes of fanciers and in the style of a character popular with them. When we do not like a certain type the posing of a bird of that type often strikes us as ludicrous, but that does not alter the fact that, to the best of its appreciation of the situation, the bird is trying to attract the approving attention of the beholder. This quality varies much in birds and animals, and is quite lacking in many, yet we see far more manifestations of it than are needed to establish the fact that birds enjoy being admired, and also often show plainly the disposition to try to find out what a person handling them wants them to do.

When creatures naturally have such traits it is reasonable to suppose that many generations of exhibiting and training for exhibition, with constant selection of the birds that respond best to efforts to make them show their points to advantage, will greatly accentuate them. Further that birds of equal beauty may be very different in apparent consciousness of it, and in the moods and attitudes they show when in favor or in disgrace, when in order or disorder, is plain to anyone who knows them as individuals. So let us take it for granted that an exhibition bird has some feelings about its successes and failures in the showroom. Let us assume that, at the least, it feels better—enjoys the experience more—when it is a center of attraction than when it gets no attention.

The Breeder and His "Handiwork"

Next let us note and remember that the bird on exhibition is not merely representative of the breed and variety to which it belongs, or the strain or family from which it comes but is, above all, representative of its owner's ability to breed, grow, select, condition, fit, train and show Standard birds. An exhibition bird is expected to win for the glory, and incidentally for the profit, of its owner—to bring him high credit for taste, judgment, knowledge and skill as a breeder and in everything relating to the production of superior specimens. The bird can show only what it has by inheritance, care and training. If a bird shows more in the exhibition pen, or coop, at the time of judging, than it showed at home, it is because the judge can show it better than the exhibitor. But few

THERE'S NO SUCH WORD AS "SHOO"

birds show as well in the exhibition coop as they do at home, among familiar and comfortable surroundings, when properly handled. As a rule, the exhibitor must allow for this and offset it as far as possible by leaving nothing undone for the bird that could increase its chances of winning. That is a simple statement that means a great deal more than it suggests. It is only as the occasion or opportunity to do the many things that enter into the production, conditioning and showing of exhibtion birds arise, one after another in continuous succession, that one appreciates all that it means.

The Heritage of Quality

The first right of an exhibition bird is to be "born in the purple"—to have a substantial inheritance of exhibition qualities. It is as easy to "gather grapes of thorns, or figs of thistles," as to breed exhibition quality in Stand-

"JIMMY" GLASGOW, FANCIER
This bird certainly likes to be handled and admired

ard poultry from strains or stocks that are not regularly producing a good proportion of offspring of exhibition quality. The breeder whose ambition does not go beyond winning at a small local show may not need the best exhibition quality to be obtained, but he does need Standard stock of the best winning lines to be had; for it is stock (directly or indirectly) from a few leading breeders, whose lines are often quite similar and more or less related, that usually wins wherever shown, being more in favor than other types. In general, certain grades of stock appear at certain grades or classes of shows, and everywhere the birds of a popular strain or style win over those of even better quality in another style.

Timely Hatching

The next right of an exhibition bird is to be hatched at the right time. This time—in each case—depends upon the time it is to be shown. A bird is at its best—in the pink of condition for exhibition—for only a short period, from the time that its plumage is full grown until the feathers begin to show wear, or the effects of laying. There are ways of hastening or retarding its development a little, but as far as possible the hatching of exhibition birds should be timed to suit the shows at which they are to be exhibited. This point is of most importance to those who pay least attention to it—the small breeders who exhibit at only one show.

The big breeder who hatches through a period of several months can find specimens fit for shows through a period of about the same length of time. The breeder who raises only a few must either time the hatching with reference to showing at a definite period, or make his decision where he will exhibit in accordance with the development of the birds he has to show. As a rule, the former is the better policy because it gives greater definiteness and directness to his plans and work throughout the season, and further because if one goes to all the trouble and expense of showing he ought to make it a point to exhibit at the accessible show where his prospects of winning are best, and the win is worth more to him.

Exhibition Stock Needs More Room Than Other Stock

The next right of an exhibition bird is to conditions of life that give a young bird the opportunity to make its best possible development, or an old one the opportunity to renew its youth as far as possible. The bird that is to be exhibited is entitled to much more in this instance than the bird that is not, for its quality is to be tested by higher and severer standards. Many novices, and some who do not consider themselves in that class, have the idea that the economical way to produce birds for exhibition is to grow as many as there is room for according to estimates of capacity of land or coops in ordinary poultry growing, and then select the show specimens among them just long enough before they are to be shown to give time for such special conditioning as may be found necessary. The economy in this practice is supposed to lie in the fuller utilization of equipment, and the great saving in time and labor by not giving exhibition birds special quarters and special care through a long season.

If the purpose of the poultry keeper is to grow utility stock, and any birds of exhibition quality secured are accepted as so much gain, that may be good practice. But if the object is to produce exhibition stock, and to develop every specimen that can be made an exhibition bird, it is the most wasteful method that could be devised, for by it no specimen attains its best, and a large proportion of those that under suitable conditions would make exhibition birds fall so far short in one or more particulars that no one wants them, either to show or to breed. In the stock of the average amateur breeder of Standard poultry there is rarely as much as one-fourth of the chicks produced in a season that even with the best of conditions and care would make good show birds. An expert who has bred closely in line for many years may have less than one-fourth of his chicks unfit for exhibition.

I have seen hundreds of ribbons on birds that were not disqualified, yet were so generally lacking in quality that they ought never to be given prizes. And again and again I have been in the yards of breeders who had hundreds and sometimes thousands of birds, whose poorest specimens (after deformed birds were removed) were far superior to many winners of the season. At least half of this superiority is in the growing, and at least half of the better results in growing is to be accounted for by the fact that the breeder of exhibition stock who thoroughly understands his business is careful not to overstock his land, or overcrowd the chickens in his coops; and unless he has culls in such large proportion that the rest get a good deal more room by the removal of culls at broiler size, he starts only about the number of chickens he can carry satisfactorily to maturity.

Good Breeders Have Few Culls

The average amateur—especially the one with rather limited space—too often starts two or three times as many chickens as he can mature, and finds it hard to cull out as fast as he should because all are backward. So he keeps them on with conditions getting more and more unfavorable every day and wonders why he has nothing that comes up to his expectations. The only wise course in such cases is to cull for the points that one can form judgments on while the chick is small, and take the chances of discarding some specimens of greater merit on the points that cannot be estimated until growth is more advanced. The alternative is the failure of any specimen to make its best development. It is better to sacrifice some possible good ones while their quality is undetermined, than to fail to produce a single one.

The breeder for exhibition, whether producing on a large scale or on a small scale, ought within a very few years to have his stock at a point where the number of culls that it will not pay to raise is so small that he can hatch just about what he wants to keep on his premises. Until he reaches that stage he must either cull early and hard, taking his chances on errors, or waste much time, room and feed, on chicks that as they mature prove disappointing. It is only the good chicks that are worth growing. The chick of exhibition quality has a right to be identified, or to have its companions of poor quality identified, and to have its needs considered first before crowding in any way affects its growth. Checks of any kind are never made up. By as much as it was set back by any unfavorable condition, a chick at maturity falls short of what it would have been if all things were favorable. The final effect of a check will be determined by the period at which it occurs. Early in life unfavorable conditions affect the size and rate of growth; later they have more influence on the plumage, and particularly upon quality of color.

Many deformities of chickens, which are not noted until the fowl is pretty well grown, or perhaps not until it is taken up and inspected carefully for faults when being

IDEAL CONDITIONS FOR GROWING CHICKS
Scene on Farm of D. W. Young, Master Breeder S. C. White Leghorns, Monroe, N. Y.

considered for exhibition, are due to overcrowding in brooders or brood coops, to rough treatment by hens unfit for mothers, to huddling in roosting coops night after night because the caretaker will not take the trouble to teach them to roost, and to various other causes operating in the period when to the unpracticed eye all chickens look very much alike. The faults of conditions in this early stage may be largely offset by extra-careful attention. After a chick reaches the weaning age it needs nearly as much house room, and more land room than an adult bird to make its best development. Also it needs cleaner land—land that has been less used for poultry.

Importance of Individual Attention

Finally, the exhibition bird at any age and stage of its development, has a right to all the time and attention the person responsible for its development can give it. Where it is one of many, its share may be limited, but the successful exhibitor handles and studies every bird of quality as much as possible. He does this not merely to make the bird show to better advantage, but because he himself takes the same kind of pleasure in working with it that a lover and expert handler of live stock of any kind does. He is on familiar terms with his birds. He knows them individually and they in him have confidence.

GENERAL VIEW OF BREEDING AND CONDITIONING HOUSES AT WILLOW TREE POULTRY FARM, S. C. WHITE LEGHORN SPECIALISTS, BEVERLY FARMS, MASS.

CHAPTER V

Early Selection and Care of Exhibition Birds

Common Faults Noticeable in Poultry at an Early Age—Faults Which Become Apparent Only As the Birds Grow and Develop—Exhibitors Must Learn to Anticipate the Changes in Type and Color in Individual Specimens That Will Come With Age and Consider Them Carefully in Selecting Birds to Show

FULL and final judgment of the quality of exhibition stock can only be made upon each individual at maturity, or upon observation at an earlier stage of growth of some fault which is not tolerated in the exhibition room. Some faults are plainly seen as soon as a chick is hatched, others appear at certain stages of its development—some early, others late. This makes the culling or weeding out of undesirables a more or less continuous process, the object of which is simply to get all specimens of no exhibition grade out of the way that the others may have all the room and attention. This culling, when well done, is much more rigid than most amateurs and commercial breeders make it. Especially is this the case with regard to birds that are undersized and lacking in vitality. Few new breeders are willing to ruthlessly weed these out regardless of the possibility that some of them might turn out to be the best colored specimens.

Undersized Chicks—Which to Cull and When

There are very few cases where the preservation of undersized specimens of good, or even phenomenally good color turns out to be of real advantage. The most of these cases are found in breeds and varieties so new, and so generally of inferior color, or faulty in a particular point of color, that until there is marked general improvement in color birds of extra-good color are needed for breeding, even if the use of them retards improvement in other respects. On the same principle a novice—especially one who cannot afford to pay the prices necessary to secure exhibition stock of the highest quality—may find it to his advantage to use birds of superior color, no matter how deficient in size. But he cannot do this systematically and regularly and build up a line noted for all-round excellence, and no matter how good the color of the weak and undersized individuals he uses as breeders may be, if he persists in using such he will soon find that all birds from such lines have the peculiar color faults that go with lack of vitality. We often find marvelously good color in a single specimen that has no other attraction, but never really fine and finished quality either in texture or color of plumage as a characteristic of a stock generally lacking in physical vitality.

The prospective exhibitor at the present time should consider further that the whole trend of judging Standard poultry now is against decisions giving undue weight to color, and that the birds that have only color quality to recommend them have little chance of high honor in good competition. Under these conditions there is little object in keeping such birds unless they are desired for the owner's own breeding. Buyers as a class object to them, and even those who can use them want them at prices too low to make it really worth while to hold them. I have seen so many such birds kept to the detriment of a whole stock all through the growing season, and then sold in the later winter at prices that made the net return on them no greater than if they had been sold as broilers, that I do not think I can do the poultry keeper disposed to keep them for their possible exhibition value a greater service than to urge him to cull out every undersized specimen at broiler size.

How Wrong Conditions Cause Deformities

Most breeders of Standard poultry cull out the crippled and badly deformed chicks found in the nests or incubators as they take them out. Slight deformities are generally not noticed until the chicks get out of the puffball downy stage and the different parts and members of the body become more conspicuous. Many of the faults that now develop are neither constitutional nor congenital, but are due to some wrong condition or to accidents which did not appear serious. Hens that are rough with their chicks, often stepping on them while small and knocking them about considerably, no doubt make a great many cripples, and their treatment of the chicks undoubtedly retards their growth through the brooding period, and by so much affects their final development. The bones of a little chicken are not easily broken, by accidents which ordinarily befall it because they have a certain pliability and toughness that prevents fractures; but they are easily bent and the joints slightly twisted or dislocated either by rough treatment by hens, or by large chicks, or by the mass of chicks of the same size when they huddle together and pile up in the corners of brooders and roosting coops at night.

LOOKING THE YOUNGSTERS OVER

From the time that his chickens begin to show their individual peculiarities the breeder of exhibition poultry should watch them as closely as he can for indications of slight malformations which may be due to crowding or accidents. If it is observed that in a lot which from their breeding would be supposed to grow very uniformly, and all straight and strong, there is more lack of uniformity than can be accounted for, some chickens being normal and standing out from the others, the probability is that some condition affecting the comfort and growth of the chicks

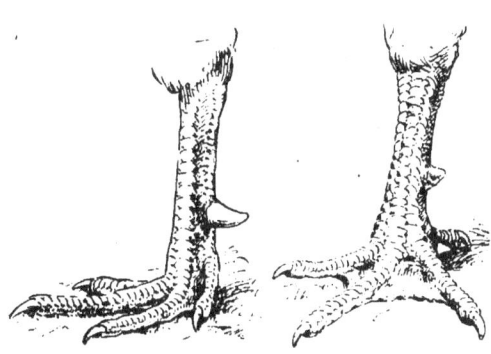

"DUCKFOOT" AND "WEBFOOT" IN FOWLS

is responsible for the difference. Or if all the chicks fall somewhat short of what was expected in the way of development, though feed is right and weather conditions favorable, it should be apparent that some condition has affected or is affecting the development of the chicks. If the effects of previous conditions are to be remedied, or causes still operating removed, the poultry keeper must recognize the fact that the cause of trouble is in every case a very simple one—easily found if all circumstances are properly taken into account; that there is nothing mysterious about it; and that it is usually to be found in the class of conditions which come under the general head of mismanagement.

The greatest source of trouble is the failure of attendants to see, night after night, that chicks at any age and under any conditions, are comfortable as they have disposed themselves for the night. This is one of the details of the care of live stock—and particularly of the care of young domestic animals in numbers—that becomes intolerable drudgery to most persons, unless they really like to work with stock. Even to these it is often tiresome at the end of a hard days work to give the time that may be necessary to train chickens to spread out around a brooder stove, or on the floor or roosts of an unheated coop. However, it has to be done if the chicks are to develop normally and with the minimum of malformations caused by crowding at night; and the grower of exhibition stock consults his own interests best when he takes the necessary precautions or actions to train the chicks right at the beginning of each stage of their care. It can be done either way—by making conditions in which the chicks will not crowd, or by training them not to crowd.

Teaching Chicks Not To Crowd

The simple methods of preventing chickens from crowding after the weaning age are: to keep them in small flocks, or in long, low and narrow shelters, or to use low roosting coops open on all sides. The latter is by far the best plan. The second answers well in emergencies and when only temporary quarters are required. Making groups so small that there is no danger at all of chicks crowding is practical only when operations are on a small scale. Where it is not practical to use any of these precautionary or preventive measures, the only thing to do is to go to the coops night after night just at dusk, make the chickens distribute over the floor or on the roosts, and see that they stay so until it becomes dark, when they will, as a rule, keep the position they are in. Many people claim that they cannot train chickens to spread out properly no matter how persistently they work with them. Failure to break the chicks of bad habits of this kind is always due either to putting off training until the habit is firmly fixed, or to spasmodic attempts—caring for the chicks properly one or two nights, then neglecting them until bad effects begin to show.

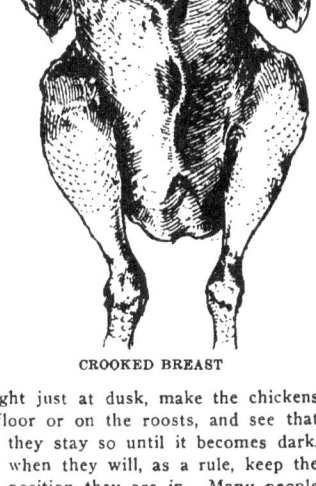

CROOKED BREAST

If the chicks are trained to spread out on the floor or the roosts as they should, most of them will develop symmetrically—unless from a mating which would not give symmetrical offspring. If attention is given to this early, any small tendencies to malformation that may be developing will be checked and may, as the bird grows, disappear

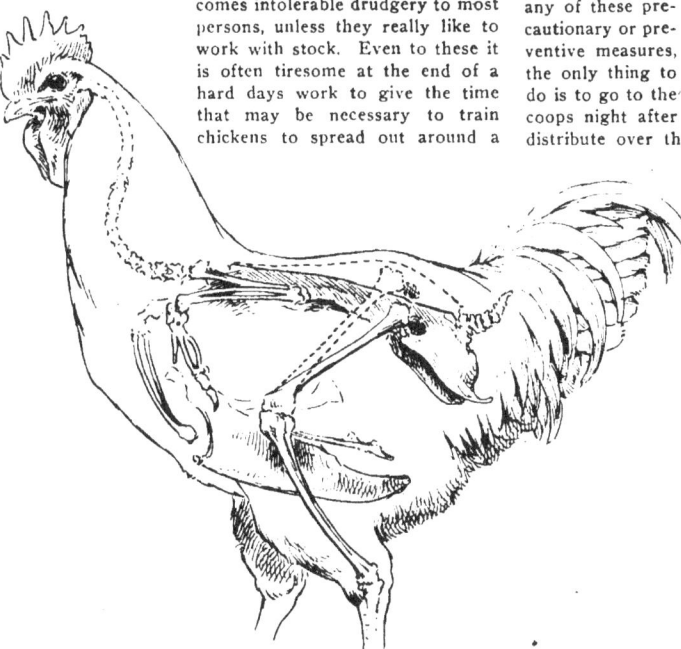

CROOKED BACK WITH OUTLINE OF SKELETON TO SHOW HOW IT IS CAUSED

entirely. After such faults become conspicuous they probably cannot be altogether corrected, but right conditions tend to improve them.

Faults Noticeable in Chicks at an Early Age

The common faults of form which can be detected early are: crooked backs, knock-knees, crooked toes, crooked breasts and crooked beaks. The bird's carriage of body and of the wings also show very early. A little later seriously faulty combs in cockerels of the large-combed breeds will show for what they are, and slipped wings, twisted flight feathers and wry tails, may be detected. Early detection of crooked backs is easiest when they are looked for at the stage when only the wing feathers in medium and large breeds, or the wing and tail feathers in small breeds, are out, and the down on the back is so scant that the position and prom-

THREE COMMON FORMS OF CROOKED BEAK IN FOWLS

by keeping chickens on smooth floors, or making them roost on flat roosts too wide for the toes to grasp. Their theory is that whenever the toes cannot take the flexed position which they naturally do on the round limbs which are the natural roosting places of fowls, or rounded perches of suitable width that are provided for them, the tendency is for either the toes, or the whole leg, or both, to take unnatural positions, leading to lateral curvature of toes, and to contraction of the position of the hocks with corresponding displacement of the feet at rest.

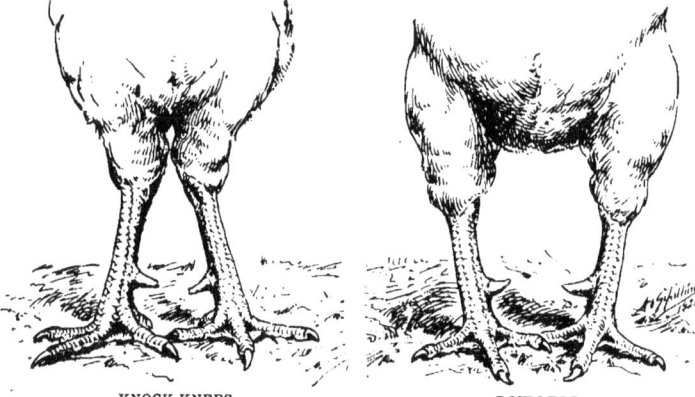

KNOCK-KNEES   BOW-LEGS

inence of the bones is easily seen. After the feathers are grown these are concealed, and when the plumage is long a considerable malformation of the back may pass unnoticed until it is discovered in handling the bird by passing the hand over the back.

Knock-knees are most common in birds with long narrow bodies and long legs, but sometimes develop in apparently sturdy chickens. Twisted toes are commonly supposed to be hereditary. Some close observers say that both knock-knees and crooked toes are very often caused

I am inclined to think this view a little far-fetched, for it appears that the majority of birds that by breeding should have straight legs and toes retain them even on smooth floors and wide roosts. It is reasonable though, to suppose that a tendency to knock-knees may be increased under the conditions mentioned, and that some cases of crooked toes may occur from such causes. Whatever the facts may be, the breeder of exhibition stock should keep floors well covered with sand or litter, and avoid the use of very wide perches except as they appear most useful when the chickens are first beginning to roost.

Crooked Breasts

Early roosting and unsuitable roosts have much more effect in making crooked breasts than anything else. Here their effects are plain and unmistakable, though by

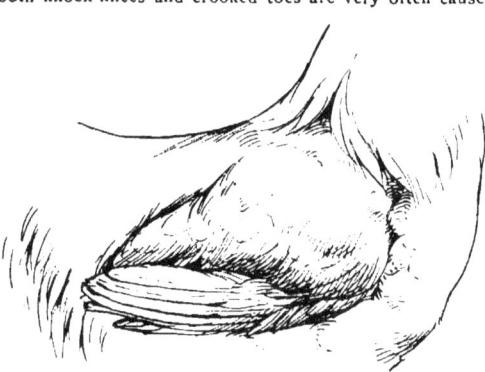

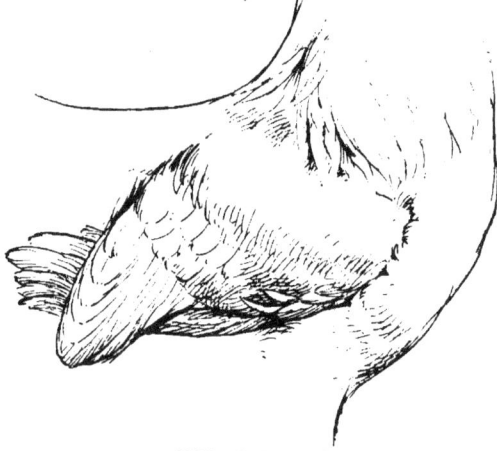

SLIPPED WING   TWISTED FLIGHTS

EARLY SELECTION AND CARE OF EXHIBITION BIRDS

LOPPED SINGLE COMB   SIDE SPRIG ON COMB

no means all birds subject to these influences develop crooked breasts. Nor, on the other hand, are these precautions more than a partial insurance against the development of the fault in birds with a hereditary tendency toward it. The place to begin the prevention of faults and tendencies that pass from one generation to another is in the breeding pen. If the occasion for special precautions to check hereditary tendencies does not impress upon the exhibitor the advantage of reducing troubles as an exhibitor, by breeding to keep all faults that can be in any degree controlled by selection of the parent stock at a minimum, other preventive measures only half serve their purpose.

Crooked Beaks

Crooked beaks are probably mostly congenital, but are often overlooked when culling chickens as they come from the nest or machine. Indeed, in a large flock which the caretaker has not time to look over carefully, a crooked beak may not be noticed until the chick is of pretty good size, for unless it is very bad the fault is not so outstanding to the eye of the experienced poultry breeder as some others.

Slipped and Twisted Wings

Slipped wings—that is, wings that are not folded and "tucked" smoothly and neatly—are most common in heavy breeds that have least power of flight. This fact suggests that the cause is partial degeneration of the wing as a result of lack of use. In Brahmas, Cochins and Pekin Ducks the tendency for slipped wings to occur is very pronounced, even when the use of birds having the fault as

ROSE COMB WITH INGROWN SPIKE   HOLLOW ROSE COMB

breeders is carefully avoided. An apparently slipped or twisted wing should be closely examined at the last joint to determine whether the wing is actually malformed, or only the flight feathers twisted. If there is any doubt on that point the bird should be kept to see how it develops. Many twisted flight feathers in young birds will come in straight in the first adult plumage.

Wrong Carriage of Wings

A very common fault in fowls is a tendency to carry the wings too high, with the points well up toward the base of the tail, and often with the points lapped over on the back. This fault is not common in well-bred exhibition stock, but the breeder who has never shown and so had his stock subjected to critical examination for all faults, not infrequently gets a lot of birds with it. If the wings are carried only a little high in the young chick while the plumage on the back is still very scanty, the feathers of the back as they grow will force the wings down a little, and the fault may not appear nearly as serious in the mature specimen as it did in the chick of two months. A very bad wing of this kind is generally beyond remedy. In many cases it is associated with crookedness of the back at the shoulders.

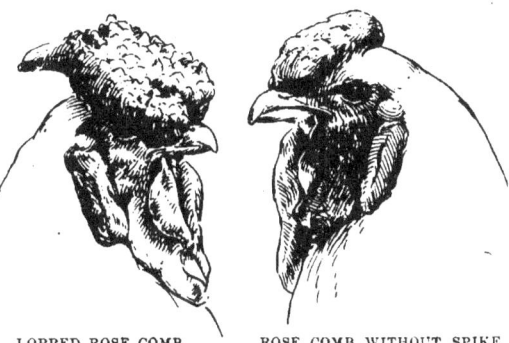

LOPPED ROSE COMB   ROSE COMB WITHOUT SPIKE

Wrong Carriage of Body

The tendency of a bird to carry the forward part of the body lower than is considered ideal for its type is usually due to the body not being properly balanced on the legs, and is apparent when the chick is quite small, though not so conspicuous that the untrained observer will notice it. If it is a prevailing tendency in a flock it may be good policy to try what can be done to remedy it by making the chicks lead as active a life as is consistent with good growth. This gives better muscular development and a more upright carriage. Where only a few birds show it, corrective measures may be postponed to the time of special fitting for the show.

Faults in Combs

Some faults in combs can usually be detected almost as soon as the combs begin to grow, and various faults in combs can easily be remedied at this time. In the rose comb varieties the rear spike is often "ingrown." A very simple operation at this time will make the spike protrude and grow out right. Side sprigs that show early can also be removed. Generally it is not worth while to go into these points carefully unless the birds are in appearance so promising that the breeder wants to be forehanded in everything relating to preparing them for exhibition.

Eyes Should Match in Color

It is well at this stage to observe whether the eyes are alike in color. Difference between them does not call for discarding a bird from consideration for a show bird, but where stock has to be reduced such points will often decide whether or not a bird shall be kept.

Uncertainty of Development of Plumage Color

Culling chicks at this age for color is limited to those which are so far from normal coloration for standard chicks of their kind that it is certain they will never have good color. A few of these may appear in chicks from well-bred stock, but they are likely to be numerous only in new varieties which produce comparatively few specimens of good color. A breeder who has intimate knowledge of his stock and of the peculiarities of color it may show at different ages, can often form a pretty good idea how chicks of certain lines with certain color characteristics in their chick feathers will develop in their adult plumage. Lacking this knowledge even a very expert breeder may be entirely wrong in his estimates. There are faults of plumage in the chick that disappear at maturity, and there are faults of the plumage of aged birds that sometimes develop in the first adult plumage immediately as it reaches completion of growth, and may come even earlier—like premature baldness or gray hair in human beings.

The bird that to anyone familiar with the appearance of good specimens of its kind in the early stages of its growth appears to have faults beyond remedy (that looks like a real cull), rarely turns out a good show bird. The only exceptions to this rule that I know are the cases where really bad faults in the chick feathers regularly disappear when the adult plumage grows. These appear to be cases where the chick plumage is inherited from one ancestor, or ancestral line, and the adult plumage from another; or where the character of the plumage at different periods of growth seems to correspond with its character at different stages of the development of the quality it now shows. Cases of the first class may be regarded as freak cases of heredity. Usually they appear in the progeny of a single hen, but occasionally they become established as characteristic of a strain, and if that strain becomes very popular their characteristic may become so common that it appears regularly.

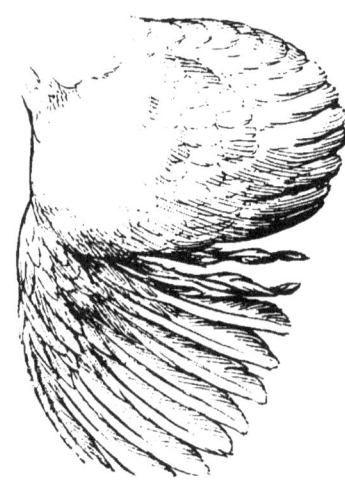

TWISTED INNER FLIGHT FEATHERS

One of the most interesting cases of this kind that I have known occurred some years ago in a popular black-red variety. In this, as in nearly all black and dark varieties, there may be a good deal of white—especially in the flights—in the chick plumage that will disappear in the adult plumage. A well-known breeder of the variety, who had for years bred to eliminate white in plumage at any stage, went to the yards of another breeder for blood to improve the color of his females. In mating this stock with his own he found the results most disappointing in that some of the pullets came with so much white in their

WRY TAIL

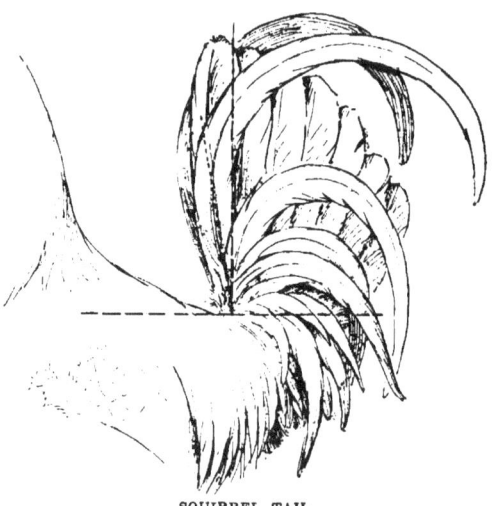

SQUIRREL TAIL

wings that he rated them as hopeless culls, and killed them at an early age. Meantime he was not getting any pullets as good at maturity as the combination of blood lines he had made ought to give. Thinking it over he came to the conclusion that he had perhaps been in too great a hurry in killing the white-winged pullets. The next season he let them grow, and found that they regularly molted the white chick feathers and replaced them with sound-colored adult feathers, and that they made his best birds.

Early Selection of Adult Birds

Theoretically it would be supposed that an exhibitor and breeder of exhibition poultry would have culled out and disposed of all specimens that were of no value for exhibition purposes shortly after they arrived at maturity; or that if one is carrying both exhibition and utility grades of stock, the separation at maturity could be final, leaving no occasion to consider at the beginning of their second year birds rejected as show birds the previous season. But a very large proportion of birds in some color varieties do not show the same quality of color in their first and second years, and in nearly all breeds there are many specimens that change their type quite a little as they age. So it is necessary in making definite selection of birds for exhibition, and also for breeding, to consider very carefully the characteristics that change with age. As the balancing of slight faults in mating to produce standard exhibition specimens requires the use in the breeding pens of many birds that are either never of real exhibition quality, or attain their best exhibition form only during a part of their lives, the occasion for carrying such birds in the seasons when they are not exhibited is not the burden on an exhibitor which a novice might suppose.

Early selection of adult birds is of most importance when birds are wanted to show at fall fairs. It takes about three months for either a young or an old bird to grow a full coat of feathers, hence to make sure that he has allowed ample time to do all the things that will contribute to the finest development and finish, the exhibitor should know three months before he expects to make an exhibit just what birds he has that are promising material for conditioning for that show. If he has been studying and observing his birds carefully he knows the old ones much better than the young ones, for he has seen them grow their first adult plumage, and has seen how well the colors in it hold. To

TWISTED HACKLE FEATHERS

a considerable extent also the changes in shape that come with advancing age are indicated by the time a bird is a year old, though because of the extent to which the development of plumage affects apparent shape, and because the second adult coat of plumage is often more abundant than the first, especially in the hackle, saddle and tail plumage of males, it may not be possible to foresee the extent of such changes very accurately.

In all such matters a thorough acquaintance with the stock, and knowledge of the characteristics of the lines bred is of very great help. The exhibitor who has been closely following the annual changes in the plumage of the birds of a few lines for a few years can know pretty near what to expect in his adult birds.

With regard to changes in the form and type of birds, the reader can appreciate the conditions with respect to these best if he will consider that the birds of different types and breeds, and to a considerable extent families, strains and individuals in each breed, present just such differences in this respect as he observes in the people of his acquaintance. In poultry, as in people, many

WRONG CARRIAGE OF BODY
Wings too low; legs too far forward

WRONG CARRIAGE OF BODY
Wing-tips too high; legs too far back

individuals keep just about the same condition and weight from the time they reach maturity until they begin to fail with age, when they become more or less emaciated. Others begin to put on flesh sometime after reaching full growth—some comparatively early, some later in life. Some of these make a progressive increase in fat until it is a burden to them, others acquire an amount not at all burdensome and keep in about this same condition while in normal health throughout life. The only class of development in human beings we cannot find paralleled in poultry is extreme fleshiness very early in life.

The breeder who has noted the changes in individuals in his stock has learned to know, partly by his knowledge of the breeding of the birds, and partly by observing, that certain characteristics of form in early life are either invariably, or in particular lines and families, associated with certain later changes in form. Such differences as this are generally too inconspicuous to be described in words or pictures. Indeed, the estimate based only on inspection is not at all dependable unless supplemented by knowledge of family history and observation of the usual changes in the individuals of a line. I cannot tell the reader how to do it, nor can anyone. Information can go no farther than to say that the cultivation of the habit of observing individual birds closely enables a breeder of exhibition poultry to judge with considerable accuracy what changes age will bring in either form or color to birds of his breeding.

Usual Changes in Type with Age

While there are some birds, both male and female, that hold for several years the form they have at first maturity, the greater number become somewhat heavier. Consequently, it is usually safe to assume that a cockerel or pullet that in its first maturity is fuller formed than is usual in birds of its age, so that except for its spurs a cockerel might pass as a cock, and a pullet would easily pass as a hen, will be too heavy in its second year to show an attractive figure and carriage. The tendency may not be plainly marked at the close of the breeding season, when both males and females are apt to be a little down in condition, but when the males are separated from the females, and the latter stop laying, and both are fed as liberally as they must be to enable them to grow new plumage free from faults that come when birds are fed too sparingly, many of them will put on flesh and fat very quickly, with disappointing results as to the preservation of the type they showed in their first season.

How to Hold Correct Type

The best way to prevent this is by giving the birds good range, and limiting their feed quite closely to hard grain. Where this is not practical much can be done by liberal feeding of green feed, by compelling a reasonable amount of exercise scratching for the grain fed, by using moist mashes sparingly, and late in the evening each day giving the birds an opportunity to fill up on either grain or dry mash. By this method good form and vigor may be preserved in many specimens that as they are usually handled through the period between the breeding and the show season, lose both form and condition to the disappointment of expectations of winnings and the diminution of their future breeding value.

As the only differences in Standard descriptions between cock and cockerel, hen and pullet, are the differences in Standard weights, and the lower weight requirements for young birds are not made primarily or properly for the purpose of establishing regular differences in weight between birds under a year and those over that age, but simply to prevent the showing of immature birds; the mature adult form, in the type within the Standard specifications which those whose ideas dominate in judging at any time approve, is really the standard of typical shape by which both old and young birds are judged. From what has been said of the variation in the changes of form with age in different individuals and families, it is plain that under these conditions the cockerel that looks most like a cock, and the pullet that looks most like a hen will, other things being equal, win in classes of birds of their own age but less full in form.

The Adult Type is the Standard Type

Considering the point of breed type independently it would appear that, inasmuch as many (possibly most)

THE SECTIONS OF A SINGLE COMB FOWL—MALE
1—crown of head; 1a—point of comb; 1b—throat; 2—beak; 3—nostril; 4—comb; 5—face; 6—eye; 7—wattle; 8—ear; 9—ear-lobe; 10—hackle; 11—front of hackle; 12—breast; 13—cape; 14—shoulder; 15—wing bow; 16—wing front; 17—wing bar (coverts); 18—wing bay (secondaries); 19—flights (wing primaries); 20—flight coverts; 21—back; 22—sweep of back; 23—saddle hangers; 24—sickles; 25—lesser sickles; 26—tail coverts; 27—main tail; 28—body feathers; 29—fluff; 30—thigh; 31—hock; 32—shank; 33—spur; 34—ball of foot; 35—toe; 36—toenail.

birds are fuller formed as cocks and hens than as cockerels and pullets, the Standard specifications should be made to show what appears as the normal difference in shape of the same birds at different ages. But the wisdom of the policy of specifying certain differences between the typical first adult form and the typical form in later periods of life is very doubtful. The probable effect of it would be quite opposite to what is desired, leading many judges to favor cockerels that had neither the correct adult type nor the promise of making it, because they felt it necessary to emphasize the difference. In addition, to split the Standard on a point like this would inevitably tend to further division on other characters where a certain ripeness of development comes with age.

The practical control of all such points lies in the fact that it is to the interest of the breeder to produce birds that will show good exhibition form for as many years as possible, and this result is best secured in the birds that hold the weight they make at their first maturity or increase it very little. The more such birds he has the easier it is for a breeder to select and fit a winning string year after year. Such birds are further valuable from the fact that they usually keep in good breeding condition too. They may not invariably be of special value as breeders, but the stock will benefit by as much of this characteristic of keeping fit to a comparatively advanced age as is distributed in it through their use in the breeding pen. In selecting birds for exhibition a breeder will take, as far as he needs them, young birds that he knows he will not want to show again, and he will also mark to hold over to show the next year some that do not show good form in their first year. But in selecting breeders he will give the preference to the birds that can qualify for strong competition at any age, and will hatch as many chicks as he can from such matings.

Color More Difficult to Hold than Shape

The proportion of individuals of high quality that in their third year and after can compete with birds of the quality they had as two-year-olds is very small, but this is due more to the failing of color in parti-colored birds than to anything else. On shape alone a good proportion of high-class Standard stock can hold good exhibition form for from three to six years. The heavier breeds usually show the effects of age earliest, for in them the tendency to put on fat at maturity is very prevalent, and is controlled only when conditions and care are directed continuously to that end. It is rare to find a Brahma or Cochin that will show good type and carriage on exhibition in its third year, except in response to motions of a hand or judging stick to make it take the desired pose. Yet occasionally hens are seen that at four years old are as smooth, and carry themselves as well as when in their prime; and there are also cocks that except for their spurs, and a little slowness of movement, appear as good as they ever were.

In the lighter breeds many specimens retain their best form, style and vigor, until four or five years of age, and an occasional bird will hold its color as long. In all breeds the method of handling has much to do with this result. The worst cause of what is really premature decay in exhibition quality is indifferent care and mismanagement between show seasons. Too many poultry exhibitors follow the policy of letting down in condition, especially after the breeding season, and then trying to bring the birds back by forcing methods. Results are far more sure, and in the long run easier and more satisfactory, if the birds are never allowed to go out of condition. It is to an exhibitor's interest to take care to insure this, and at the same time to breed to the lines that give birds that are easily kept in condition without constant adjustment of the diet and habits to the tendencies toward undesired developments.

Normal Differences in Young and Old Birds

What may be regarded as the normal difference in form between a well-developed cockerel and cock, or pullet and hen, is simply a very little more apparent substance and solidity of form. There is usually a similar increase in the character of comb and wattles, and the hackle, saddle and tail of the male are more fully developed. In color points the general tendency is just the other way, and both the rate and the extent of divergence from characteristics of the first adult plumage are commonly much greater than the changes in form which age brings. In general, as to all colored birds, we may say that the expectation is that every bird will lose color to some extent, and that

THE SECTIONS OF A SINGLE COMB FOWL—FEMALE
1—crown of head; 2—beak; 3—nostril; 4—comb; 5—point of comb; 6—face; 7—eye; 8—wattle; 9—ear; 10—ear-lobe; 11—throat; 12—hackle; 13—front of hackle; 14—cape; 15—breast; 16—shoulder; 17—wing-bow; 18—wing front; 19—wing bar; 20—wing bay (secondaries); 21—flights (wing primaries); 22—flight coverts; 23—back; 24—sweep of back; 25—cushion; 26—main tail; 27—tail coverts; 28—lesser tail coverts; 29—body feathers; 30—fluff; 31—thigh; 32—hock; 33—shank; 34—spur; 35—ball of foot; 36—toe; 37—toenail.

most birds will lose a noticeable amount or degree of color, and some a great deal. These losses have the general tendency to make a bird that is good Standard color in its first adult plumage too light after its first annual molt; and a bird that in its first adult plumage has faults due to excess of pigment nearer to Standard after each succeeding molt, until which it may be a very good Standard-colored bird, after which it begins to fall below the requirements for quality in color. With the general tendency to loss of pigment with age, there are some birds that remain practically sound in color for four or five years. These however, are exceptional.

In looking over birds at the end of the breeding season for the purpose of deciding which are to be kept for their possible exhibition value, a breeder needs to be able to remember each as it appeared when at its best. Unless he can do this he has only half the knowledge needed to make wise choices. Besides the loss of pigment in the new plumage each succeeding molt, there is the fading in each coat of feathers to consider. This varies as much in individual birds as the fading from year to year. In many cases it is far greater than the reduction of pigment in the new coat, and it is quite invariably some greater, as shown by the fact that the new coat, though lighter than the old originally was, is deeper in color as well as cleaner and fresher than the old coat just before molting. As to brightness and freshness of appearance the new coat will almost certainly look better than the old one did previous to the molt. But the fading of the coat after it has grown does not eliminate any of its original faults. The undesired appearances of pigment may become less conspicuous because of the dulling of the color, but that does not give clean surface or sharply defined markings.

Examples of Common Changes in Color with Age

In the new coat of feathers many small faults may disappear; and the opposite fault may appear in a section which before was free from it. Thus, mossiness—the prevalence of small dark flecks—in the ground color of penciled and laced varieties may be molted out in sections where it appeared, and the same reduction of pigment that gives this result may diminish the intensity of the dark markings. The breeder not only needs to be able to recall to his mind's eye the picture of each specimen as it was, but he needs also to be able to see in his mind's eye the picture of the bird as it will be when a degree of loss of color that eliminates faults which kept it from qualifying as a show specimen in the preceding season, has worked its effect also on the sections that were right at that time. With respect to the birds that were right in color the preceding season, he needs to be able to picture them as they will be after losing some of their pigment. This is the easier of the two processes, because the change will be all loss and no gain.

In the Ermine Color Pattern

. The changes that take place in the ermine color pattern (white with black points) afford the simplest and best examples of the effects of color changes with age, putting birds that in their first plumage were good Standard color out of consideration for exhibition in following years, and making Standard-colored specimens of birds that in their first adult plumage were so full of color faults that, to one not aware of the changes that time would bring, they appeared hopeless culls. This pattern is seen in the Light Brahma, the Columbian Plymouth Rock and Wyandotte and the Light Sussex. Ideally this pattern shows a white surface, as the birds stand, in all but the neck and tail sections, the saddle of the male, and the foot feathering in both male and female. The main tail is solid black; the coverts and the hackles are black edged with white, or white with black stripe in centers (according to the form of description adopted); the saddle of the male carries a more or less black striping; and a part of the foot feathering in the Brahma is black edged with white; the flights of the wings are predominantly black, but the secondaries must have enough white so that the folded wing will be white.

To get the desired combination regularly, with strong color in the black sections and parts of feathers, and no black showing in the surface of white sections, is next to impossible. Good quality in black parts is regularly associated with the presence of undesirable black markings on the surface adjoining black or partly black sections, and with much black—in the form of slate—in the undercolor. Especially when the black in the wing in both sexes and in the saddle of the male is emphasized, there is a pronounced outcropping of black feathers, or feathers with small black spots prevalent in the web, at the front of the neck and shoulders, on the back, and on the thighs. The style of bird in this color pattern that is favored changes from time to time. Twenty years or more ago, before the other varieties with the color came out, Light Brahma breeders regarded the purity of the white sections as of cardinal importance, and tried to get other sections as good as posisble without sacrificing white surfaces. The modern style demands extra good black in wings and strength in black sections, and if these points are secured, is tolerant of the outcroppings of black in sections which the Standard still describes as white.

Whatever may be the style in favor, few specimens that in their first adult plumage show good black markings in the sections containing black, and little or no black in white sections, can in their next year win against the darker birds which originally had so much smut on back, neck, at shoulders and on thighs, that they could not in that form be considered as show birds, but which when they molt these faults out make the richest and handsomest colored birds. In neither case, as a rule, is the white surface perfectly clean. Nearly always a good deal of plucking of feathers with a little black or gray in the web is necessary before the bird is ready to show. But the exhibitor who is out to win finds year after year that while now and then a specimen of phenomenal quality and great stability of color may win year after year in its class, most of his best selections for strong competition are birds not equally valuable in the old and young classes.

In the Black and White Mottled Varieties

In the black and white mottled varieties—as the Ancona and the Houdan—we have a similar case, but with the difference that both the number and the size of the white mottled tips increase with age. It is plain that where such changes take place the same bird cannot win in the young class one year, and in the old class the next, unless in one or both years there is some condition other than color determining the award. It could not win as a young bird and then also win as an old bird against a bird that duplicated the form in which it won as a young bird. It may, and frequently does, happen that a young bird that is too dark, or an old one that is too light, will win over a bird nearer the Standard color. An exhibitor will often take chances on such a bird if it is particularly good in other respects, and (or if) notwithstanding its failure fully to meet Standard color requirements, it appears the most promising he has for a certain show. Also an occasional bird does hold color so that it can win year after year on its color quality; but such birds are so rare that their existence does not affect the general necessity for holding birds too dark to show in their first year to show when molting has given them better Standard color. An exhibitor who fails to do this will find that he is very short of birds that meet Standard color requirements when in the cock or hen plumage.

In Buff and Red Varieties

Practically all buff and red birds fade to a noticeable degree in the spring and early summer, and a large proportion of the ordinary stock of varieties of these colors fade badly and very unevenly. Most birds will molt a little lighter than their original color, but an occasional one will be much lighter.

Until a bird has gone through an annual molt the probable degree of loss of color can only be conjectured, birds are practically laced or spangled, with markings so little darker than the ground that the unevenness of color is apparent only on very close observation. At a little distance the color appears perfectly sound. Besides regular markings in such variable colors there are often irregular, blotchy markings on each feather. All these faults tend to develop in more pronounced form in each coat of feathers as it fades with exposure, and also to increase in succeeding molts.

What has been said of the vagaries of buff color applies to red, with the difference that as some black is required in the plumage of all red varieties the presence of black where it is not wanted is more troublesome.

In general, birds that are in their first adult plumage of a rich red color, whether of the lighter red formerly considered most desirable, or of the rich wine color now fashionable, hold their color better than the light bird with a "bricky" cast in the plumage, or in some sections

THREE-FOURTHS GROWN COCKERELS PLAINLY SHOWING WHAT THEIR ADULT TYPE WILL BE

These three White Plymouth Rock cockerels were photographed at Owen Farms in June. No. 1, at left hand, is an attractive bird at his age, but will be small and too fine. No. 2, in center, is a model bird for his age, one of the kind that make good size and Standard type. No. 3, at right hand, will make a large bird, but not with either the substance or finish of the bird in the center.

though it is safe to assume that most birds will lose about what is usual in their line and family. After observation of one molt a breeder can judge pretty accurately about how much lighter a bird will be with each succeeding molt. Occasionally his judgment will be at fault because the condition of the bird or some circumstances affecting the quality of color of plumage is different in two years compared.

The lightest even-colored buff cockerels and pullets can rarely be shown in the following year as old cocks and hens with any hope of winning against sound-colored birds, for they will present too much of a washed out appearance even when even in surface, and if the buff color was anywhere a little weak the first year it is likely to be more or less white after the molt. This shows most in the stiff feathers of wings and tail. It is among the birds that were a little dark in their first year that the best colored cocks and hens are to be found, but the reduction in intensity of pigment does not always give a better-colored bird. It may bring out the faults of color, especially where the feathers are uneven in shade, through more reduction in one shade than in the other. Many buff a decided yellow cast; or the dark ones that are brown, rather than red.

Parti-Colors with Buff, Bay or Red Ground

In all varieties where the ground color is buff, bay, red or light brown, the behavior of this color is the same in fading as the behavior of buff. Usually the darker markings of brown in these patterns lose color in about the same degree, the result being a general and quite uniform softening of the whole surface. Sometimes the light color fades much more than the dark, giving rather harsh, contrasty effects. In all parti-colored patterns the general tendency is for light grounds to extend a little with each succeeding molt, and to become clearer, giving the general surface color of the bird an even lighter tone than would result from fading without any extension of the light areas. Perhaps the nicest illustrations of this are seen in the barred patterns—the Barred Plymouth Rock, Campine and Penciled Hamburg. In these, the widening of the white or bay bar by a hardly perceptible margin on either side, the dark bar being reduced in width to the same measure, makes quite a difference in the appearance of the bird. Whether the change is for better or worse

COCKEREL HOUSE AT WILLOW TREE POULTRY FARM, BEVERLY FARMS, MASS.

depends on how near the desired shade the color was originally.

In Blacks and Whites

Plain white and black change less with age than any other colors. Old cocks in white varieties often show increasing tendency to brassiness year after year, but these are mostly birds of indifferent breeding that have not purity of color well established. The black flecks sometimes present in young white birds become less troublesome as the birds advance in years: many will completely disappear in the first annual molt. Under suitable conditions for growing good plumage, a bird that is good black in its first adult coat should be as good or better in after years, until age begins to bring in white in the wings and tail. Some birds are as good as ever in color at three and four years.

Where Undesirable White Appears First

In all color varieties the most accurate forerunner of weakness in color is the appearance of white at the base of the hackle, at the juncture of neck and back, and at the base of the main tail feathers, and at the base of primary wing feathers. The persistence of white in these sections appears to be a survival of an original natural tendency to white undercolor in dark birds. It is only since 1915 that the Standard did not contain provisions especially lenient to weakness here in so old and long-established a variety as the Brown Leghorn. The serious objection to it in places where it is only seen by "digging into" the bird to look for it, is that it has a pronounced tendency to extend until it shows quite plainly. An otherwise good bird is not to be discarded for a little weakness in this point. If it has sufficient merit elsewhere it may win in spite of this weakness. But a bird that before its first annual molt begins to show white in the places mentioned is altogether unlikely to molt into a top-notcher.

Birds do sometimes come better here the second year than the first, because they are in better condition or given better care and feed. These are exceptional cases.

Whenever a breeder feels doubtful about a bird molting into good exhibition plumage, it is wise to give the bird the benefit of the doubt, and at the same time to give it the benefit of the best of conditions and care for the few months required to show just what it will do. It is especially desirable to do this in the case of birds that he has reason to think did not develop as good plumage as they should in the first year. It is also well to bear in mind that everyone else has in some measure the same troubles in deciding what birds to keep over, and that it is always particularly difficult to get cocks in good show plumage as wanted.

COLONY HOUSES IN YOUNG ORCHARD
A fine place for either old or young exhibition stock

CHAPTER VI

General Care of Exhibition Stock in Summer

Special Color Feeding—Cayenne and Carbonate of Iron—Scientific Observations on the Effects of Aniline and Other Dyes in the Feed—Some Effects of Common Feeds on Color—Effects of Heat on Growth of Feathers and on General Condition and Development—Effect of Lice on Development of Exhibition Stock—Effects of Sun and Shade on Plumage—Relation of Character of Soil and Climate to Exhibition Quality— General Sanitary Conditions

THE points of greatest importance in the handling of exhibition stock have been mentioned, and to some extent discussed incidentally in considering matters relating to the attitude of the exhibitor toward his birds, and the selection of the specimens giving most promise of winning quality. Many matters relating to the care and handling of exhibition birds are also treated more fully and in greater detail in books devoted especially to such matters as housing, breeding and feeding. It is presumed that the exhibitors, and prospective exhibitors, who read this book are sufficiently acquainted with special works on such subjects, to appreciate the necessity for limiting discussion of them in a book like this to the points in which the conditioning and fitting of exhibition stock involves the use of appliances and methods little used except by those who undertake to do all that can be done to make the birds they exhibit win.

Special "Color Feeding"

The writer has painstakingly read many an article—including some very long and elaborate treatises on the color feeding of exhibition birds—which at the outset professed to set forth the methods of obtaining color tones in plumage superior to those produced on birds fed only common poultry feeds, only to find that except as to the use of a very few articles long commonly used they had no specific information to give. Some of the most labored and pretentious articles by men professing a thorough knowledge of chemistry especially qualifying them as authorities on this subject are wholly speculative and suggestive and deal with theories not results.

The two special ingredients that have been quite extensively used are cayenne pepper and carbonate of iron. The effect of these is to (sometimes) intensify buff, bay and red shades. Carbonate of iron is also said to bring out more prominently the black flecks in plumage of these ground colors. Very little has been done or attempted in this line in America for two reasons: first, "the fancy" in this country did not tend to the extremes of cultivation of particular characteristics which in England led to experiments on this line; second, because the reports of results, and the accounts of the methods used in that country

WHITE WYANDOTTE HENS ON RANGE IN SUMMER
Scene at Sabrina Farm, Wellesley, Mass.

were not of a nature to stimulate interest in the practice among exhibitors here.

The practice of color feeding to improve or intensify buff color would appear to have been suggested by the effects on the plumage of white and very pale yellow canaries of cayenne and iron fed for their tonic effects. Given in only the moderate quantities appropriate for that purpose they were observed to have some slight effects in staining plumage as well as in contributing to the finish and luster. Heavier feeding of cayenne in particular, gives some canaries very high color. According to Lewis Wright, the application of this method to poultry began about 1897 in England when a very rich, deep buff became fashionable in the showroom. Knowledge of the use of cayenne and iron to intensify the color of canaries then led a number of exhibitors of buff fowls to try feeding them to poultry.

The results were admittedly far from uniform. Some birds would not respond to this kind of color feeding at all. In others the effects were visible only in patches. In still others there appeared to be a deepening of the buff color which made winners of specimens which but for this treatment would have been too light in color to be considered. The extent to which it is successful and the extent to which it has been and is practiced are very much in doubt. How difficult it is to get at the facts in such matters is shown by the situation at the present time in the United States in regard to the "processing" of red fowls to improve their color.

EARLY WHITE LEGHORN COCKERELS AT OWEN FARMS, VINEYARD HAVEN, MASS.
Photographed in June

Basing the belief on a peculiar, unnatural appearance of some Rhode Island Reds of remarkably uniform color, and on the testimony of persons who claimed to recognize in certain "processed" birds specimens which natur-

EARLIEST WHITE ROCK COCKERELS IN PERMANENT HOUSE AT OWEN FARMS
The photograph shows only a few of the trees in the yard in foreground

ally had neither the quality nor the uniformity of color they had when seen at certain exhibitions where they won, a considerable number of Rhode Island Red breeders are fully convinced that by one or more processes unknown to any save the few who devised and practiced them, the faults of red plumage can be remedied as effectually as creaminess is removed from white plumage by washing and bleaching. The presumption is that the change is made by some similar process. Further discussion of that is deferred to the proper place in a succeeding chapter. The point of interest here is the existing doubt both as to the nature of the process and the extent of its use. While some judges say that the peculiarity of the surface finish—the unnatural tone of color—led them to pass in certain cases birds which developments brought under suspicion of being processed, none pretend to be able by inspection of birds to determine whether they have been thus faked or not. No one except those who use whatever process or processes are employed can say whether all specimens that undergo a process to improve their red color show an unnatural finish in the surface, or only those cases in which the operation did not give perfect results.

With regard to color feeding the situation was precisely the same when it first came into use; and since the process ceased to be a secret the uncertainty as to the extent of its use remains, though common-sense consideration of all aspects of the matter tends to the conclusion that it is of much less consequence and value than those who are looking for some secret as the clue to special excellence suppose is the case.

The process and results of color feeding to improve buff color as given by Wright in "The Book of Poultry" 1902 are: "Color feeding for enriching buffs, if carried on at all, must be from the first beginning of the growth of the plumage to the very end of that growth, whether in chicks or molting adults. The regimen usually recommended is half a teaspoonful of cayenne, of which the cool kind is just as good as the hot, given every day in the soft food, along with two grains carbonate, or three grains saccharated carbonate of iron. A little fat should also be mixed with the cayenne. Merely for enriching bays and crimsons, as in a Partridge Cochin cockerel, there can be nothing gained by anything more than saccharated carbonate of iron; plain carbonate is cheaper, but saccharated is more readily assimilated by the animal system. On the other hand, there is some reason to think that iron may occasionally be injurious to a buff bird, in accentuating the slightest difference of color. While a uniform color would probably be slightly deepened in tone, any deeper patches, or the tendency to black specks, would probably be brought into stronger relief by iron, while cayenne would be less likely to have this effect. * * However, if by careful breeding an even and rich buff has been produced, there is not the slightest reason to believe it will ever be surpassed by color feeding; the sole question is as to how nearly inferior color may be made to equal or approach it."

Again Wright says in regard to the feeding of cayenne in particular: "In a long controversy upon the subject during 1898, several large breeders of buff varieties, whose word there is no reason to doubt, stated that they had experimented extensively, and given it up as yielding no result. It is absolutely certain that no better buffs have been produced in general since 'feeding' was practiced than were bred before it was known; but, of course, this does not prove that their number has not been increased by specimens which would only have been inferior otherwise. Also in the earlier experiments alluded to cayenne alone was used, whereas it is now believed that iron and some amount of fat are also advisable. Our own opinion is that in a certain number of cases there may probably be appreciable gain, but that it has been greatly exaggerated. * * * Attentive scrutiny of buffs generally at exhibitions, since publication of the color feeding process, has led us to surmise that the more usual effect when marked (for in many birds none at all is admittedly produced) may probably be to deepen the color in localized patches rather than all over—in pullets usually at the sides of the breast and of the cushion near the tail, some-

FAVORITE STYLE OF SUMMER COOP FOR GROWING STOCK AT OWEN FARMS
The chickens are cockerels of the second size on this farm in June, when this and other photographs used here were taken

times on the flat of the wing, the deepened color being a peculiar 'bricky' tint by no means attractive. It appears probable, however, that in individual cases color may be so gained without this patchy effect."

In further discussion of the effects of color feeding, Wright cites the experiments of Dr. Sauermann as reported to the Vienna Ornithological Association, published in German in 1890, and afterwards translated into English and published in the "Feathered World." These were made first by feeding cayenne to White Leghorns, and then by feeding aniline dyes combined with oil to canaries. In the first case cayenne was fed to twelve White Leghorns, in what amount is not stated. The effects began to show after eight or ten days, but only two out of the twelve birds were affected by it. In these there was some red color imparted to all soft feathers on the surface, but more on the hackle and breast than on the body, while the flights and tail remained white. The color appeared only on the surface where exposed to the light. The legs and feet were colored orange red. The same birds when fed cayenne in the next molting season became darker and more brownish where the color took effect. The yolks of their eggs were red, in some cases—it is said—bright blood red. Chickens hatched from the eggs showed no red in the down, but had some red in the first feathers. This faded out quickly if the chickens were not themselves fed with cayenne. All the chickens hatched from the two hens that absorbed the red color from cayenne inherited the capacity to do so. In this we can see how there would be a marked difference in the results obtained by color feeding by different breeders. Certain lines as well as certain individuals would respond to color feeding, and if a stock was predominantly of a line or lines that would do so, the beneficial effects might appear quite marked. Otherwise, so few individuals might be affected that there would be little if anything gained by the practice.

In feeding aniline dyes to canaries it was found that the effects lasted only for a short time. It appeared in this experiment that while the dye might have a desired effect to color the feathers, the oil combined with it seemed to check molting. When large quantities of cayenne were fed to fowls the process of molting also seemed to be delayed. When instead of oil, albumen was combined with a few dyes, some striking results were obtained. The dyes used were at first boiled with grain, but afterward it was found that the result could be obtained by mixing them in bread before baking and feeding the dyed bread. This is said to make the dyes so much less unpalatable that birds will readily eat the bread containing them. Only two results are reported: Fawn-colored pigeons fed methyl eosin became red. Yellow Budgerigars fed methyl violet became blue.

About fifteen years after Sauermann's results were given to the public, Dr. Oscar Riddle of the University of Chicago, made some experiments in feeding a dye known as Sudan III to fowls. While these were made primarily for the purpose of following certain food elements through the processes of digestion and assimilation, some of the results observed are of value in connection with the subject of color feeding to improve plumage. It should be kept in mind in considering these that the Sudan III has not a color that would give white or buff birds color a breeder would want. What effect it might have on dark-red birds is conjectural. Its use in the experiments of Riddle and others was simply to stain food elements in a manner that would make it possible to identify them in different parts of an animal organism. For the purpose of this experiment constant feeding of dye was not necessary. On the contrary, varying the length of the periods between doses, as well as the size of the dose, gave more definite observations as to a number of the effects.

Among the interesting results noted was that the red dye stained the yolks of eggs that were growing rapidly for a time proportionate to the size of the dose. As long as the dye was available the new growth of the yolk would absorb it, but what yolk was formed before the dye was distributed through the system would remain yellow, and what grew after the supply in the system had all been absorbed in the tissues would also be yellow until a new supply of color became available. The result was alter-

EARLY RHODE ISLAND RED COCKERELS AT OWEN FARMS IN JUNE
These birds had to be "herded" into the small treeless corner of their large yard to be photographed. No picture of them could be made in the shade they frequented most

nate layers of red and yellow yolk. This shows clearly why in color feeding it is necessary to continue the process steadily through the whole period of the growth of the feathers, with the doses so close together that the system will always have a supply for the growing feathers. Failing that, parts of feathers would be of their natural color.

The deposit of Sudan III in the body tissues was limited to the fat. As Riddle puts it: "It seems certain that the dye is bound to the fatty constituents, cannot loosen from them, and is dragged with them mechanically, so to speak, wherever they may go." He points out that in fowls the fat in different parts of the body differs in its tendency to take the dye. This seems to accord with Wright's observation that the feathers in different parts of the fowl differed in their response to color feeding. Riddle does not mention the staining of the plumage of hens to which the dye was fed, but says that young chickens fed Sudan III while their feathers were growing became distinctly red. His failure to observe a like result in mature fowls is readily explained by the apparent fact that it was not fed to them while molting. In the

COOP FOR GROWING CHICKENS USED BY JOHN S. MARTIN, PORT DOVER, ONTARIO
View with front open

experiments with this pigment it was found necessary to clean the floors of pens and brooders very often to prevent the staining of bills, feet and feathers, by contact with the pigment voided with the droppings. The bills and feet of the birds were very highly colored, but how much of the color was due to internal and how much to external applications could not be determined.

Of very great importance for its possible significance in other cases of color feeding is Riddle's observation that stained fat did not appear as available for reabsorption by fowls as the natural fat, and that chicks fed this dye "ate much more than those not fed the stain; they seemed always hungry and did not grow as well as the other birds of the same age and breed." He adds: "This, of course, may easily have another explanation than the one suggested"; but bearing in mind observation of others that aniline dye interfered with the molt of canaries, and that cayenne in some cases interfered with the molt, it is reasonable to conclude that Riddle's explanation is correct. Sauermann is said to have found that the aniline dyes when pure were not poisonous. It would seem though that, properly qualified, the statement of his finding is that, if not adulterated, the dyes, in such quantities as he used them, were not violent poisons. Anything that taken into the system interferes with its normal functions is a poison in practical effects.

On the whole, the cautious poultryman must conclude that whatever results attempts to dye the growing plumage by internal applications of more intense pigments than are found in the natural feeds of fowls may have toward the object of the practice, the possible bad effects are both too numerous and too serious to make it advisable to subject valuable birds to such processes. Before leaving the subject it is in order to say something more of the value of testimony to the effect that color feeding was in certain instances very successful. While poultry exhibitors as a class are quite willing to freely discuss all their practices verbally among themselves and with all actively interested in matters relating to exhibiting, there are always some who, either from a natural secretiveness, or the desire to appear to have more knowledge of unusual matters than others, like to make it appear that their results are accomplished by methods known only to themselves. There is also a common, and perfectly natural, tendency to "string" competitors or reporters who go after information too precipitately—not having acquired the art of taking it as it comes. On a bad day, when the attendance is light at a big show, and the exhibitors and regular attendants find things becoming too dull and monotonous, a gullible person who wants inside information can pick up from the banter of exhibitors and their friends a lot of misinformation about the extent of practices of this kind and the identity of those engaged in it. It is simply amazing how seriously some people take things of this kind that they hear, or happen to overhear.

There are also occasional plain cases of an exhibitor, who is getting a little ahead of his competitors in his breeding, taking pleasure in letting them get the idea that his advantage lies in greater skill at some other phase of the production of exhibition birds, simply to divert their attention from their own breeding work—"keeping them guessing" about his methods of fitting, so that they will not apply their minds too intently either to learn what they can of his breeding, or to make experiments of their own which close observation of his birds might suggest.

Some Effects of Common Feeds on Color

In the use of regular feeds these points chiefly concern an exhibitor: the general effects of substantial rations abundantly supplied; the effects of yellow corn on the color and finish of plumage; the influence of green feed; the results of feeding seeds especially rich in oil, or preparations of the oil from such seeds.

One cannot study and compare many practices and statements of advocates of special feeding for exhibition fowls without coming to the conclusion that, as a rule, there is little occasion for these when the birds have had the opportunity to make the best development they can upon good, substantial, ordinary feeds, under good living conditions. A comparison of American and English practices, with particular note of the greater importance English exhibitors appear to attach to the use of sunflower seeds, linseed-oil preparations, and the inclusion in special rations of articles which furnish more fat than the common feeds, also suggests that the absence of corn from many rations in England is the prime cause of the occasion for an apparently much wider use of special feeds, stimulants, etc., in preparing birds for exhibition than obtains in America. It would appear further that a larger proportion of exhibitors in England grow birds under less favorable conditions than most of our exhibition birds get, and so would in any case have a larger proportion of specimens that would need special treatment to bring them up. The most discriminate English writers on this sub-

SAME COOP AS ABOVE, CLOSED FOR THE NIGHT
Thorough ventilation and perfect security

ject are usually careful to caution their readers against more than a very moderate use of forcing and highly stimulating things for birds that are in good condition, observing that with such high feeding often overreaches its purpose.

The exhibitor should make it a rule to do all that can be done with ordinary feeds, good conditions and good

TWO-COMPARTMENT COOP FOR COCKS
Used at Owen Farms, Vineyard Haven, Mass.

care, before using any special feeds for special results. In general, corn is the principal grain used in American rations. The tendency of the meal and cracked corn—especially when made from soft corn—to heat and mold in warm weather makes it necessary for all poultry keepers to be careful about the quality of corn feeds they use, and the grower of exhibition stock needs to use more care than those growing stock for other purposes, because he has more at stake in the development of each individual. When good, clean, bright corn cannot be obtained, the proportion of corn in a ration should be reduced, or possibly it should be cut out entirely. The proper course must be determined by consideration of the quality of the corn to be obtained, of the conditions under which it is fed, and the results with the grains that might be substituted for it. A slight inferiority in the quality of corn fed makes little difference to birds living under natural conditions with abundance of green feed and fresh animal feed to make them resistant to the bad effects of the corn, but birds in bare yards, or in yards where the grass is so soiled that they do not eat it, and birds fed heavily of dried meat and bone by-products, are apt to be unfavorably affected by a steady diet of slightly moldy corn.

On the other hand, good corn is not only usually our cheapest grain feed, but is the grain most relished by poultry, and upon which they thrive best; and the writer observed long ago, and has repeatedly pointed out, that our best poultry keepers generally, in all lines, use corn much more liberally than the writers who reflect largely foreign ideas about the use of corn recommend. He has noted further that poultrymen's appreciation of the need and value of other articles of feed that provide fat is most marked where the use of corn is most limited. When good corn feeds are available the exhibitor may use all that the birds will eat with a relish and without getting over-

fat; except that he must be careful about the use of yellow corn he feeds. Birds differ in respect to the capacity to take pigment from corn, just as it has been shown that they do in susceptibility to the effects of cayenne and of dyes. Unless a poultry keeper knows that the color of the white stock he has is not affected by the amount of yellow corn he feeds, it is wisest either to use white corn, or use a ration containing little or no corn, for birds he wishes to exhibit; at the same time trying out some of the other specimens of the same breeding with yellow corn.

In considering and testing the effects of yellow corn on white plumage, one must also bear in mind that birds on good pastures, where they eat large quantities of tender succulent green grass and weeds, get a great deal of yellow color from these. Hence it will not infrequently happen that white birds given yellow corn, but not much high-colored green feed, do not appear to have the color affected by the corn, while those on range do. Conversely, when little or no yellow corn is fed, the coloring influence of the green feed may not be appreciable.

For buff, red and all dark colors, the feeding of yellow corn tends to give greater depth and lustre to the color. This may be further increased by using also sunflower seed, hemp, linseed meal and oil, and cut bone and other animal feeds containing fat. With the exception of the meat and bone feeds the advantage of such things is not so marked with birds getting liberal supplies of corn that there will appear to be any real advantage in using them. There is no gain in excluding corn from a ration because of its fattening properties and then adding these usually higher priced fats to supply the deficiency. In using meat feeds freely their tendency to give coarseness to comb and wattles, to add to the length and profusion of plumage, and also to increase the growth of frame in

REAR VIEW OF COOP SHOWN ABOVE
With hinged roof raised

young birds, must be taken into consideration. This result is desirable or objectionable according to the characteristics of breeds, or of particular lines or individuals. Where fine, small, neat combs and wattles are required meat feeds must be used sparingly. Wherever their use can have no objectionable effect, they may be used as freely as in feeding for growth or for egg production. To grow a full coat of feathers free from all faults due to undernourishment of the feathers, the bird should have

SUMMER COOP FOR EXHIBITION COCKS
Used at Willow Tree Poultry Farm, Beverly Farms, Mass.

some fat always in reserve throughout the whole period, thus insuring that all other feed elements are available for their specific purposes.

Effect of Heat on Growth of Feathers and General Condition

It has long and often been observed that early-hatched chickens which are growing their first adult coat of plumage in the summer rarely show the quality and finish of the later-hatched birds timed for the winter shows. The prime cause of this difference is in the fact that all the principal kinds of poultry are better adapted to cool climates than to warm climates, and that all the functions of growth are in a measure checked by extreme hot weather. This may affect both the shape of a bird and the texture and quality of its feathers. The chick coming to maturity in hot weather tends to be rather coarse boned and weedy, and to grow a comparatively light coat of feathers. If the weather is variable, hot and cool spells alternating, and the breeder fails to keep the birds in the best possible appetite and condition by the careful use of green feed and milk, and milk by-products, the plumage will grow very unevenly, often showing faults corresponding to periods when the bird was off its feed and out of condition.

To make the best development of which it is capable a chicken needs to have the last three months of its growth at a season when there is no weather hot enough to be at all uncomfortable for it. That is why our northern states produce so many more of the birds that win than the southern states. The southern breeder, and to some extent breeders everywhere that extreme hot weather may extend well into the fall, have much more to contend with in growing exhibition stock than those living where the summers are short, and hot weather rare after August.

Effect of Lice on Development of Exhibition Stock

Birds intended for exhibition should be kept wholly free from lice during the period of the growth of their plumage. On the not impossible theory that the body parasites of birds serve a useful purpose in stimulating the bird to take dust baths for the elimination of the lice, with the incidental result that the feathers are cleaned and scoured, a few lice on a fowl are not objectionable. Some lice are said to be beneficial also in consuming dead matter of skin and feathers. However that may be, it is easily apparent to those who examine birds for quality of plumage that lice are directly responsible for many faults in feathers.

They damage them in two ways: by eating a part of the web of the feather as it comes from the quill in the process of growth, and by so lowering the vitality of the fowls that both the growth and the pigmentation of feathers will show the effects. The direct damage is often seen in straight lines—about a thirty-second part of an inch in width—running in both directions from the quill across the web of the feather, and looking much as if something had scratched the web without disturbing its arrangement.

ROW OF COOPS LIKE ABOVE ON NORTH SIDE OF LONG HOUSE
Airiness of coop shows better here than in the other picture— you see right through them

That is the first impression it gives one not familiar with the phenomenon and acquainted with the cause of it. A moment's consideration is all that is necessary to convince any discerning mind that the damage could not be caused after the web expanded, and was undoubtedly caused by injury as it was folded in the quill. Observation of the lines and of the fact that while not always the same distance apart they are always parallel soon leads to the conclusion that they are produced by lice, each line representing a meal made of that feather. Nothing can be done to repair such damage. If it is limited to a few feathers the consequences are not serious, but too often toleration of lice results in so much of it that the plumage is badly marred.

The indirect damage by lice produces just the same effects as undernourishment. Undesired white in the plumage of wings, tail and neck, is often due entirely to one or both of these causes. In that case a bird that as a cockerel or pullet had such faults, may—by good care—be made perfectly sound in color at its first annual molt. A breeder who knows the characteristic of his stock can tell pretty well what birds that showed faults due to these causes in their first year

EXHIBITION S. C. WHITE LEGHORN HENS' SUMMER HOME AT WILLOW TREE POULTRY FARM
Roomy, airy house and large, grassy, shady yard

may be expected to come right in the following year with proper treatment.

Effects of Sun and Shade on Plumage

The ordinary changes which take place in a coat of feathers after it has grown are due partly to wear, but also conspicuously to the action of the sun and wind. All colors are in some measure affected by light and especially by constant exposure to strong sunlight. White birds that have been bred to a pure white are less affected by sunlight than any others. A few such birds can stand a great deal of sunlight without any marked deterioration of their pure white color. And in all colors we find a few birds strongly resistant to the effects of sunlight, but the number is so small that general practice in handling birds to grow good-colored plumage can take no account of them. Provision must be made to protect birds that are growing the plumage in which it is desired to exhibit them from the sun as much as is necessary to insure that their color will be as good as it can be. Experienced breeders think it wise to give this protection to birds that have the least need of it, the same as to the others, for though a "nonfading" bird may stand a lot of sun and still have better color than one that fades more easily has when given full protection from sunlight, it will not have its best color,

COOPS USED FOR R. C. BLACK BANTAMS AT WILLOW TREE POULTRY FARM

for comfort. The few exceptions to this general statement are where delicate colors like buff are best preserved if the birds are never allowed to remain in strong sunlight, except for occasional brief periods.

As a rule, poultry having access to suitable growing shade will keep in it as much as is necessary to prevent the fading of plumage while growing. The shade that serves this purpose however, is not a limited shaded space with perhaps just enough room for all the birds to get under cover when they have nothing else to do and no place else to go except to a house not particularly comfortable on a warm day.

The shade that meets the requirements is the shade of a roomy yard with one or more fair-sized trees giving ample shade in some part of it at any time of day; or the shade of an orchard or a wooded lot, or a cornfield or raspberry patch. There must be room in the shade for the birds to live and something more than just the shade to attract them to it. Many exhibitors plant corn or sunflowers in their yards so that practically the whole yard is shaded. Many who have not time or inclination to cultivate the yards allow them to grow tall weeds which make splendid shade, though they may give the place a rather unkempt look.

LEGHORN CHICKS AT WILLOW TREE POULTRY FARM

and may easily be beaten in the showroom by a bird not naturally quite so good in color, but at its best.

Buff and red show the effects of sunlight most plainly on the solid-colored birds; but protection from sun is just as necessary to hold the full tone of ground color, and the richness of dark markings on all varieties of the black-red color patterns—Brown Leghorns, Partridge Cochins, Plymouth Rocks and Wyandottes, Dark Cornish, Golden Campines, Polish and Hamburgs, etc. Blue birds generally fade easily and rather unevenly. Some black birds fade badly, becoming when much exposed to the sun a dingy, reddish brown, which may not look near as bad as it is to an exhibitor whose birds all fade until he shows them beside some good black specimens. In all the black-white varieties—the Dark Brahma, and all the silver laced, penciled, spangled and barred color varieties, both the purity of the white and the intensity of the black color are best preserved when the birds live much in the shade.

Natural shade, or as it is sometimes called "green shade," is much better for exhibition stock than artificial shelters. Wherever it can be obtained it should be used. The advantages of natural shade are in the conditions that go with it. The birds are outdoors, on grass or cultivated ground, and they get enough of the drip of dew and rain from the leaves over them materially to benefit the growing feathers. The shade requirements for exhibition stock are on the whole no more than all poultry ought to have

Where artificial shade must be provided it is a good plan to use coops with wire sides all around, so that the circulation of air is always good. Shades of a single thickness of burlap or cotton cloth over open spaces are often used. The sufficiency of these depends on the circumstances. Sometimes a yard is so located that it gets natural shade, or the shade of a building a part of the day, and the artificial shade is needed for only a short time on bright days. Late in the season when the sun's rays have lost much of their strength such shade may be sufficient. But

A GOOD COLONY COOP FOR EITHER YOUNG OR OLD BIRDS

for ample protection against strong sunlight something more substantial is needed. Where it can be obtained a piece of old sail or tent cloth of medium to heavy weight will answer, especially if it is of good size. For shade alone it is not essential that a roof or cover be waterproof. In fact, until it gets cool in the fall, a chicken coop with plenty of cracks in the sides and a leaky roof is more the ideal place for poultry than the substantial wind and water-tight structure necessary for the comfort of most breeds in winter. Old boards, not worth using for any other purpose, can often be made to serve purposes of shade for several years by nailing them on light supports.

Even with such special provision for shade it is a good plan to make the roosting quarters as comfortable as possible on hot days, and where it is practical to do so to use space under a poultry house for a shady loafing place for the birds. It is quite the common thing throughout all our northern states to find poultry houses built especially to suit winter conditions and not admitting ventilation enough in extreme hot weather. Nearly all the permanent poultry houses in the North would be better sum-

The advantage of this depends upon the houses and sheds being roomy and comfortable, and light enough for the birds to see well anywhere in them when the sun is excluded by curtains or awnings. As shading to produce the desired results must be carried on through about three months birds kept indoors for the most of the time must be made very comfortable, and all their wants fully provided for, or loss of condition in other ways may more than offset the improvement in color. Keeping birds in the shades also helps the color of white ear lobes which generally tan more or less, even on the most naturally white, if too much exposed to sun and wind. Much exposure to wind also has the effect of roughening the surface of the ear lobe.

While the effects of wind on the plumage are not usually of much consequence, exhibition stock should not be kept in exposed wind-swept situations unless amply provided with shelters and windbreaks. Young chickens in particular, will fail of their best development if they have constantly to battle with the wind. The natural home of the fowl is in and near thickets where the wind usually

J. V. McCONNELL'S COCKEREL HOUSE, GARDEN GROVE, CALIF.

mer houses if built so that a considerable part of the back —enough to give very good circulation of air through the house at any time—could be opened in the summer as a large proportion of the good poultry houses in the South, where they must provide for long periods of hot weather, are. Space under a house is oftenest available when colony coops are used, which can easily be raised a foot or more from the ground. This makes a much cooler spot than the space under a single thickness of wood or other material.

Breeders of buff birds in particular, who wish to insure that the plumage of the birds is not in the least affected by exposure to the sun, make a practice of keeping the birds in their houses or in open-front sheds, except for a little while each morning and evening, and on cloudy days.

has little sweep and force, and though birds kept for utility purposes may adapt themselves to rough conditions of life, such conditions are not favorable to the finish in form and feather desired in an exhibition fowl. With reasonable protection from the elements, exhibition birds should not be coddled. The idea is to have them live as natural a life as is consistent with protection from accidents or conditions that would mar their perfection.

Character of Soil and Exhibition Form

Birds that are kept on dry porous soils—even when these bear some vegetation—do not, as a rule, have the smooth surface and good color of shanks and feet which appear so attractive in birds on grass and on moderately moist or even wet land. Alkali soils are especially bad for the feet, absorbing much of the oil from the skin.

Birds differ greatly in ability to keep good condition of the shanks and feet under unfavorable circumstances, but few retain really good color and surface in very dry places, or when constantly on dry floors and in yards containing much coal ashes.

The damage caused by too much contact with dry earth, ashes and the like, is not irreparable. External applications of oil will greatly improve their condition and appearance, but the legs that have to be so treated never have the natural finish. This may not be a serious disadvantage in competition. In local shows, where practically all the local birds are on a parity in this respect, a bird from outside that is good in shanks and feet will have no advantage on that account except in close decisions. But in sending a bird that is poor in color and condition of shanks and feet to a large show, where most of the birds exhibited are good in that section, an exhibitor will often find that, unless it has outstanding merit somewhere else to catch attention, its lack of color in shanks and feet so detracts from its appearance in a class generally good there that the judge is prepossessed against it at first sight.

Fine condition of the shanks and feet is a matter of prime importance entirely apart from appearances. When the skin is allowed to become dry the scaly leg mite establishes itself much more easily under the scales which are then slightly loose. Also, dirt of all kinds works farther under the scales and is more difficult to remove. Poultry keepers generally, unless the point has been especially brought to their attention, do not realize that the scales on the shanks of fowls are modified feathers, and are normally molted like feathers. In many cases they do not molt—as seems apparent by the fact that a rough scaly shank does not have smooth new scales at the end of the molt. The reason for the failure to shed the old scales and grow new ones would appear to be that the condition of the shank—and perhaps also the drain on the juices in it made by the mites—interferes with the processes.

There ought not to be any scaly leg among exhibition stock but, if there is, the time to begin systematic work to eradicate it is before the birds begin to grow the feathers in which they are to be shown. Where there is much of it on the premises and the young chickens become infected early, traces of roughness will begin to show at the junction of the foot and shank by the time they are three or four months old. For both chicks and old birds ordinary cases of scaly leg are easily cured by simply dipping the legs to the hock in a mixture of raw linseed oil and kerosene, using equal parts of these ingredients for ordinary cases, less kerosene for very mild cases, and more for severe cases, and repeating the application at intervals of a week or so as long as seems necessary. When done before the new plumage grows it is not necessary to take pains to prevent the oil from coming in contact with the feathers at and above the hock. What little may get on them will not give the birds any particular discomfort and the feathers will soon be shed.

General Sanitary Conditions

The coops, houses and yards, into which growing and molting exhibition stock go, should be in good sanitary condition at the beginning of their several months' occupation for this special purpose. Most breeders are careful to have them so for birds that are to be shown in winter, because unless the renovation before the birds go in is thorough there is little chance of its being done at all until the next spring. Many exhibitors, however, are inclined to give very indifferent attention to the matter of providing thoroughly clean quarters for lots of birds that will come out of them before winter, thus giving the opportunity for a complete overhauling at that time. Carelessness of this kind is much more likely to affect old birds and young in permanent yards than young birds colonized on the range, for the coops of the latter are generally well cleaned before the birds go into them, and are either placed on fresh ground or remain on ground which has had no poultry in the early part of the season.

Where the birds are in permanent houses and yards the most thorough clean-up of the year should be made at this time. If the houses have wood or cement floors they should be made as clean as shovel and broom can make them. Where the floors are of earth what is plainly mixed droppings should be removed and new earth or sand put in. If there is going to be opportunity to clean up again before winter, it may be as well not to put in more new earth or sand than is needed to give good conditions at this time. Then with removal of the droppings and the filling of the floor to the desired level, the floor is in good condition for the birds that use the house through the winter.

The ceiling, or underside of the roof—as the case may be—and the walls should be brushed down, all movable nests and roosts-parts having first been removed; and if the house is old, or has been at any time infested with parasites, the whole interior should be sprayed with a suitable disinfectant. The floor of cement or wood should be thoroughly disinfected, and special attention should be given to saturating any cracks or joints about roosts and nests that have harbored mites or could do so.

After the floor has had time to dry out thoroughly an inch or two of dry sand or loam should be spread over it, and over this a litter of clean straw, hay or leaves, if these are available. The advantage of these materials over shavings and other short, fine material is that when not broken up so fine that they pack their light contact with the plumage helps to polish the feathers and also to keep them clean. To get this benefit the litter must be reasonably fresh, loose, and deep enough so that the birds working in it are nearly half buried. Many exhibitors use deep litter only in the few weeks of special preparation just before showing, and many do not use it at all. Practice is determined largely by what is available, and by the ordinary practice of a poultry keeper in the use of litter throughout the year.

Yards and runs that are to be used for exhibition stock, and that are not in grass, should be freshened up as much as possible. It is a good plan, wherever practical, to have such land turned over early in the spring, and planted to a crop that will make either shade or green feed for the birds when they are put on it. Small yards, or any land where the droppings are seen accumulated on the surface, should have what can be taken up removed before the soil is turned over. Growing crops of itself does not sufficiently purify land that carries considerable poultry every season. If the droppings remain on the land and are turned under no deeper than the ordinary spading, forking or plowing of poultry yards puts them, several seasons of crop growing may be necessary to purify the soil.

CHAPTER VII

Preliminary Work in Fitting Exhibition Birds

Gentleness in Catching Poultry—The Right Way to Catch a Fowl—How to Catch Turkeys and Waterfowl—How to Hold Birds for Inspection—Taking Birds from Coops—Returning Birds to Coops—Penning Birds in Small Space to Catch—Supervising the Molt—Partial Molting—When Premature Molting Is an Advantage— When to Remove Unmolted and Damaged Feathers—The Humane Way to Pull Feathers

THE practices and points discussed in the last chapter generally relate either to things which a poultry keeper can see without handling the bird and closely examining it, or to things that come up in poultry practice in other connections as well as in the special consideration of exhibition birds. We cannot draw clear lines between the practices of exhibitors and of others, and sometimes convenience dictates consideration of matters of peculiar interest to exhibitors with equal interest to the breeder whether he exhibits or not. Again it may seem preferable to emphasize a matter that ought to be of common interest by treating it in a sense of special interest to exhibitors.

Gentleness in Catching Poultry

Such a matter is the method of catching and handling exhibition poultry. All poultry ought to be handled just as carefully, but the penalties of mishandling specimens of only ordinary market value are usually in each case so light that the consequences are not brought home to the offender as they are when rough treatment either seriously injures an exhibition bird or so impairs its confidence in men that it will never show to good advantage. In counseling gentle handling under all conditions, I am not unmindful of the fact that some birds do not respond at all to the most gentle, skillful and persistent efforts to train them to allow themselves to be caught readily, and to stand and pose as the person handling them wishes. Many specimens that have exhibition quality never make good show birds for this very reason. As stated elsewhere, submission to handling and posing tends to become second nature with birds descended from long lines of prize-winning ancestors. But even with these the response to the advances of the person handling them is influenced very much by his attitude and method. Only the most docile birds become submissive under handling that makes them uncomfortable or injures them.

The three points to be kept in mind especially, when catching exhibition birds, are "USE NO HOOKS;" take your time; and, handle only one bird at a time.

The long wire hook much favored for catching fowls should be absolutely prohibited from the premises where exhibition stock is kept. It has been for some years a question in my mind whether the hook ought not to be barred from all poultry yards, for the reason that the use of it seems to be always accompanied by the abuse of it. Still there are occasions when it is a great convenience and the damage it may do has no serious consequences because the injured bird is destined to be killed shortly. But with exhibition stock the use of the hook is so entirely out of place that there is no room for an argument for its use in emergencies. The objection to the hook is that even with the greatest care, and when the wire is wrapped with soft material where it comes in contact with the shank of the bird, many birds have shanks badly bruised, or the leg strained at the hock or hip.

The Right Way to Catch a Fowl

One should not attempt to catch a bird until he has it where he knows that it cannot get away from him, and where with a little patience he can maneuver it into a position where he can take hold of it securely without danger of twisting a leg, or loosening a lot of feathers. In pens and coops up to about twenty feet square birds can easily be driven into a corner. Then by advancing straight toward the corner, stooping a little, and with hands outstretched to stop the bird if it tries to run out at either side, advancing cautiously and quietly, one can get close enough to touch the bird. From the time one starts to catch the bird he should keep his eye on it, and be ready with a movement of the hand to turn it back as it starts to one side, and as quickly to sign with the other hand when it tries, as it usually will, to make a

THE RIGHT WAY TO CATCH A BIRD IN THE YARD
Stopping the Bird — Taking Hold With Both Hands

break in the opposite direction. After being turned and checked a few times, and finding the pursuer always has his eye fixed on it and is coming nearer, the bird of ordinary natural docility will stand still and wait developments.

Now an expert in handling birds will simply advance one hand until it touches or almost touches the bird on one side, the hand being open with palm toward the side of the bird and the movement directed with the idea of holding the wing close to the side when the other hand advances the fraction of a second later from the other side. In this way the bird is caught and held firmly between the two hands, its head being toward the catcher. Now by quietly and easily slipping one hand down and under the bird, slipping one finger between the legs and so grasping them both just above the hocks, the bird is held so that it is secure and realizes that struggles are useless, and at the same time being comfortable is not alarmed and disposed to struggle to free itself.

The point in catching a bird in this way is to hold its attention and anticipate its every move until it realizes that it is cornered, and then to handle it so gently yet firmly that it has no occasion to feel frightened. In general, a bird that is cornered keeps its head toward the person trying to catch it. Often as a bird moves about the floor or between corners if it slips past the catcher's guard, one has a chance to catch it as it is going from him, in the same manner as described—that is, between the hands. The surest way in such cases is to advance one hand over the bird's back and a little forward, with the thumb turned toward you, instead of out, and at the instant of touching the bird wheel it about so that it faces you, and just as it comes around bring the other hand into play and catch it as previously described. With a little practice it is much easier to do this than to catch the bird wrong end first and then turn it in the hands to get proper hold.

After a bird has been caught and handled a number of times it is often easy to catch between the hands from any position when the catcher is near enough to touch it, but comparatively few birds will submit to capture without making the pursuer do a little work. Poultry are just like cows and horses in this, and some of the most docile seem to enjoy dodging capture when they have the opportunity to do so.

After one becomes accustomed to catching and handling birds there are other methods that may be employed when the bird is in a position that admits of using them to better advantage than catching between the hands. Thus one may have a bird in such a position that it is easier to get hold of it by the legs in much the same position as when the bird is held in the hand; or he may be able to catch it by the wings, close up to the shoulder. A novice in either of these holds is very apt to pull feathers out. One who becomes somewhat expert in them, or either of them, not only is surer of getting a quick firm hold, but knows by the sense of touch at the instant of contact with the bird whether he has such a hold, and if he has not does not tighten his grip, but lets the bird slip away, and tries again.

NEAT MARKET SHIPPING COOPS USED WHEN SORTING AND MOVING BIRDS AT OWEN FARMS

After handling certain birds a number of times and getting an idea what they will do one may sometimes with very little risk take liberties in the way of holds that would bring disastrous results with a frightened bird. Thus a bird that is well broken to handling, but likes to play its master, may be caught by the tail, or even by a handful of soft feathers without pulling any of them out, if the man and the bird understand each other. I have seen a bird caught by the neck submit gracefully, though a bird not used to handling may injure himself or his captor by struggling if the latter does not quickly let so insecure a hold go.

One who has little experience or skill in handling birds without injuring them, and particularly without loosening feathers, is often deceived when he sees an expert handling them by the apparent roughness of some of the latter's movements. I would not say that all persons who have handled many exhibition fowls are not rough when they appear so. I have known men who exhibited for years and never learned to handle birds right. They were rough even in their application of right methods. But the greater number of seasoned exhibitors rarely do anything that injures a bird because even in their most strenuous efforts to catch a bird they always have vividly before their minds the necessity of catching it without doing any damage. I have seen a breeder (probably with some foot-ball experience) "fall on" a bird that instead of trying to slip past him at one side made a dive forward

JOHN S. MARTIN'S CARRYING COOP

under his guard, and come up with a proper hold and scarcely a feather on the bird ruffled.

How to Catch Turkeys and Waterfowl

Turkeys are caught and handled as fowls are but with the difference that in catching large birds it is a matter of using the whole arm, not merely the hand. In handling ducks and geese it is often easiest to catch the bird by the neck near the head, and then as quickly as possible a body or leg hold. The necks of these waterfowl are so strong that it does not hurt the smaller birds to be carried by the neck, though the heaviest birds seem to be a little inconvenienced by that treament—even if not actually injured. Ducks should never be caught by the feet for their legs are very tender and easily broken. Geese have stronger legs than ducks, yet it is not a good plan to catch those that may struggle by a leg. In holding waterfowl in the hand the shortness of the

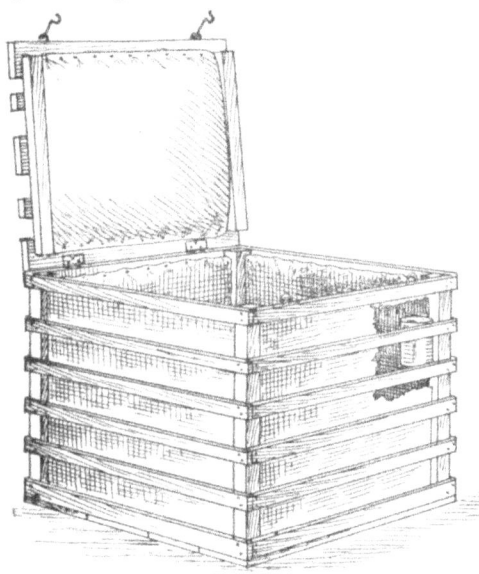

A GOOD STYLE OF SHIPPING COOP
Very handy in sorting and moving birds

thighs, and the width between them is so great that one cannot get a firm hold of both legs instantly as he may with chickens, and often has to use both hands to adjust the legs so that they can be firmly held in one.

How to Hold Birds for Inspection

A small, light bird can be held in one hand while being examined. When the weight of a bird is such as to tire the hand, or when constant handling of medium and light birds tires the hand and arm, one handling and examining birds either sits down and holds the bird in his lap, or stands with one foot on a box or chair, or perhaps on the edge of the platform on which low coops stand at a show, with his arm and the weight of the bird resting on his knee. The bird held by the legs as described can be examined all over, turned in any desired position, and passed from hand to hand as desired, so that whichever hand is most convenient may be used in opening the wings or parting the feathers. An excitable bird will sometimes struggle a little, and is more apt to do so if the grip in which it is held is unsteady—alternately tight and relaxed.

Taking Birds from Coops

In taking birds from an exhibition coop the best way is to turn the bird facing you, then slip the hand under it and secure the hold on both thighs well down toward the hock; then with the weight of the bird resting on the hand and wrist draw it gently toward you and through the door of the coop. A bird that is broken to handle well will submit to this readily. An ill-broken or shy bird will often crowd into the corner of the coop, or may fly and strike the top of it. In the first case a little coaxing will usually get a bird in position where he can be properly caught; but if a bird is stubborn it may be necessary to get both hands in the coop, using one to bring him in position where the other can catch and hold him securely. Also in case of a very large bird, or one that spreads its wings, both hands must be used. The person handling must use judgment and while in a general way proceeding as described must always keep cool, keep his wits about him, and remember that the all-important part of the process is to do what is necessary with the bird without injuring it or its plumage.

Returning Birds to Coops

In returning a bird to an exhibition coop it is usually passed in head foremost. The method of holding it while doing this will depend somewhat on the bird. In some instances if a person holding the bird will turn it a little on his arm and stand in such a position that it is just before and facing the door of the coop, it may simply be released and then will step quietly into the coop. Others it is necessary to take between the hands with the head away from the person holding, and pass through the opening and let them feel their footing on the coop floor before releasing them.

In handling birds it is sometimes convenient and sometimes necessary to put them in and take them from coops having the opening in the top, as the ordinary shipping coops and carrying coops have. In all cases, the opening should be large enough to admit of easy passage of the bird, when held between the hands.

Penning Birds in Small Space to Catch

What has been said so far applies to the catching of one bird or a few birds from a flock, and to their handling in a leisurely way for inspection, etc. It often is necessary to go over the birds in a pen, or to move a lot to new quarters. Wherever many fowls are kept much such work must be done by daylight, and arrangements should be made so the birds can be caught and handled easily.

In the ordinary small colony house as used for growing chickens the person catching them can often pick them up one by one as they crowd to one end of the coop. In wider coops, where the birds have the opportunity to dodge to either side of him, the best plan is to have a wire-covered frame six or eight feet long and from two and a half to four feet wide, according to length or to the kind of fowls to be caught, which can be put across a corner of the pen to keep fowls in it while being caught. A person working alone should have a small panel so that standing in front of it, or standing astride of it, he can easily reach any bird in the space it encloses. Then with the coop in which the birds are to be put conveniently beside him, he can catch the birds leisurely and without exciting them.

When there are two or more persons at the work a higher and longer panel may be used and the person catching can pass the birds as caught to one outside. The work should always be done deliberately. No more birds should be cornered at one time than can crowd into the corner, as they usually do at the first moment, without some being trampled. The more birds there are in a group the greater the tendency to crowd and stampede. The first lot of birds cornered usually shows most excitement. As the work proceeds and the birds become accustomed to the movements of those at work they take the proceedings in a matter-of-fact way. Some poultrymen use a box or coop into which the birds are driven until the coop is full. Where there is room to handle such a coop and two or more men are working together it may work very well, but in general a panel is more conveniently handled and more satisfactory.

In all the work of catching, handling and moving birds they should be handled carefully, one by one. As a person has two hands, he may hold or carry two birds—if the birds are small enough to be held securely with one hand—but that is as far as he should go. Trying to carry as many birds as can be carried by the legs down, and some by only one leg, is a practice that ought never to be used with high-class stock. Either a coop should be provided for carrying or the person moving the birds should take one or two (according to whether the birds can be absolutely controlled with one hand if they struggle) at a time.

Supervising the Molt

In discussing the effects of "color feeding" mention was made of the fact that in many instances it appeared that the use of such ingredients interfered with normal molting. Extended observations on conditions associating with molting and with failure to molt normally or completely have never been made at all systematically. The most thorough study of the subject of molting we have was made at Cornell University Experiment Station some years ago, at a time when forcing the molting of hens by "starving," feeding very light for several weeks, and then following with a period of very heavy feeding of rich rations, was being widely recommended as a means of getting old hens to lay at the season when eggs are most scarce—in the late fall and early winter. Many reports had been published by individual poultrymen claiming to have been perfectly successful in efforts to

COCKEREL HOUSE OF L. J. DEMBERGER, STEWARTSVILLE, IND.

control the molt in this way. Others had tried it without apparent results. The experiment did not show uniform results, nor any final advantage in measures to hasten the molt of laying hens. In connection with it, however, a number of observations on the growth of feathers were made which are of considerable interest to exhibitors, giving more definitely various facts which old exhibitors knew in a general way, but not so fully as to details and the time required for the growth of feathers.

In the Cornell experiments an effort was made to stain the first feathers that appeared on any part of the body, and when these dropped out to stain with another color those that succeeded them. The first set of feathers were stained red, those that followed them black. The red-stained feathers in the wings and tails of White Leghorn chicks had all disappeared, molted out, by the time the chicks were eight weeks old. At thirteen weeks the black-stained feathers which succeeded them had all been molted and white feathers, which were allowed to remain unstained, were growing in their places. From about thirteen weeks to between five and six months no feathers were molted out. Then the feathers were all gradually shed and the first mature plumage grown. Hence, it appears that a bird in its first year grows in succession four coats of feathers.

It was observed that in general the oldest feathers molted first, the rotation in the molt following closely the rotation of the appearance of the first feathers on the chick, except that in growing the first adult coat of feathers the stiff wing and tail feathers, which in the earlier coats were grown first, were in this coat grown last.

This peculiar fact explains why so often young birds, particularly pullets, that seem to have quite completed their molt as far as the body plumage is concerned, are often found to have the chick feathers still in the wings and tail. If these are bright, clean and good in color birds are sometimes shown with them in, especially if it happens that the situation is not discovered until so near the date of showing that it is not possible for

INTERIOR OF A LONG COCKEREL HOUSE

the new ones to grow out in time if these are removed. There is always the possibility at this stage that the feathers may drop out without assistance, leaving the appearance of the bird more or less marred. However, there will be little trouble on that score if the exhibitor takes the precautions shortly to be described.

In cockerels the "furnishings"—that is the decorative male feathers—usually finish last and the cockerel is from one to two months later (according to breed development) than his sisters in completing the growth of his first adult plumage, so the tail and wing feathers are not so likely to remain unmolted when the bird seems otherwise fit to show. The time required to grow this first adult plumage is about three months, but it appears to vary considerably in individuals, as also does the time of the first annual and later annual molts. No definite observations on the time of growing the first adult coat have been made. We only know that it is approximately three months, which brings the age at which fowls are ready to show in full adult (first) plumage at from eight to ten or eleven months, according to rate of growth and size and sex.

In the Cornell experiment it was found that the average time required by 215 White Leghorn hens to molt and grow a new coat of feathers was 95 days. But this was for flocks containing hens up to their fourth year. The average time for the yearling hens was 82 days, for the two-year-olds, 101 days, and for the three-year-olds, 104 days. The report states that there is "a wide variation in the length of time it requires individuals to complete the molt," but does give an average for a lot of about forty hens at 65.5 days. From this it would appear that the common observation of exhibitors that hens frequently made a complete molt in six weeks or a little over is borne out by the experiment. It was found further that the latest molting hens usually molted the fastest. No positive conclusion was drawn from this observation, for the late molt appeared to be in a sense an abnormal result of the success of attempts to get heavy egg production. The statement is made that "When fowls molt naturally and well, one should scarcely be able to notice that the flock is molting, except that the shed feathers are found in large quantities about the place."

Whatever attitude the breeder for heavy egg production may take in regard to the advisability of prolonging laying as long as possible with the result that the hens are made "late molters," the breeder of exhibition fowls is interested first of all in having his birds lead a normal life with all functions operating normally and seasonably. With respect to molting, as in the matter of feeding, the wise general policy is to use special methods only in emergencies — and by proper forethought to avoid emergencies. As will appear in many instances to be cited in succeeding chapters, most special processes are more or less uncertain (in the nature of chances) and of temporary effect—and, sometimes, undesirable after-effects. An exhibitor who has a large stock of high exhibition quality from which to select usually has yearlings and older birds beginning to molt all the time from the beginning of summer till very near winter, with the average normal bird beginning to molt about midsummer, and having the plumage full grown about the first of November.

FRONT OF COOPS IN COCKEREL HOUSE
Feed and water cups outside

Partial Molting

Any change in the life of the fowl at the beginning of summer, about the time when breeding pens are usually broken up, and the exhibition quality birds that are to be kept over are given new conditions of life—most often good range in a large flock—seems to stimulate further egg production, or to check production and start molting, according to the tendency of the individual, or perhaps also according to its condition. It is not possible to say in advance how hens approaching their first annual molt will be affected, but after a bird has been seen through one molt the person attending it knows about what to expect of it in any subsequent year. The birds that start molting in June and early July

ONE WAY OF ARRANGING EXHIBITION COOPS FOR CONDITIONING BIRDS

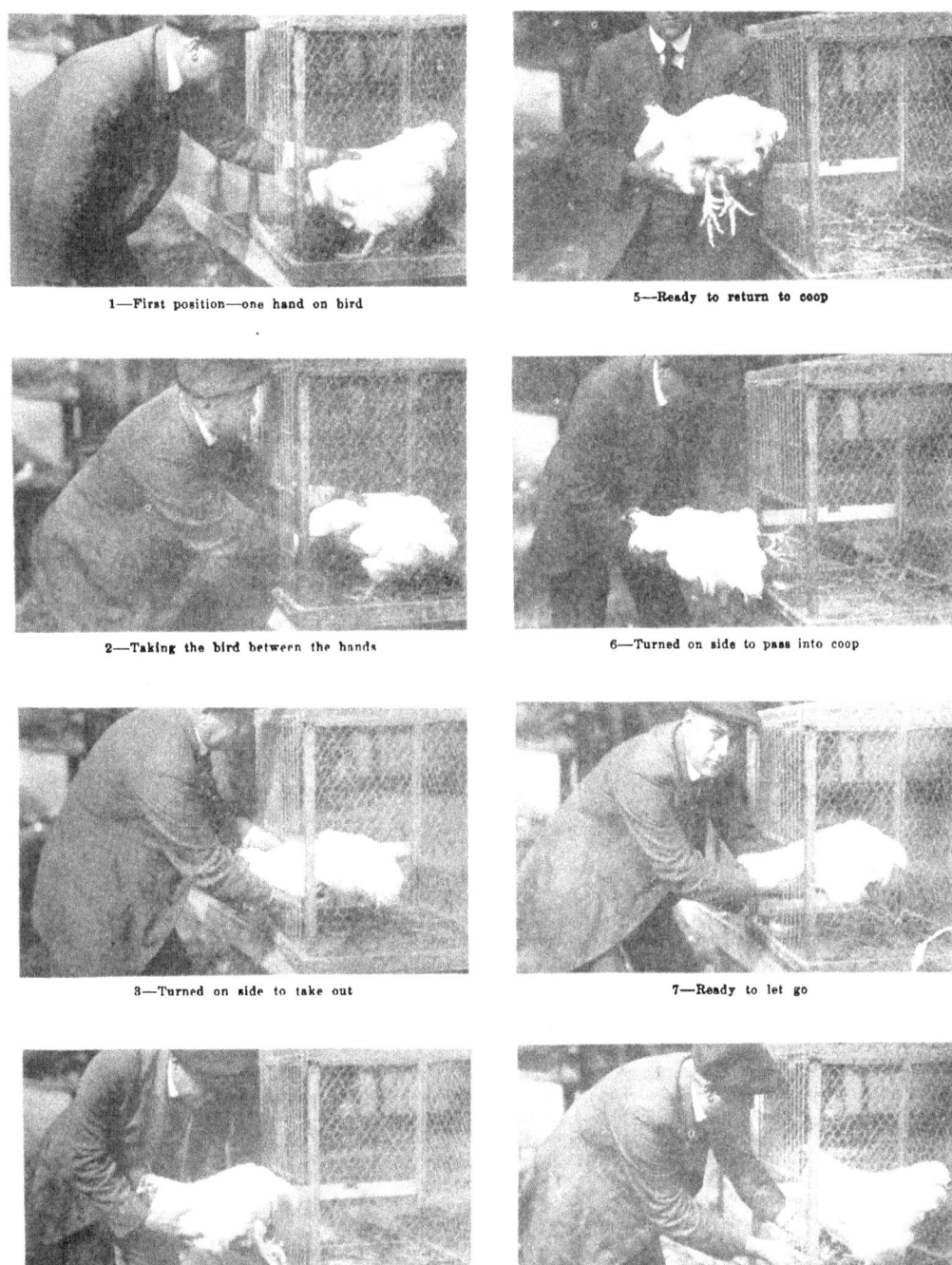

JOHN S. MARTIN DEMONSTRATING METHOD OF TAKING BIRD FROM AND RETURNING TO COOP WITH HEAD TOWARD HANDLER

will be in fair feather for the fairs in the last of August and September.

Many hens that have been allowed to incubate late in the spring, whether they run with chicks after hatching them or not, will then begin to molt. To superficial and never critical observation these birds may appear to make a complete molt, yet it is doubtful that many of them do so, and the common observation that hens that molted early in the summer often go through a partial molt again in the fall is probably due to failure to note that the first molt is incomplete, and what appears as a second partial molt is in reality the renewal of the feathers not molted earlier. Not only is there this somewhat conspicuous division of the molt in the early molters, but many later molting hens do the same thing, as far as the incompleteness of the main molt goes, and may either complete this molt in winter or carry the old feathers until the next annual molt. Besides the occasional unmolted old feathers scattered through the soft plumage, large patches or sections of the plumage may be retained when the greater part is molted.

When Premature Molting Is an Advantage

While it is possible that in any case an early molt might be started by an abrupt change in the life of the bird, the only case where there seems to be any real advantage to an exhibitor in a premature molt is when it is of prime importance to him to win at a fall show, and he cannot get the birds he needs in condition any other way. For reasons already pointed out, the plumage that is grown in warm weather is not usually as good as that grown in the cool fall and early winter months, hence, it is a disadvantage to an exhibitor to have a bird that he wants to exhibit at a late show molt early, for not only will the new plumage begin to show the effects of wear and weather long before the show season is over, but if a partial molt occurs afterward the new feathers will be brighter than the rest of the plumage and it may be impossible to get the bird in good show condition.

Old cocks are especially troublesome about molting. I am of the opinion that in the greater number of cases the cause of this is using too many hens with males in the breeding pens, the drain on the vitality of the bird being so great that it neither molts normally nor is fit for breeding service early in its second season.

When to Remove Unmolted and Damaged Feathers

Estimating that a young immature bird or a yearling will take about three months to grow a full coat of feathers, and an older bird will take somewhat longer, the time a bird begins to drop feathers freely should be noted and, about two months before a date at which it is wanted to show, a thorough examination of the whole plumage should be made and all the unmolted and damaged feathers removed. If the bird is backward and slow in molting and there are many such feathers, they may not be fully grown when the bird is to be shown but they should be near enough to it so that their lack of development will not be a serious handicap. The only way that birds timed to the day for particular shows can be secured with any regularity or certainty is by having enough of them that will be ready about that time so that among them some will be right at show time. The supposed wizardry of some uncommonly successful exhibitors in always having birds ready for shows at which they exhibit is about half in timely preparation and half in numbers of birds from which to select.

When only an occasional feather is to be plucked, all the old and damaged feathers on a bird that has commenced to molt may be removed when the examination for such is made. When all the stiff feathers of both wings and tail have to be pulled, or when most of the feathers in a section that is backward in molting need to be removed, the number to be taken at one time should be regulated by the ease with which they can be drawn. It is not usually advisable to take all the large stiff feathers at once, even if they pull with little resistance, because that is too much of a shock to the bird and may affect the other feathers that are in various stages of growth, making many of them faulty. The better way is to pull a few every day, taking one or two of the corresponding feathers in each wing, or corresponding feathers from the opposite sides of the tail. In the same way, if the hackle or the back feathers hang in molt they should be removed a few at a time.

The Humane Way to Pull Feathers

In pulling feathers the most effective and the right movement—the one causing the bird least pain and discomfort—is a strong quick pull, made with the bird or the part from which the feather is to be removed held firmly and steadily, and the pull at the angle at which the feather leaves the body, so that it is a direct, straight pull. The feather should be grasped between the thumb and finger of one hand, close enough to the body to avoid bending or breaking the quill. One not expert in the operation should then give a very easy test pull to determine whether he has such control of the bird or the part that it will not yield when the real pull is made. Having satisfied himself of this he may remove a small feather with one quick movement. In the case of a large feather it is better to pull strongly but slowly for an instant and then, after relaxing a very little, with as strong a quick pull as may be necessary, draw the feather out. One can get the idea best by practicing a little dry-picking birds that have been killed for the table, taking one feather at a time between the thumb and finger instead of taking them by the handful as in ordinary plucking. Except while the bird is in the stunned condition, in which the feathers relax and are easily drawn, the feathers on a dead bird are much harder to draw than on a living one. In either case, the deliberate pulling of a feather takes much more strength than would appear necessary when we consider how easily feathers sometimes come out—almost at a touch—when they are not wanted out.

In pulling large wing feathers the wing must be held close to the root of the quill so that there is no strain on the bird or movement of the wing to give a slant to the pull at the moment pressure is applied. An expert can hold a bird between his arm and side or in his lap with the arm keeping it in place, leaving both hands free to hold the wing with one and draw the feather with the other. A novice usually finds himself rather clumsy and apt to hurt the bird when he tries this, but it is probably on the whole better for him to practice doing the whole thing himself until he is proficient, rather than to have someone hold the bird for him. When the person holding the bird performs the operation he is able—after a little practice—to judge accurately of the state of resistance or relaxation of the bird and to apply the pull in the instant of relaxation when it is easiest for both the bird and the operator. In pulling tail feathers the bird is much easier held, for its wings are both folded and all its own weight is pulling in the opposite direction from the force drawing on the feather.

CHAPTER VIII

Definite Selection of Birds for Exhibition

Methods of Systematic Examination for Faults Affecting Show Value—Discussion of Disqualifications and Most Common Defects of All Breeds and Varieties—Distinctions Between Removable and Permanent Faults—Points to be Considered in Selecting Single Birds—Importance of Studying Each Specimen from Many Angles and Under Varied Conditions of Light—Matching Pens—Why Birds do not Always Look Equally Well When Shown Single and When Shown in Pens

About a month before a string of birds is wanted for exhibition the intending exhibitor should begin the work of definite selection of the specimens which then show—as far as they are developed and finished—exhibition quality and the possibility of being fit for that particular show, or of being forced to readiness for it if a little backward. This selection should take nothing for granted on the strength of former observations as noted or remembered, but should be searching and thorough, and especially attentive to trivial faults. It might be sup-

Conditioning and Training Coops

Before beginning the work of selection provision must be made for the individual cooping of all male birds during the periods of fitting and training, and for the cooping of the females singly or in small groups. For fitting small strings of birds a room of appropriate size may be fitted up wherever is most convenient. Most small exhibitors have rather makeshift arrangements, often using a fairly well-lighted basement or cellar room in the house, or any available space in an outbuilding. Those who

CONDITIONING ROOM OF E. B. THOMPSON, AMENIA, N. Y. MR. THOMPSON LOOKING A BIRD OVER

posed that in previous examinations, and especially in handling when removing dead and broken feathers, an observant person would discover every fault. But it is surprising how many of the little faults in birds will escape observation through several examinations, and occasionally through all handling prior to the show. The reason for this is fairly obvious after one gets a little practice and is continually finding new faults which he had previously overlooked. In such examinations we, as a rule, really see only what we are looking for—except as the eye may happen to note something else in passing. So it is only by the most careful, systematic examination that one can be sure that he knows all the faults of a bird.

have to condition many birds find it necessary to arrange and equip a good-sized building for this purpose. In rooms devoted especially to birds being prepared for show or shipment the coops used are mostly the regular exhibition coops. Where it is desired to keep the birds confined separately or in small numbers, with more room than such coops afford, quarters are often fitted up with larger stationary coops, both on the ground and in tiers above. Houses arranged for small breeding pens are also sometimes used in the show season for small lots of birds in various stages of preparation for exhibition. The important thing is to have all the birds that have to be handled placed where they can be caught and handled

TOP ROW—MODEL HEADS OF SINGLE COMB MALES
Second and third rows—faulty heads of prize-winning birds*

*Mr. Schilling has selected for this and the groups of heads on following pages only the heads of birds that were otherwise good enough to win—winning in spite of their heads. The reader will find it a useful exercise to identify the faults in these heads. He should examine and compare every character—comb, wattles, lobes, face, crown, neck, beak, eye and general expression.

easily at any time, and where there is no danger of their being injured in any way.

If the entry is to be limited to a certain number coops should be provided for enough reserve birds in excess of that number to insure that if anything goes wrong with a bird a substitute will be available. If it is the intention to show all available birds, final arrangements for cooping must of course wait until selections have been made. If one is short on regular exhibition coops, or the substitutes for them in the room where the work of selection is done, shipping coops as used for high-class stock may be used to advantage to hold birds whose final disposition it may be desirable to postpone until full account of available stock has been taken.

Order in Systematic Examination

For convenience in presenting the subject and to avoid unnecessary repetition the points to be considered will be taken up in the following order:

1. General disqualifications and defects—applying to many breeds.

2. Disqualifying and serious defects in each breed.

3. Color and color pattern faults, each color pattern being discussed for the varieties of the several breeds which have it.

In following this outline each reader should have a bird in hand for examination. It is of little use for a novice to read such matter as this right along with the idea that he can retain it or any considerable part of it accurately in his mind and go out to the poultry house and apply it without a book. Considering only the matter of speed in making selections it is often an advantage to look for some of the most common disqualifications and for faults that cannot be remedied in other sections before beginning with the head and going systematically through the bird section by section. For practice to acquire knowledge of the faults of birds and proficiency in detecting them, one should examine fully all the specimens he has which to general observation appear to have Standard quality, and try to make an accurate estimate of the exhibition value of each.

TOP ROW—MODEL HEADS OF SINGLE COMB FEMALES
Second and third rows—faulty heads of prize-winning birds

SINGLE COMBS

Disqualifications—Lopped comb, except in females where the Standard calls for comb falling to one side; side sprigs; and split comb.

A lopped comb is one that leans badly to one side. Such cannot be remedied. A comb that leans only a little may be straightened. A side sprig may be removed. Very expert surgery might fix a split comb so that the scar would not be noticeable, but unless the bird is uncommonly good and the fault slight the chances of detection make the risk greater than cautious exhibitors care to take.

Serious Defects—Twisted combs—any lateral variation from a straight, symmetrical form.

A twist in the front of a comb, over the beak, may be near the base, in which case it appears as a "thumbmark," or may cause adjoining spikes to point in different directions. This irregularity of the spikes may occur at other points. A twist at the rear takes the form of turning the blade to one side. All these faults, if not too serious, may be remedied by surgical operations, or by manipulation, or by the two processes combined.

All single combs (except Minorca combs which have six points) are described in the Standard as having five points. Too many or too few points call for a cut of $\frac{1}{4}$ for each point more or less than required. On this basis a four or a six-point comb, if neat and symmetrical, may be better than a five-point comb. Number of points alone does not make quality. But two points more or less than the required number makes a comb so different from the Standard in appearance and so likely to be heavily cut otherwise, that the one point cut for irregular number of points becomes serious. Coarseness of comb is also a serious defect, whether the coarseness be in size and outline or in texture. It is of particular importance to observe this point because coarseness of comb is often associated with general coarseness of structure and a loose unsymmetrical carriage.

ROSE COMBS

Disqualifications—Lopped combs; combs so large as to obstruct the sight. (This is due to the comb spreading out and down while symmetrically balanced on a strong base); absence of rear spike.

TOP ROW—MODEL HEADS OF ROSE COMB FOWLS, MALE AND FEMALE
Second row—faulty heads of prize-winning males.   Third row—faulty heads of prize-winning females

The first two conditions cannot be remedied. In some cases the absence of spikes is really failure of the spike to protrude normally as the comb grows, and this if taken in time can be remedied.

Serious Defects—The common faults of rose combs all come under the general description of irregularity of form and excessive size. The Standard in all cases calls for a neat symmetrical comb with even, smoothly pebbled surface. There is a general tendency to extreme coarseness of pebbling with deep indentations between the rounded points which make the pebbled surface. There is also a common tendency to angularity of the comb, both in front and toward the rear where there may be quite a pronounced "shoulder" on either side of the base of the spike. In addition, many rose combs that follow the skull, where the Standard requires that, turn down with almost an angular bend in the upper surface which should curve smoothly; while others fail to follow the head and have a straight upper surface with the spike in the same line. This last type of comb is Standard for some breeds, but the comb with an angular turn above the crown is one of the most common and also one of the ugliest faults in good birds of the popular rose-combed breeds and varieties. Another almost as common fault is the rose comb with a depression in the center, sometimes coming well down forward giving a slightly cleft appearance to the front of the comb. At the present time these faults of irregularity and coarseness in rose combs do not appear to handicap a rose-combed bird as heavily as the corresponding faults in single-combed birds do. Hence, in general, an exhibitor of rose combs does not have to be as particular about quality of comb as an exhibitor of single combs.

PEA COMBS

Serious Defects—Lack of clear definition of the longitudinal sections which give this comb its "triple" character, and of the serrations in each section.

Originally the disqualification for lopped comb applied to this as to the preceding forms, but for a long time pea combs have generally been too small for the lopping when present, as it frequently is, to be very unsightly. For his own satisfaction a breeder of pea comb fowls should cultivate a good type of comb, but in selecting birds for exhibition he can be quite indifferent about

DEFINITE SELECTION OF BIRDS FOR EXHIBITION

irregularities in the form of the comb, provided it is not large and coarse enough to make the faults conspicuous.

V-SHAPED COMBS

In Standard breeds this type of comb is associated with large crests and beards and no refinement of form is demanded.

CUP COMBS

Serious Defects—Lack of cup-like formation and irregularity of the leaf-like points which form the sides of the cup.

WATTLES

In a measure, the character and texture of wattles correspond to the comb though the form is the same whether the comb be single, rose or pea.

Serious Defects—Coarseness of texture is the only defect in wattles which has so far been listed in the Standard as demanding a specific cut. In selection it is of most importance for its association with coarseness of comb and of structure generally.

MODEL HEADS—CORNISH AND BRAHMA PEA COMBS—MALAY STRAWBERRY COMB

EAR LOBES

Disqualifications—Ideally an ear lobe should be either all red, or all white or creamy white. In nearly all breeds the Standard disqualifies for a proportion of the undesirable color, the exact proportion being different for different breeds, and being gradually reduced as careful breeding increases the number of specimens free from foreign color in lobe. The precise requirements will be given in the lists of breed faults. White in red ear lobes appears in two forms—as positive enamel white in spots or patches on the surface of the lobe, and as mere absence of color or paleness of the lobe. The first is permanent and cannot be eradicated permanently without damaging the lobe. A small spot may be cut out and the lobe fixed so that it appears to have a scar which might result from some other cause. Small spots may be temporarily concealed by the use of cosmetics or dyes. Pale whitish ear lobes may be remedied by any treatment that brings the blood to the surface, or concealed by cosmetics or dyes. The latter treatment is quite easily discovered if the judge suspects it, but will often pass unnoticed. The redness of lobes may often be increased by giving the birds an abundance of fresh air. Confinement and inactivity are not conducive to high color in the skin.

Red in white ear lobes is sometimes due to exposure and disappears when the birds are confined under cover. Permanent red in a white enamel lobe cannot be removed but may be concealed by cosmetics and powders. A judge who is suspicious or who has the habit of examining lobes closely will easily detect faking of this kind. Yellow-white ear lobes are made white by many exhibitors by the process described in the next chapter.

FACE

Disqualifications—The only important point to consider in this section is enamel white in fowls of the breeds having large white ear lobes. This may be concealed in the same manner as in lobes—subject to the same risks.

CRESTS

Disqualifications—The Standard disqualifications relating to crests are practically obsolete, their appropriateness was really limited to the days before Standards were well distributed. The absence of crest in a crested breed is rated a disqualification, but no one thinks of accepting such a bird as representative of a crested breed.

Serious Defects—The most serious shape defects in crests are irregularity of form partly due to the fact that the feathers in male and female crests, differing in structure as the feathers in male and female hackle do, corresponding degrees of fullness in the front of the crest do not give the same effects in the different sexes. The male crest corresponding to that of the best female crest tends to spread, the feathers turning sideways and forward instead of falling back smoothly. The female crest corresponding to the finest type of male crest is likely to lack fullness in front. As crest is the most valued feature in these ornamental breeds, good form in it must be given precedence in selecting birds for exhibition.

BEARDS AND MUFFS

Disqualifications—Absence of these where required constitutes a disqualification.

BEAKS AND BILLS

Disqualifications — Crooked or deformed mandibles. The most common form is the point of the upper beak twisting to one side and coming down over the edge

MODEL HEAD OF HOUDAN FEMALE

FAULTY CREST OF HOUDAN MALE
Dotted line shows size and outline of a model crest.

of the lower mandible, but sometimes both mandibles are twisted. There are also cases where one mandible is undeveloped. If the malformation is very slight it is immaterial, but if it is considerable a judge may construe it as a deformity within the meaning of the Standard. In ducks and geese the large bill makes deformities so conspicuous that they can hardly escape ordinary observation as slight deformities in the beaks of fowls often do.

HEADS AND EYES

Serious Defects—It may seem that after all the points so far considered there could be little room left for serious defects about the head. There are no more disqualifications, and the defects for which specific cuts have been prescribed in the Standard are in color of the eyes; but the character of the head and the expression of the eyes should be given close attention when selecting birds for exhibition.

The most common objectionable type of head to be avoided is the snaky head often called "crow head." If one has looked the head adjuncts over, and thinks he has found that comb and other appendages are free from serious faults, and are well developed and about right, and still it appears to him that the head is rather fine drawn, or rather coarse, he may reasonably assume that the impression the head gives him is correct, and that he needs to revise his judgments about the other parts to the extent of concluding that though he can find no serious fault in them they are not good. If the head impresses him as too fine, he should in further examination of the bird look particularly for corresponding lack of development of the whole frame. The head is not an unfailing index of this, but it is a fairly reliable one. If the head impresses him as too coarse, he should be on the alert for coarseness everywhere and for indications of the lack of symmetry and of the ability to pose gracefully, which are the usual results and accompaniments of coarseness. In a very few breeds, which will be especially mentioned in course, massiveness of structure, including massiveness of head, is desired; but on the whole the best heads, the most appropriate to the size of the bird, are those of which one never thinks as especially showing any character except as that is shown in the expression of the eyes.

The color of the eyes should be especially noted, and as eyes frequently differ in color it is not enough to examine one eye and take it for granted that the other is of the same color. Nothing can be done to change the color of eyes. Color faults in them may subject a bird to a cut of as much as 1½ points. Particular attention should also be given to the shape and expression of the eye, for in a measure that is a check on one's judgment of the standard of the head. In nearly all cases the Standard calls for a round, full eye. Sometimes the phrase "prominent" or "bold" is added. This character and expression of eye are rarely found in a poor head or in a bird lacking vitality. The tendency in such cases is to contracted and sunken eyes. In a few breeds with very broad crowns a god eye will be overhung by the brow above but it will have the appearance of a sunken eye. If a specimen at this stage of preparation for exhibition shows lack of character and expression in the head and eyes, its future value as an exhibition bird is very doubtful. If it has good quality in other respects it is worth while for the owner to do what he can to bring it on; but he need not feel disappointed if it fails him before he has finished it for the show or in the showroom.

DAMAGED HEAD PARTS

Birds are often injured in ways that mar their appearance, sometimes very much so. Parts of combs and wattles are lost by the frost or in fighting, or by cutting on wire fences and partitions, or the raw edges of pails or hoppers. Ear lobes are torn, and white ones in particular have their surface damaged so that it will never come smooth again. The novice is always in doubt about exhibiting such birds. No advice applicable to all cases can be given. The extent of the injury, the grade of competition, and the attitude of the judge, all enter into calculations as to the wisdom of putting a bird that obviously is permanently out of exhibition form in competition. The sex and age of the bird also are to be considered. An old cock that is in excellent condition apart from such an injury may come through creditably where a cockerel would not have the ghost of a chance, for the simple reason that nearly all exhibitors have fewer old cocks to show and usually have a proportion of the cocks damaged more or less about the head, while well supplied with cockerels in prime condition. For strong competition an injured bird is very poor dependence.

NECK

The serious faults in this section are principally color faults. Many birds in all breeds have rather poor form and carriage of the neck, but unless the bird is very unsymmetrical, and unresponsive to training as well, it can be taught to pose so that deficiency in this section will count little if at all against it. The form and habitual carriage of the neck are of most importance in the Cornish, Malays, Exhibition Games and Game Bantams, in all of which the length of the neck itself and the shortness of the plumage give its actual lines and carriage an importance they do not have to anything like the same degree when the hackle is abundant and long.

BACK

Disqualifications—In all kinds of poultry a crooked back disqualifies for exhibition. The most common forms of crooked back are what is known as "roach back," a sort of hump back; elevation of one hip; and crookedness at the shoulders. Roach back is often apparent at sight, but may not be in short-backed birds having a profusion of feathers. There are of course all degrees of this, as of the other two faults. By passing the palm of the hand from the neck to the tail, holding it flat on the bird's back, one can feel any irregularity of formation. If this leaves him in doubt as to whether the fault is so serious that the bird would unquestionably be disqualified, or is light enough to give the specimen the benefit of the doubt, he should observe closely the carriage of the bird as it stands or responds to posing, noting whether it stands squarely and balances itself well, and turns easily and gracefully. If it evidently has a "limp" he may conclude that there is deformity at the hips. If it does not carry its wings alike or carries both wrong, or holds one shoulder plainly higher than the other, it may be concluded that the unevenness noted by touch at the shoulders is great enough to attract the notice of the judge and to cause him to set it back, if he does not qualify. In general, a back that is not a good, straight, strong, symmetrical back so affects the carriage and shape of a fowl that the lack of symmetry affects nearly every section.

In considering shape of back as noted by the eye, account must be taken of the effect of length and profusion of plumage on apparent shape of back. Thus the convexity of the lines of the back noted in some short-feathered breeds of poultry, in turkeys, and some breeds of geese and ducks, is due to the shortness of the plumage at shoulders and rump, and to some extent to the high carriage of the wings. Nothing can be done to remedy a crooked back. If a novice can discover it, he can hardly expect to get by an expert, for very few judges fail to look for it, and most judges are not inclined to be lenient with faults in shape in this section.

TAIL

Disqualifications—Wry tail in all poultry; squirrel tail in all kinds except Japanese Bantams; absence of main tail feathers.

A wry tail that disqualifies is one that is carried permanently to one side to a degree that is noticeable. A tail that is very slightly to one side is not considered a wry tail, nor is one that the bird carries straight a part of the time. Decidedly wry tails are usually due to deformity or injury of the fleshy protuberance out of which the tail feathers grow. It cannot be remedied. A slightly wry tail or occasionally wry tail may be caused by a minor injury to the muscles or tendons, or in some cases be a habit contracted as a result of roosting or sitting where the tail is held to one side. This form of wry tail may be remedied by a simple surgical operation, or by bending the feathers.

Nearly all high-tailed birds carry their tails at times so high that if judged with the tail at highest angle they would be disqualified for squirrel tail. A judge in passing on this point is supposed to take into consideration the tendency of birds to throw the tail up when excited and to make an effort to have the bird pose with the tail at a proper angle. Unless a bird persists in throwing the tail past the allowed limit for disqualification a judge does not disqualify it for this fault, but he does discount as a serious fault the carrying of the tail at any angle higher than that described in the Standard for a breed. A high tail makes a sharp angle with the back and parts the saddle so that the lesser coverts and saddle feathers on the back, at the root of the tail, fall sharply to the sides instead of flowing back as they should to cover the main tail feathers and sickles at the base.

The tails of fowls may be bent in any manner desired, and the saddle feathers bent and raised to give breadth and fullness in front of the tail, and the desired curve to the back. The method of doing this is given in detail in the next chapter. It is in order to observe here however, that to do this well is an art requiring some practice and the beginner would do well to acquire some proficiency with birds he does not intend to show before trying it on one he is depending on for an approaching exhibition.

Entire absence of main tail feathers is a condition presumed to be due to their removal because of badly defective color. The occasion for inserting it in the Standard however, arose in connection with the custom of plucking the tails of Cochins to have them grown by the time the bird was to be shown just enough to give the desired outline and hold the abundant soft feathers of the small sickles and coverts in place. Repeated pluckings for this purpose gave in some cases very old tails on comparatively young birds, for each time a tail was plucked and grew in, it came with the qualities of a tail a year older than the one before it. The result was that occasionally a new tail came so poor in color that the exhibitor pulled it out and showed a bird either without a main tail, or with the feathers of it just starting and not yet showing their faults. The usual explanation was that the tail had been pulled by accident.

Serious Defects—The most common faults other than those already mentioned, seen in the tails of birds at shows, are tails only partly grown out and tails of full length and development, or nearly so, with one or more large feathers removed or broken off. In general, the presumption is that such feathers have been removed or broken to conceal color faults. It is, of course, possible and probable that occasionally such feathers were broken or lost by accident.

Partly grown tails are frequently seen in the show room in all breeds. From what has been said of the frequent delay in the molting of tail feathers, it is apparent that many such cases are due to failure to remove the old feathers in time for the new ones to grow out before the show. Punishment for this fault of condition is varied according to the degree of development of the tail, on the theory that the nearer full length it is without showing a fault the greater the probability that it is free from fault. Thus, a tail that is only one-fourth grown is cut 3 points, one that is half grown is cut 2 points, and one that is three-fourths grown is cut only 1 point.

Broken or missing main tail and sickle feathers are in some breeds cut so much for each feather, and in others made a disqualification if the damage is in a section where foreign color is a disqualification. Where a feather is accidently broken off the loss may be concealed by splicing a perfect feather to the quill, if one that matches the others can be obtained. Feathers having color faults are sometimes cut off near the base and perfect feathers spliced to the quills. This practice appears to be more common in England, where the birds are not handled and examined as closely as in our showrooms, but occasionally a specimen turns up at an American show with a spliced tail that may not excite any suspicion when seen from the aisle, but is easily discovered when the judge goes through the bird.

In all breeds where a short full tail is desired there are many specimens that have such a tail only when it is about three-fourths grown. Later the main tail feathers are too long to look well, and the sickles also straggle out beyond the tail in a manner not at all pleasing. The principal cause of this characteristic is probably the use of long-tailed hens as breeders. Most breeders of exhibition stock of the breeds in which what may be called modified bushy tails are desired are quite careful to avoid the use of males with this fault in the breeding pen, taking them only when they are so good in other respects that this fault must be overlooked; but few breeders are as careful as they should be to avoid the use of long-tailed hens, and above all their use with males having longish tails. The result is a continuous outcropping in the flock of males that show really good form in this section a little before fully developed; the constant necessity of making careful computations as to the probable size of a tail when fully developed; and the need of pulling a growing tail in order to have the next one the appropriate length when the bird is to be shown. With the most careful calculations and the most expert management the matter is difficult to adjust, and the wise exhibitor always has enough males in various stages of feather development to avoid being caught at the time of exhibition with none that are near right in this section.

WINGS

Disqualifications—Clipped flights or secondaries; white in same, in various color patterns.

This point concerns chiefly new exhibitors at new shows. In all localities where no shows have been held for some time there are often amateur breeders with some very good stock, who have not been showing anywhere, and who have thought it more economical to clip the wings of birds disposed to fly out of their yards than to make fences high enough to keep them in. Plans are made perhaps not more than four or five weeks ahead for a show near them. They decide to exhibit and when they come to select birds find a part or perhaps all of them have clipped wings. The breeder thinks that he knows those wings were good in color. In clipping them he had no thought of anything but the quickest way to keep the birds in bounds. It seems an unfortunate thing in the light of the situation that has developed, but he thinks perhaps considering the circumstances an exception might be made in this case. The local show managers might take a similar view if new to the game, for there are likely to be prospective exhibitors in the same fix, and they want to bring in as many local birds as possible. I would not say that no judge would connive at such an understanding, but I never knew one to do so. The judge has to consider both his own reputation and the interest of competitors whose birds are shown in perfect condition, and who have the right to insist, and if present certainly would insist, on the application of the rule provided for such cases. There is nothing to do with the birds having clipped wings but to leave them at home, or show them "not for competition." Color faults are so common in the flights and secondaries that the rule is a wise one.

Broken and missing feathers in the flights are a disqualification or defect according to whether a serious defect which might have been in the missing feather or part of a feather is a disqualification. Cases where foreign color disqualifies will be considered undercolor.

Serious Defects—Twisted flights; broken or missing primaries or secondaries.

The two wings of a bird usually have the same faults, but occasionally a fault will be found in one and not in the other. Hence, each wing should be fully examined. Twisted feathers in one wing only would be supposed to be caused by an injury to the wing which prevented the feather or feathers from growing in the normal position. Twisted feathers in both wings might possibly result from such an accident, but are more likely to be due to congenital slight deformity of the last joint of the wing.

A slipped wing may or may not be permanent. Many birds that normally fold the wing right and tuck it close will slip the wing more or less when excited or tired. Almost any bird that is being handled and examined, the wings opened and perhaps held open for some time, may fail to tuck its wing immediately when put on its feet again. In that case a light touch at the lower edge of the folded wing may prompt the bird to tuck it closely. Or, if this is ineffective, to simply take hold lightly with the thumb and finger and draw the wing out and down a little bit, and then release it, may result in the bird tucking it right. The natural thing is for the bird to fold the wing properly when it has been moved out of position—if it can. If repeated efforts to make the bird fold the wing right are unavailing, it is still possible that if the bird is let alone for a while it will fold the wing right. Of one thing an exhibitor may be quite sure —that a bird that slips its wings with a very little handling at home will be as bad or worse in the showroom. He will also find that some birds that were never known to fail to fold the wings right after handling at home will show slipped wings after a day or two on exhibition.

This cannot be guarded against except so far as good care and sufficient exercise while the birds are being conditioned gives them strong vitality. Wright describes a method of helping slipped wings while the birds are molting by carefully tying the flight feathers in the position they should have when properly folded. The efficiency of this would appear to be much greater where the birds are not handled than where the wings are opened as much, both by judges and by exhibitors themselves to show to others, as is usual in American shows. In many color patterns the wings and tails are the parts most difficult to keep free from color faults, and the places where weakness in color shows first. In some, excellence of wing color is sought to an extent that gives it undue importance. So birds generally at American shows have wings opened and handled a great deal, and it is of great importance that they should be trained to fold the wings promptly if they can. The judge will generally give a bird that otherwise would be in the awards every opportunity in reason to show that the fault is not permanent. I have known a judge to delay awards in a class overnight simply to see whether the cockerel which, but for a slipped wing, was the best in the class could show perfect form.

Broken and missing feathers are always assumed to represent removed defects, and the effort of an exhibitor to at least raise a doubt whether the fault existed. A feather that is partly broken—that is, the broken part not completely detached—is given a lighter cut than one that is broken off or missing. Usually it is inadvisable to show birds in strong competition if they must submit to the heavy cuts for damaged wings, but in a small show or in weak competition the damaged bird might win in spite of this handicap. Were it not that the handling wings get is apt to expose anything of the kind, it would not be a very hard matter to cut off slightly twisted flights and splice them on straight. As it is there is little temptation to faking in this particular.

BREAST, BODY AND FLUFF

In these sections there are no disqualifications. It is a curious fact, however, that "crooked breast" or keel bone was for many years disqualified by many judges, by some on the supposition that the Standard expressly mentioned it as a general disqualification, and by others on the supposition that the Standard contained a specification calling for disqualification for "any positive deformity" which should apply to it. As far as I have been able to discover, the term crooked breast never appeared in the Standard until 1905, when that defect was added to the list for which specific cuts were recommended; nor did the Standard ever contain a general provision that birds should be disqualified for any deformity—either positive or doubtful. The nearest approach to this was the provision in the case of one or two breeds in some of the older editions of the Standard, that birds should be disqualified for wry tails, crooked backs, crooked beaks, "and any other deformity." The idea that crooked breast was a disqualification appears to have been due to confusion of standards in the mind of judges, and to the fact that in all times since the judging of poultry by written standards began, a great many judges with no accurate knowledge of the provisions for the different breeds and varieties have undertaken to judge "all varieties" by a sort of general standard formula made up by themselves. "Any deformity" is a very convenient formula, and most people would say—a good one; but it has never been "Standard."

Serious Defects—Crooked breast; flat breast; narrowness and shallowness of body; short fluff—or apparent short fluff.

Crooked breast is a generally preventable defect which has been considered in connection with the care of exhibition chickens to insure normal development. Detected at a very early age, it might be largely if not wholly corrected by keeping the birds on thick, soft bedding so that there would be no pressure on the keel and it might grow straight again. Few breeders can watch and examine small chickens closely enough always to find it early. When found at the time of selecting birds for exhibition nothing can be done for it farther than to have the bird in as good flesh as possible so that the keel will not be prominent and its irregular form noticeable at first touch. How far this may be successful depends a great deal on the structure of the bird and of the keel. A prominent crooked keel cannot be helped much, while a rather low keel may be so concealed by fullness of flesh on either side that a slight crookedness in it is overlooked. Unless very bad, a crooked breast does not seriously handicap an exhibition bird, for it does not show, nor does it affect the carriage of the bird as deformities of the back do.

The other faults mentioned are of less consequence from the exhibitor's standpoint for most judges do not give the same attention to them as to crooked breast; but a few judges have always been severe on lack of development in these sections, and of late years much more attention is given to them than formerly. In nearly all breeds the Standard calls for fullness of breast and body and moderate fullness of fluff. Fullness of breast is evinced in breadth, depth and roundness of breast as the bird stands. Depth of body may also be observed in the standing bird, but width and fullness of body are better determined by touch as the bird is held in the hand. Practically the first thing that one accustomed to handling birds in the method used in examining exhibition specimens notes when he catches a bird is the characteristic of the body, whether full or poorly developed, as the bird rests on his hand.

Shortness of fluff is desired in some breeds but not in others. The general point to note in regard to it is the relation between apparent length and fullness of fluff and the shape of the body. Looking at the Standard illustrations, or at illustrations of almost any superior Standard specimens, one may observe that the well-balanced bird, unless very short in fluff, shows almost as much length of body back of the leg as before it. Apparent length, depth and fullness of the posterior part of the body depend a good deal upon the condition of the bird. In a female there is, as a rule, more fullness here at any time than in the male, and in the laying hen or pullet the difference is increased. In males, and in females not laying, any apparent lack of development here can be greatly helped by putting on all the flesh the bird can carry and keep in good condition.

This method can be applied to some extent to improve the lines of breast and body, but if the posterior parts have fair development to start with an amount of fleshing that would fill out a thin breast might make them baggy, especially in old birds of the heaviest breeds.

LEGS AND TOES

Disqualifications On All Breeds—Legs and toes of color not conforming to Standard requirements for the breed; toes not conforming in number to the Standard requirement for the breed; web feet.

In nearly all breeds and stocks some specimens are found which do not conform to the Standard in color of beak, legs and skin of body. Thus, in yellow-legged breeds, we find some birds with flesh-colored legs and a pale skin, and some with very dark or willow legs and rich yellow skin; in breeds required to have flesh-colored legs we find some decidedly yellow; and so on. Between the color that undoubtedly meets the Standard requirement and one that does not any grade may be found. The question is where to draw the line in disqualification. Usually it is decided according to the preponderant color. That is—a yellow leg may be a weak, poor yellow, with a fleshy or whitish tinge, but as long as it is yellowish it would not be disqualified, though the color would be discounted.

In the case of a dark leg on a yellow-skinned bird, the color is mostly in the scale and is caused by pigmentation of the scales corresponding to pigmentation of the feathers. Usually this is more pronounced in females than in males, and most prevalent in pullets before they mature. Another point of color often noted in yellow and light-colored legs is a reddish tinge, sometimes quite a bright red on the outside of the shank on males. This high color on the shank is a common phenomenon on male birds of many kinds, and is usually most intense in the breeding season. It is not considered "foreign" color, but a bird having a bright leg of the desired color all over is preferred when a decision comes down to the point of turning on color of legs. In score-card judging, practice as to cutting here for color varies considerably; usually the cut is not made unless the red impression is most prominent. All color faults on shanks can be concealed by use of the appropriate coloring pigment, but where very heavy applications are required to give the effect desired a judge is quite apt to rub enough of the color off to satisfy him that he can justify himself for leaving the bird out of an award it might be entitled to were its legs naturally of good color.

The disqualification for wrong number of toes is of most interest to breeders of the few breeds required to have five toes, and to breeders of Cochin Bantams, into some strains of which the fifth toe appears to have been introduced by crossing with Silkies. Occasionally, one of the common four-toed breeds will have a fifth toe on one or both feet. If the toe is well developed, an observant poultryman is apt to note this as the chick runs about before its gets very large, but if it is quite small he may not. Occasionally a bird has only three toes on one or both feet. Usually the rear toe is the missing one and its absence may not be noted until the bird is caught and examined carefully. In five-toed breeds specimens often appear with only four toes, and not infrequently with six toes.

Web foot is not very common. When pronounced it is so conspicuous that it will hardly escape the notice of the amateur long before selection for exhibition begins.

Serious Defects—Crooked toes; scaly legs; bumblefoot; corns.

The most common form of crooked toe is the positively deformed, badly twisted toe. Very few birds in the yards of exhibitors get to maturity with bad faults in this respect because it is the common practice to kill them when hatched, but one who has not culled early may find an otherwise fine specimen with a badly crippled foot—possibly all the toes on one foot twisted. In a score-card show the most that can be done to a crooked toe (no matter how bad), when one does appear, is to give it a cut of 1 point. In a comparison show the judge can set the

SOME COMMON EXAMPLES OF BAD FEET ON PRIZE-WINNING BIRDS

bird back for the fault as much as he wants to. The instructions to judges using the comparison system say that the judge must give all defects their proper valuation, but as there is manifestly no way to determine whether a judge has done that except by a score card, the judge can do as he pleases. In cases of this kind, at both score-card and comparison shows, where the Standard instructions do not cover an unsightly fault, a judge may apply the first of the general disqualifications and simply declare that the bird is unworthy of an award and should be, and is disqualified. The opinion of exhibitors will usually support the judge who does this.

The ordinary case of crooked toe to which the specified cut of ½ to 1 in the Standard lists of cuts for defects is intended to apply is any toe that is not straight. In the three forward toes crookedness usually is in curvature to one side, the curve starting at one of the two last joints of the toe. Curvature of the hind toe is more likely to be at the junction with the foot, giving in extreme cases the formation called duck foot, and in others only a little displacement which may pass unnoticed unless one is on the lookout for it. Most judges are not at all lenient toward crooked toes except in cases where the defect is very slight. Hence, it is in general not good policy to show a bird that has two or more toes subject to cuts for this fault, unless it is so good otherwise that it can stand these cuts and still rank among possible candidates for high honors.

Scaly leg, bumblefoot, and corns are common faults of condition which the judge can treat as he pleases. Most judges take the position that there is no excuse for showing birds with any of these faults; and that the fact that a bird has uncommon merit, instead of justifying the owner in showing it in such a condition, aggravates his shortcoming. If any of these troubles exist and are found a month before the show there is ample time to cure any but bad cases of bumblefoot.

Disqualifications on Smooth Legs—Feathers, stubs or down on shanks or feet; evidence of removal of feathers, stubs or down from shanks or feet.

These growths are different grades of the same fault. They are most troublesome in the smooth-legged breeds of Asiatic ancestry, on one side, but no breed is free from them. Feathers and stubs grow on the outside of the shank, usually on the upper part near the hock. There may be only one or two of them, or there may be a line halfway down the shank or more. Technically, a feather is a growth having, however small, quill, shaft and web; a stub has quill and web, but no shaft; down is a minute fuzzy growth without visible quill or shaft. Down may come anywhere along the line on which feathers and stubs come on the outside of the shank or at the junction of the shank or foot in front, or between the toes—usually in the angle at the junction of the web and toe, but frequently out in the surface of the web. Small stubs and down may also come occasionally on the side of the toe where the feathers grow in feather-legged breeds.

At this stage of selection and preparation the exhibitor need not concern himself about down. If he finds some feathers or stubs, the question is first whether they can be removed and traces concealed, and if so, whether he intends to do that. If any of the feathers are of considerable size—large enough to leave quite a hole when removed—or if stubs are numerous, the prospects of faking the legs and escaping detection are poor. If there are just a few small stubs they can be fixed up so that

CORRECT AND INCORRECT FOOT FEATHERING
The feathers on the shanks and toes of fowls should grow nearly at right angles from the part

the chances of the traces being detected are small. If the exhibitor is not disposed to fix them up, there is no use working further with the bird having them. It is appropriate, however, to say in this connection that any bird, no matter how free from this fault when examined at one time, may develop down or stubs within a few days and sometimes overnight and the novice in exhibiting should bear in mind that this is a point that may come up with any bird at any time. If the faults are to be removed the time to do it is at the latest possible moment before the birds are judged.

Disqualifications on Feathered Legs—Shanks not feathered for their full length on the outer side; outer toes not feathered to the last joint (except that in Langshans the feathering may stop at the middle joint, and in Silkies and Sultans a bare outer toe does not disqualify); vulture hocks; evidence of the removal of vulture hocks. In Cochins and Cochin Bantams a bare middle toe disqualifies.

The disqualifications for absence of feathers are made to suit the grade of feathering required on the feet in each

ENGLISH-TYPE LIGHT BRAHMA
Note the extremely heavy foot feathering. Also note the bird is not short legged. This type is correctly described as "cochinny" but that description is not appropriate for a Brahma that has Standard foot feathering, though it should be very short on the leg and appear dumpy.

case. It might me said that in general the disqualifying point is halfway to the extreme point of the toe which the Standard requires should be feathered. Bad cases of vulture hocks cannot be remedied. Slight cases of vulture hock may be fixed by plucking the undesirable feathers so that they will be partly grown when the bird is shown.

A good judge of the heavy-feathered breeds in which this fault most frequently occurs can generally determine easily whether there is occasion to look for evidence that very bad hocks have been fixed, for the foot feathering of vulture-hocked birds shows, as a rule, much larger, coarser and stiffer feathers than are found on birds with soft hocks.

Serious Defects—Bare toes where feathering is required, but absence does not disqualify; scanty and broken feathering on legs and feet.

Bare middle toes in Brahmas and Brahma Bantams are cut 1 for each toe. A bird that is good in other respects can stand being discounted here but one that has any other pronounced fault has little chance in good competition with the additional handicap that the bare toes impose. While it is not invariably so, it commonly happens that birds with bare middle toes are subject to a further cut for scantiness and shortness of the whole foot feathering.

Broken foot feathering is mostly due to wear. Birds with feathered feet cannot be allowed to do much scratching, or to run in long grass or weeds while the plumage in which they are to be shown is growing. Exhibitors and breeders of this type of fowl take special care to have the foot feathering in good condition until after showing and then to compel the birds to take exercise as others do. While it is desirable to have the foot feathers perfect when shown, a little raggedness at the ends is not a fatal fault. Other things being equal the bird in best condition would win, but a bird that has the right amount of foot feathering, and in good color, will stand better in this section than one a little scant in feather, and perhaps a little weak in color that has the feathers perfect.

Weight Disqualifications—The margin between the required Standard weights and the disqualifying weights is uniform at two pounds below for old and young, male and female, in all kinds of poultry but turkeys, where it is six pounds below, and bantams, where it is four ounces above Standard weight. As the disqualifying weights for chickens and poultry do not apply until December 1, the allowance is in most cases so liberal that there is no excuse for anyone having birds disqualified for weight. The weight requirements are on the whole of less importance at comparison shows than at score-card shows, for at the former neither disqualifications nor cuts in this section are regularly and systematically applied. The judge may or may not give weight and size due consideration, making them the factors in his decision that they necessarily are when the birds are scored according to Standard.

Weight Defects—These are intended by the Standard to apply at all times whether disqualifications are in force or not. Hence it is always to the advantage of the exhibitor to have his birds approximate Standard weights as closely as possible. It does not particularly profit an exhibitor to have a bird escape disqualification for lack of weight and be cut from 4 to 6 points for the shortage that just misses disqualification. In selecting birds a month before the show ample time is given to bring all that are healthy and rugged well up to Standard weight when shown. The exhibitor should weigh every bird, make a note of the weight, and systematically work to have all at the nearest to Standard weight they make and remain in good condition when shown, preferring to give all that can stand it a little margin over Standard weight as they enter the show.

BREED DISQUALIFICATIONS AND SERIOUS DEFECTS

NOTE—Under this head are given the disqualifications, additional to general disqualifications previously mentioned, peculiar to certain breeds, and the defects of breed character to which the exhibitor needs to give special attention in selecting, and which are too commonly overlooked.

PLYMOUTH ROCKS

Disqualifications—Any positive white in ear lobes.
This a most important point because in all other breeds of the same class a considerable amount of enamel white in lobes passes without disqualification.

Serious Defects—Lack of size, weight and type; coarseness in large birds; knock-knees; pinched tails.

Center—Standard-shape Plymouth Rock male. Left—A common poor type in Plymouth Rock males, a rangy bird, lacking substance, too much bone and too little flesh. Right—Another common poor type, coarse and decidedly "baggy" in the rear. It is males like this mated to hens with the same fault that produce so many Plymouth Rock hens that after their first laying period store up excessive amounts of abdominal fat.

Undersized birds are always most numerous in the large and medium-large breeds, because poor management in growing the chickens affects them more. Faulty type in small birds is likely to be an overrefined type, which is not in itself unattractive to one who has not well fixed in his mind the correct Standard Plymouth Rock type. Where such off-type birds are selected, though birds of much better type are in the flock, it is usually because the person making the selection is following his own ideas without reference to the Standard and model illustrations. The selection of birds faulty in an opposite direction is usually due to misconceptions arising from attempts to follow the letter of the Standard description without referring to model illustrations to get the true interpretation.

Thus the neck and back of the Plymouth Rock male are described as "rather long." The Standard makers and revisers in formulating this description had in mind the question of differentiation in the type of Plymouth Rocks and other breeds of the same class, and particularly the Wyandotte. The breeder who is considering only Plymouth Rocks, and probably has never paid any attention to other types, quite naturally interprets it as meaning that the bird should appear to be rather long in both neck and back. Another breeder new to the selection of birds for exhibition, noting that the Standard calls for a "broad, deep and full" body, for large thighs, and shanks "stout, set well apart," chooses a heavy, low-down bird in which these are the conspicuous characteristics of shape. The only way to avoid such errors is to learn what the words of the Standard mean by observing how they apply to Standard models and to the best specimens one has an opportunity to see.

Different varieties of Plymouth Rocks sometimes have certain peculiarity of form more pronounced than in other varieties. Thus White Plymouth Rocks at the present time show in many flocks an undue proportion of rangy, extremely long-legged and long-backed specimens resulting from the reaction in breeding to make the White Rock conspicuously unlike the White Wyandotte in shape. Buff Rocks, on the other hand, are apt to be rather short. In all other varieties—Partridge, Silver Penciled and Columbian—specimens of either sex having the size, weight and type of the birds that win in good competition in Barred and White varieties are comparatively scarce. So in making selection in these varieties one who finds

Center—Standard Plymouth Rock female shape. Left—A common faulty type, high on the legs, body not properly balanced. Right—Very heavy and coarse Plymouth Rock shape

Center—Standard Wyandotte male shape. Left—A common type of undersized, ill-formed male, too low on the legs, and plainly lacking in vitality. Right—High-stationed, rangy Wyandotte male, a type especially common in the Silver Laced variety, because of the English blood introduced years ago, but frequently found in all varieties.

his birds rather deficient in these respects can make allowance for that, knowing that the condition is general in the variety.

WYANDOTTES

Disqualifications—Positive white covering more than one-fourth of the surface of the ear lobe.

Serious Defects—Extreme shortness and blockiness (most often found in White Wyandottes); absence of blockiness of type (giving in some birds ranginess and a stilted carriage, and in other undergrown birds an over-refined Wyandotte type, having the Wyandotte lines to some extent but in too small size and too finely drawn to give true type); long, loose feathering (giving bulkiness of appearance out of all proportion to the actual weight, and what is called "cochinny" type).

The extremes of fault in shape are found most often in White Wyandottes, and by far the greater proportion of birds of good type are also found in this variety. In most of the other varieties variation from Standard type is principally in one direction—toward spareness of form and too great length in all sections. This characteristic is most marked in the silver laced variety, having been introduced in it some years ago by the use of English birds of superior color but very much drawn out perpendicularly and wholly lacking in Wyandotte type. It is no use to show in good competition White Wyandottes that are not distinctly typical in form, for the class is one of the most popular, and strong classes come out in a great many of the small shows. In the other varieties good color will greatly help a bird that is somewhat off in shape, and the classes, as a rule, are not so large that there are good-shaped birds good enough also in color to take all the honors. Wherever the class is large, however, it usually takes a bird of good type to win high honors.

Extreme length and looseness of plumage is oftener found in White Wyandottes than in other varieties. It came in vogue first through the preference of a leading judge of this variety for that type of feathering, and was afterward cultivated to a considerable extent to exaggerate the difference between the White Wyandotte and the White Rock—to get away from Rock type. The proper length of plumage can be estimated best by observing how much of the thigh and hock show and whether the outline surface of the plumage about the thigh follows to a degree the lines of body and thigh, or conceals these, and possibly also a part of the shank, and fluffs out around the thighs. The lines should always show as in the Standard illustrations—rather more than less.

Center—Standard Wyandotte female shape. Left—A high-stationed, weedy hen; fine color but totally lacking in Wyandotte type. Right—Large, coarse, loose-feathered Wyandotte female—a type to be avoided

Center—Standard Rhode Island Red male shape. Left—Common type of male accepted by many breeders and some judges as meeting Standard requirements. His faults are plainly seen when he is placed beside a bird of correct type. Right—Coarse type of Rhode Island Red male often accepted as Standard.

JAVAS

Disqualifications—Positive white in ear lobes; absence of yellow color on bottoms of feet.

Most Common Defects—Lack of size and type. So few Javas are shown that, as a rule, anything that has the other breed and variety qualifications can get a place regardless of these faults. There are however, a few Java breeders whose stock is of uncommonly good type, and in competition with these birds lacking in type will get only the leavings in the awards.

AMERICAN DOMINIQUES

Disqualifications—Any positive white in ear lobes.

Most Common Defects—Lack of distinctive type. This is a handicap in a class where birds of correct type are present, but does not count against a bird under other conditions. The most common type of Dominique suggests a poor Rose-Combed Plymouth Rock. The typical Dominique as shown in the Standard is a very different looking bird.

RHODE ISLAND REDS

Disqualifications—Positive white covering more than one-fourth of the ear lobe.

Most Common Defects—Lack of distinctive type—Outside of a few good strains anything and everything but correct type is common in this breed. The reasons for this are: first, that the breed is comparatively new and contains an extraordinary mixture of breeds in its ancestry; and, next, that the type adopted for the breed is a very difficult one to get—at least at this stage of breeding. In trying to produce a long body with level line of back the tendency is to an attenuated, slab-sided style of bird, and when such birds are mated with more compact specimens to remedy that fault all kinds of undesired combinations in form appear. The difference in type which the Standard shows between Plymouth Rock and Rhode Island Red is one that I think it will be found must be made principally by the difference in the character of the plumage, and the carriage of the body, head and tail. The feathers of the Red should be shorter and harder, and it should habitually carry its head rather more forward, and its tail lower than the Plymouth Rock does.

BUCKEYES

Disqualifications—Positive white covering more than one-fourth of ear lobe.

Center—Standard Rhode Island Red female shape. Left—Underdeveloped Red type. Right—Coarse type. The peculiar position of the R. I. Red in the general-purpose class of fowls depends upon maintaining stocks free from the "beefy" tendency of birds of this character.

LIGHT BRAHMA COCK, HEN, COCKEREL AND PULLET OF GOOD TYPE

The males here are both good Standard models, though the cock has a tendency common in large fowls that are not perfectly well-knit to stand with the shoulders low. The hen is obviously one of the big birds well over Standard weight that Brahma breeders like—when they meet all other requirements. The pullet appears a little fine to meet the requirement for massiveness in Brahmas, but she has more of it than the average Brahma pullet that wins at the winter shows.

Most Common Defect—Lack of breed type. The Buckeye is in origin a pea-combed Rhode Island Red. Because of the opposition of influential breeders of the other varieties of the breed admission to the Standard as a variety of the Reds was refused it. Then, in order to come in as a distinct breed, it had to have a Standard description for shape different from that of the Red. The form selected may be described as a gamey Rhode Island Red type—that is, one suggestive of a combination with the Pit Game. The type in a somewhat crude state is quite common among Reds as well as in Buckeyes, but well-finished models with good size and substance are comparatively rare. Most birds are rather small and lacking in style.

BRAHMAS

Disqualifications—None for breed shape other than given under general disqualifications.

Most Common Defects—Lack of size and breed type and character; coarseness in large birds; long, loose, "cochin" feathering in stock bred to the modern style in color; short, close feathering in old-style-colored birds.

Lack of style in Brahmas is mostly due to poor work in growing them, and is most noticeable in the young birds, and the shorter feathered ones. Many loose-feathered birds appear to have weight enough but to be rather short legged. In a class of Light Brahmas that draws from the best stocks, old birds must usually have good size and type to win, but young birds lacking in these respects may stand well if they have good color and their lines are good though not well filled out. In ordinary classes very poor birds are often placed, and in the absence of good ones may win high prizes. The observation above does not invariably apply, yet there are comparatively few Light Brahmas seen that are "antique" in one respect but not in the other.

In Dark Brahmas few specimens of good size and good Brahma type are found. A good-colored bird of this variety is likely to win over anything else unless wholly lacking in type.

COCHINS

Disqualifications—None for breed shape other than given under general disqualifications.

Most Common Defects—Lack of size and type, and of profusion of plumage.

In the stocks of our best breeders of Buff and Partridge Cochins, as bred and grown by them, these faults are rare. In White and Black Cochins, and in stocks of the two varieties named that are not highly bred and well grown, they are so prevalent that a really good bird seldom comes from such sources. At the few shows that quite regularly have good classes of Buff and Partridge Cochins it takes good birds to win. Elsewhere there is little competition and very ordinary or inferior birds may win blue ribbons. The White and Black Cochins at our best shows have for years been almost without exception deficient in size and poor in type. There is practically no competition in them, and anything shown that is free from disqualification usually gets a ribbon, though occasionally a judge will either withhold all awards from an unworthy

STANDARD AND EXTREME TYPES IN COCHINS

Above—Standard Buff Cochin male and female shape. Below—A type very commonly accepted as ideal, but too low on the legs. It is this type chiefly that discredits the Cochins.

First Left—Standard Black Langshan female. Second—Standard Black Langshan male. Third and Fourth—Average good Langshan winners. Birds like these are commonly accepted as correct in type except when they come in competition with birds like the pair at the left.

class or give low prizes. If a bird is not vulture hocked it is hardly possible to get too much feathers on a Cochin.

LANGSHANS

Disqualifications—Yellow skin; yellow on bottom of feet.

This fault is a relic of the period when the Black Langshan and Black Cochin had much more in common than they now have. It is not as frequent as in earlier days, but the exhibitor of Langshans should always look for it, because it may appear at any time in stock regarded as free from it, and the judge always looks for it.

Most Common Defects—Lack of American Standard type.

In this breed the type is on a little different footing from that of most breeds in the Standard, because in England two very different types have been maintained, and while the American type is unlike either, English ideals and stock have had some influence here, mostly in directions which the Langshan exhibitor in this country must avoid if he would win in good company. The Langshans first known in England had very abundant flowing tails, and were rather low-set birds. The exhibition type that became popular there was extremely tall and stilted in appearance. The American Standard type is intermediate—a tall bird without angularity, with a good deal of substance and stateliness, and at the same time more active and energetic than the average Brahma.

LEGHORNS

Disqualifications—Red in lobes of cockerels and pullets over more than one-third their surface.

Most Common Defects—Lack of size; wrong carriage of tail; wrong shape of back at saddle; too low station; red in ear lobes. In some varieties narrow sickles and coverts.

The general lack of size in Leghorns is due to the absence of weight requirements for them until 1915, and to the popularity here of a very fine graceful type of Leghorn. The reaction from this came because, as a result of the preference for this type in the showroom, the Leghorn was being damaged in utility qualities and was losing prestige on that account. Many specimens under Standard weight are still shown, and with quality in other sections win, but judges pay much more attention to size than they used to, and dainty little birds do not win if birds approximating Standard weight are at all representative of good Leghorn type.

High carriage of the tail is very persistent in Leghorns, because it is "natural," and until comparatively recent times was considered a desirable characteristic. Comparatively few Leghorns habitually carry their tails as low as the Standard requires. Consequently the practice of bending the tail feathers of exhibition birds to give the tail the desired position has become common. The method of doing this is described in the next chapter.

The shape of the back of a bird at the saddle, or cushion, is much influenced by the carriage of the tail, and the fullness of the plumage in front of it. A high carriage of the tail makes some angle here, and if the plumage lacks fullness the appearance of angularity in the line of the back and tail is increased, and the back also appears much narrower at the rear than farther forward. All these faults can be improved by bending the feathers in the manner described in the next chapter.

Too low and too high station are probably equally common in Leghorns, but as an extremely high-stationed

Left—A pair of Black Minorcas of good Standard type. Right—Pair of Minorcas of general good quality; winners at a good show, but lacking the symmetry and finish in form of the pair beside them

Center—Standard Leghorn male shape. Left—Low-set, undersized, poorly developed Leghorn male, yet capable of winning in some competitions. Right—A common Leghorn type that, in the Whites especially, brings disappointment to hopeful exhibitors—too high stationed, coarse, lacking the neat, symmetrical curves of the model bird.

bird is apt to lack the gracefulness sought in this breed, while those a little low on the legs generally have some grace of their own, if not in just the type the Standard demands, we commonly find low-set birds more numerous in shows than leggy ones. It is often stated that the head and appurtenances count for one-third (approximately) in a Leghorn, and great emphasis is placed on quality in that section. As Leghorns are judged now at the best shows, I do not think it would be possible to demonstrate from the awards of judges that head counted as much as type. A number of our best judges of this breed frequently show that they are reluctant to give a high prize to a bird that does not at sight impress one who knows as of good Leghorn type, and observation of their awards when birds faulty in type but good in other respects are competing indicates that type carries more weight with them than anything else.

Superior Leghorn type is quite general in the best stocks of Brown and White Leghorns, but not common in other varieties. Black Leghorns are usually uniform in type but old style. In Buff Leghorns there is considerable diversity of type and rarely the finish in males that is seen in good birds of the other three varieties. Any good-colored bird with a fair comb has a chance of being well placed in good competition in this variety.

Narrow sickles and coverts are common in both White and Buff Leghorns. Few males in either of these varieties have as well-furnished tails as the average Brown or Black Leghorn. Some exhibitors pull the sickles of cockerels when they are too narrow, expecting the next pair to have more width, as the tendency is to a wider and longer feather in each succeeding molt.

Red in the lobes and white in the faces of old birds, particularly old cocks, are very common. As a rule, experienced exhibitors do not show birds in which such faults are present to a degree that calls for a heavy cut, and makes their removal or concealment difficult, though perhaps possible if they choose to take the trouble necessary to do a highly finished job. Old males with a great deal of red in the ear lobes are sometimes shown with no effort made to remove or conceal it, if they are otherwise good enough to have a chance of a place. Exhibitors will take chances on this if they are short on old males in every way fit, because that situation is so likely to affect their competitors too, that the chances of winning with a bird of this description are much better in a cock class than in any other.

MINORCAS

Disqualifications—More than one-third of the surface of the ear lobe red.

Center—Standard Leghorn female shape. Left—Underdeveloped Leghorn female, lacking in vitality. Right—Coarse Leghorn type, approaching Minorca lines, yet without the symmetry of a Minorca of finished type

Left—Pair of Standard Blue Andalusians. Right—Average Blue Andalusian winners at minor shows. In many cases the trouble with such birds as these is in the growing

Most Common Defects—Lack of size; Leghorn rather than Minorca lines in medium to small birds; coarseness in large birds; wrong carriage; excessively large combs with a pronounced tendency to lop and twist.

The typical Minorca is a big, strong, rugged bird; conspicuous for length of body and back; rather angular looking as compared with a Leghorn, yet very symmetrical. The line of the back should be nearly straight, but not level—instead the shoulders should be carried well up, giving quite a slant to the tail. Large combs are required, but it should be understood that this means large as combs run in this country where the general tendency is to a small, neat comb. The Minorca comb should be larger in proportion to the size of the bird than that of the Leghorn, yet not so large that it is obviously a burden to the bird, and with size it must have fineness of texture, symmetrical outline, and be straight and strong and firm on the head of the single-combed male, and in the female have the loop over the beak and the turn to one side of the head smooth and even, and in rose combs must have better shape and texture than is found in large combs in other rose-combed varieties.

There are few breeds in the Standard that require as skillful handling to fit for exhibition and to keep in condition as the Minorca. The comb is so large that high temperatures and rich feed—as perhaps required for conditioning in other respects—tend to grow it beyond the limits at which it retains good shape. A bird going a little out of condition will have the comb wilt in a manner most discouraging to the exhibitor. The result of the difficulty of showing Minorcas in perfect condition is that the leading exhibitors are past masters in fitting and showing them, and the novice who lacks their expertness will do well to look for shows where he can compete with men little more expert in fitting than himself until he has acquired some proficiency in the art.

BLACK SPANISH

Disqualifications—Red in the white face, except in small amount above the eye; puffing of the white face which obstructs the sight.

Most Common Defects—Puffing, wrinkling and discoloration of the white face, due to exposure, to injuries and to birds going out of condition.

The white face which is the characteristic feature of the Spanish is an abnormal development of the enamel-like skin of the ear lobe, and is subject to all the faults and defects of white ear lobes in exaggerated forms. In addition, the face contains more or less numerously, feathers and down which must be removed before the bird is exhibited.

For nearly twenty years beginning about the middle nineties the Black Spanish were almost unknown in American shows. Of late years interest in them appears to be reviving, but many of the best specimens shown here are brought from England. Outside of the stocks that are maintained by frequent importations, the Spanish shown in this country generally have moderate face development, and very ordinary quality in type and color. One can show mediocre birds and win anywhere but at a first-class show, and sometimes even at the leading shows.

BLUE ANDALUSIANS

Disqualifications—Red covering one-third of the ear lobe.

Left—Standard Ancona male and female. Right—Old-style Ancona male and female. These birds might have won in the best shows a few years ago, but are now out of place in good competition

Most Common Defects—Lack of distinctive type. The Andalusian is in type intermediate between the Leghorn and the Minorca. Small birds tend to Leghorn, large birds to Minorca type. In ordinary competition weakness of type does not seriously handicap a good-colored bird. In large and strong classes type counts for as much as color. Birds lacking in type have little chance for a place.

ANCONAS

Disqualifications—Ear lobes more than one-half red.

Most Common Defects—The shape description in the Standard is practically identical with the description of shape of Leghorns; but the ideal illustrations for the breeds are not identical, the Ancona being somewhat lacking in the finish and style of the Leghorn. The illustrations show much of the old-style characteristics mentioned as common in Black Leghorns. The situation is perhaps more aptly described as a case where improvement in color has delayed refinement in shape. The result is that at the present time we can hardly say that lack of Leghorn style is a fault in an Ancona, yet in view of what we see in the development of other breeds, it would appear that when an Ancona breeder shows birds of first quality in color, having the style of the best Leghorns,

STANDARD SPECKLED SUSSEX

that will become the style for Anconas and the type that now is accepted as correct then will be regarded as defective.

DORKINGS

Disqualifications—None for shape other than those mentioned under general disqualifications.

Most Common Defects—Lack of size and type. Outside of the stocks of a few leading exhibitors, most of the Dorkings seen here are comparatively poor to very poor type, more suggestive of beefy, low-set Leghorn type than of the typical Dorking. True type, always accompanied by good size, is the most important thing to consider in Dorking competition. It is the first point the judge considers, and a bird of good type has to be very bad indeed, in other characteristics, to lose to one of inferior type under a good Dorking judge. In the most popular variety, the Silver Gray, one who has not good typical Dorkings should keep away from shows where the breeders of high repute exhibit. So few of the other varieties are shown that almost anything that will pass for a Dorking can win unless the exhibitor happens to run into one of the small strings of good-colored Dorkings occasionally appearing at a large show. White Dorkings of good size, type and vigor are rarely seen.

SUSSEX

Disqualifications—More than one-third of the surface of the ear lobes white; yellow legs or skin.

STANDARD SILVER GRAY DORKINGS

Most Common Defects—Sussex as introduced here cannot be said to have any conspicuous shape defects. Few birds not of good size and type have been imported and the breed is largely in the hands of skilled breeders.

HOUDANS

Disqualifications—None except as noted under general disqualifications.

Most Common Defects—Shortness of body, with type and carriage too much resembling the Polish. The body of the Houdan should be very much like that of the Dorking; a little shorter in the back and fuller in the breast.

CREVECOEURS AND LA FLECHE

So few of these French breeds are shown that there is practically never any competition in them, and anything shown competes in the "Any other variety class," which, as a rule, is made up of breeds in a crude state.

FAVEROLLES

Disqualifications—None in addition to those given under general disqualifications.

Most Serious Defects—Lack of "finish." The greater number of Faverolles seen here are of fair to good size and type, but look rather "rough hewn" as compared with breeds that have been longer bred for exhibition. To some extent this is in lack of conditioning and fitting.

ORPINGTONS

Disqualifications—White covering more than one-third of the ear lobes; yellow beaks, or skin.

The general disqualifications for wrong color of shanks and feet include yellow shanks in this breed. Consistency would require that, on the same principle that wrong color of beak and skin disqualifies in an Orpington, beaks and skin other than yellow should disqualify in the American breeds. The Standard is not always consistent

STANDARD MOTTLED HOUDANS

Center—Standard Orpington male shape. Left—Short-legged Orpington male, most common in Blacks but frequently found in all varieties. Right—Big-boned, high-stationed Orpington. Many specimens of both these off-types win, but that does not contribute to the popularity of the breed.

about such matters because of differences in circumstances relating to the breeds, or difference in the occasion for emphasizing a characteristic when the Standard was adopted. When the standards for Plymouth Rocks and Wyandottes were made, it was commonly assumed that the color of beak, skin and shanks, was invariably uniform. The fact is that they are commonly but not invariably the same and, in particular, a leg that will pass as yellow, though not a good rich yellow, is quite frequently found with a flesh-colored skin and beak. The beak in an American breed would be cut for color, but the Standard paid no attention to color of skin. When the Orpingtons were first introduced many of the Buffs were much better color than most of our Buff Rocks, and in consequence they were freely used to improve Rock color, and crossbred birds were exhibited in whichever class the color of legs qualified them for. Many birds were of course not of good yellow or white or pinkish white. So to make the distinction as emphatic as possible Orpington breeders made the requirements for beak and skin color stronger than in any other breed.

Most Common Defects — Extreme variations from Orpington type.

Outside of the breeders whom constant keen competition keeps close to type, breeders of Orpingtons may be divided into two groups: those who pay no attention to type, and produce undersized, weedy-looking birds, and those who exaggerate the type, making a very short-legged, loose-feathered bird. The Orpington is a neater type than the Plymouth Rock, and both shorter on the leg and more profusely feathered, but the standard weights are only a half pound more—except in pullets, where the difference is a pound. If one will consider that this difference in weight is a little over five per cent he will see that even allowing for longer plumage, ten per cent more fullness of form than is seen in the typical Rock is ample. Many breeders seeking to make the types distinctly different get Orpingtons with lines going far beyond this.

CORNISH

Disqualifications—None except as given under general disqualifications.

Most Common Defects—Lack of type, the lines generally failing to reach the type, but in the Dark variety where good type is oftenest found a considerable number of birds appear of exaggerated type. This fault is far less objectionable than the other. In fact, many judges favor the exaggerated type, and the disadvantage of it is not in the showroom, but in the fact that the very coarse

Center—Standard Orpington female shape. Left—Extreme short-legged, long feathered type. Right—High-stationed, rangy type. Birds of this character have no Orpington quality but color of skin, shanks and beak

Left—Standard Dark Cornish male and female. Right—Dark Cornish male and female of a type widely accepted as Standard. Many such birds are very good in color and may be described as fine in type when they win, but they have little chance of a place when birds of the other type and even fair color are present.

birds as in the other large breeds are apt to be indifferent producers and inferior in quality of flesh.

POLISH

Disqualifications—None except as given in general disqualifications.

Most Common Defects—Faults in crests as already noted.

HAMBURGS

Disqualifications—Red covering more than one-third of the surface of the ear lobes; absence of distinctive male plumage in males, making them "hen feathered" like Sebright Bantams.

Most Common Defects—Lack of size and poor type. These faults are largely due to the fact that Hamburgs are largely in the hands of people who are not so situated that they can give them the range conditions that would give vitality and vigor. The preference for refined type also seeems to work against the development of good size, especially in the penciled varieties. A Hamburg that is poor in color has little chance of a place in competition, and in large classes it takes nice type to win. But where competition is weak, and really elegant birds not present, almost anything in the general range of type found in the breed will pass as good shape, provided the bird carries itself well and has attractive manner.

REDCAPS

Disqualifications—Ear lobes all white.

Most Common Defects—Lack of size and of quality in general. Redcaps have for nearly a quarter of a century been unknown in most American shows.

BUTTERCUPS

This is a non-Standard breed in which there has been a great deal of interest for the last ten years. Its recognition in the Standard has been delayed by the confusion of ideas of breeders as to the desired pattern in color. The breed is of the general Leghorn type, probably running a little larger than average Leghorns, though there is not much difference in size as observed. The original breed peculiarity, from which the name is in part derived, is the cup-shaped comb.

Most Common Defects—Poor combs; undersize with rather delicate type. The last fault appears to come from the mixture of Golden Penciled Hamburg blood in an effort to introduce precise color markings.

CAMPINES

Disqualifications—More than one-half of the ear lobes red.

Most Common Defects—Lack of alertness in carriage and manner, due to lack of vitality.

EXHIBITION GAMES AND GAME BANTAMS

Disqualifications—Combs and wattles in cocks not dubbed; duckfoot.

Most Common Defects—Oversize, with coarseness, lack of good carriage and "station."

BLACK SUMATRAS

Disqualifications—All white ear lobes.

Most Common Defects—No really marked faults of shape are observed in the few Sumatras seen; on the contrary, most of them are conspicuously nice in type, and a good proportion remarkably attractive in that respect.

Left—Standard Silver Campine male and female. Right—Coarse type of Campine quite prevalent a few years ago and still shown by some breeders

MALAYS AND MALAY BANTAMS

Disqualifications—None except as noted under general disqualifications.

Most Common Defects—Lack of breed character—if that is to be considered a defect. The Standard calls for a head with a "fierce, cruel expression." The ideal type is a bird of great height and reach, very muscular, and with plumage so hard and close that the muscularity is made very conspicuous. The greatest development along these lines is obtained by the English breeders of Malays. With the slight interest taken in them in this country a bird that in England is regarded as quite deficient in type and characteristic expression appears here as meeting the requirements very well.

SEBRIGHT BANTAMS

Disqualifications—Neck hackles in males coming down on the shoulders; sickles extending more than an inch and a half beyond the main tail feathers.

Most Common Defects—The above faults in less degree than that which calls for disqualification. The hen-feathered character in males is abnormal, and is apparently attended in the greater number of cases by a degree of sterility. Hence, comparatively few males are perfectly hen feathered, and many of those which appear so when shown have had the growth of hackle and tail feathers carefully regulated to make them as near right as possible when the birds are shown.

ROSE COMB BANTAMS

Disqualifications—None except as under general disqualifications.

Most Common Defects—None. The Blacks which are bred principally are largely in the hands of fanciers who appreciate and cultivate the type. Consequently, while there is a normal amount of variation from Standard type, few birds that do not appear quite typical in the absence of better specimens are seen.

BOOTED BANTAMS

Disqualifications—Absence of vulture hocks.

Most Common Defects—Lack of "booting"—The feathering on the feet is one of somewhat different character from that of the large Asiatics, the largest feathers resembling the flight feathers of the wing.

BRAHMA BANTAMS

Disqualifications—Same as for large Brahmas.

Most Common Defects—Oversize; lack of Brahma type. This breed of bantams is comparatively new and a large proportion run considerably over Standard weight. In the birds of Standard weight it is difficult to secure at the same time diminutive size and the proportions of the massive large Brahma. Few specimens show the desired combination, and as nearly all are deficient in shape the awards turn more on color. As a breeder, one interested in Brahma Bantams should work constantly for typical form, but as an exhibitor he is not under present conditions at all handicapped by a lack of it. Birds that are approximately of Standard weight and fair in color as compared with the large varieties will hold their own in any competition.

COCHIN BANTAMS

Disqualifications—The same as for large Cochins except that stiff hock feathers do not disqualify.

Most Common Defects—Oversize; lack of Cochin type; stiff hock feathers. The first two faults are less common and less marked in Cochin than in Brahma Bantams, for the breed is much older, and many breeders have produced good miniature Cochins. But the Buff and White varieties in particular are quite popular with amateurs, and as a result there is always considerable stock bred and shown that is badly lacking in type. Many breeders who have shown with some success, and produce pretty good colored birds, seem to have little appreciation of correct type. So we have here again one of the cases where it takes good type generally to win in strong classes, but very ordinary birds may win at good shows if a first-class breeder does not happen to be showing.

JAPANESE BANTAMS

Disqualifications—None except as given under general disqualifications.

Most Common Defects—None of importance. The type was well fixed when the breed was introduced and it has been kept mostly by fanciers.

POLISH BANTAMS

Disqualifications and defects as for large Polish.

MILLE FLEUR BOOTED BANTAMS

Disqualifications and defects as for bearded and booted breeds.

SILKIES

Disqualifications—Lack of the "silky" character in plumage; lack of crest; lack of fifth toe; vulture hocks.

Common Defects—Wrong character of feathering (tendency of the plumage to web as in the normal feather), the silky character not being absent yet not being well developed. Interest in Silkies is so limited, that provided a bird is not disqualified it can get a place in the winnings in any ordinary class.

SULTANS

Disqualifications—Beak not white nor light flesh color; red in face; lack of vulture hocks; shanks not feathered their full length.

This breed is an eccentric combination of superficial feather developments.

Most Common Defects—Poor development in the superficial feather characters, absence of which constitutes a disqualification. Sultans are so rare that, as a rule, any bird not disqualified is placed if classification is given the breed.

FRIZZLES

Disqualifications—None additional to general disqualifications which may apply.

Most Common Defects—Limited extent and degree of the outward curvature of the feathers, near the tip, which gives the frizzled appearance.

COLOR FAULTS—WHITE VARIETIES

There are white varieties in the following breeds: Plymouth Rock, Wyandotte, Rhode Island (non-Standard), Cochin, Langshan, Leghorn, Minorca, Dorking, Orpington, Polish, Hamburg, Houdan, Faverolles (non-Standard), Exhibition Game and Game Bantam, Rose Comb Bantam, Booted Bantam, Cochin Bantam, Japanese Bantam and Polish Bantam.

These varieties upon a critical examination of color are found to differ considerably in purity of color, though the Standard specification is the same for all—pure white. In selecting white birds for exhibition the same faults are to be sought in all varieties, but the prevalence and the degree or extent of fault may differ greatly. In general, color quality is determined by two things. First, by the keenness of competition in a variety. Second, by the time that it has been carefully bred for purity of color. Without competition color does not become highly improved, no matter how long the variety is bred.

In examining the birds for color faults it must be assumed that any fault which occurs in that color may appear in any bird of that color. The common faults in white plumage are the appearance of traces of black and red, and the existence at some times of a superabundance (from the standpoint of purity of white) of oil in the plumage. The oil is removable by bleaching processes, the color is not.

Traces or splashes of black or gray are oftenest found in the hackles and tails of white birds, but feathers more

COMMON FAULTS IN WHITE PLUMAGE

or less marked with black may appear anywhere in the plumage. Red or yellow-splashed feathers are not as common but are often found. A residue of red pigment in the plumage seems to have a tendency to give a faint tinge to the surface of the feathers of a section, and particularly on the hackles, backs and saddles of males, while black tends more to appear in ticks, faint stripes or patches. What is on the surface can, of course, be seen without "digging into" the feathers, but it may happen that the faults are all in the undercolor. The fact can only be ascertained by systematically working all through the plumage with the fingers in much the same way that the head of a person infested with head lice is examined for those parasites. The illustration is not an elegant one but the motion employed is somewhat similar and the same thoroughness of search is necessary. An expert holds the bird in one hand and systematically parts the feathers with the fingers of the other. One green at the work will often find it better to have someone hold the bird for him.

Disqualifications—The common formula in the Standard for the statement of disqualifications in white color is "feathers other than white." This is used for the white varieties of all breeds but the Plymouth Rock and Cornish, for which the formula is "red, buff or positive black in any part of the plumage." The latter form is apparently a survival of the period before the attempt was made to give the standards as much uniformity as possible. The form more generally used is better as it leaves no doubt about the intent of the Standard to disqualify for any other color than white and brassiness. In a strictly literal interpretation of the Standard, White Plymouth Rocks could not be disqualified for gray or blue or slate color, or any appearance of black but positive black; nor would any of the color disqualifications apply to the Polish and Games where the disqualification is omitted—erroneously, no doubt. It can hardly be questioned however, by anyone familiar with the methods of Standard making and the history of the revisions since 1898, that the intention is to apply the rule "feathers other than white" to all white varieties. The judges commonly so construe it, and the wise exhibitor removes foreign color before sending his white birds to a show.

Most Serious Defects—Brassiness. This is most frequently found in males, and in the characteristic male plumage—the hackle, the back, extending also to the wing bow and wing front, the saddle and perhaps also the sickles and tail coverts. Brassiness is usually strongest across the middle of the back and wings and may be quite marked there when little is present elsewhere.

The instructions to judges prescribe a cut of 1 to 2 points for brassiness in each section where it appears. A section in scoring a bird is an item or group of items which are cut as one in scoring. Thus the comb is a section and the eyes are a section. The hackle is a section and the back including the saddle is a section. The wings are a section—not each wing a separate section. If a bird is brassy from neck to tail the minimum cuts would set a bird so far back that it would have to be about perfect in every other respect to have any chance of a place. At a comparison show a plainly brassy bird gets no consideration at all.

Creaminess—This refers to the natural tint given feathers by the oil in them. Feathers may appear as yellow from this cause as from brassiness, but creaminess can be remedied, and usually disappears at some time after the feathers are grown. Creaminess does not disqualify, but pronounced creaminess may be cut to 1½ points in each section, and as it is usually found all over the bird it is quite as much of a handicap as brassiness. In the statement in the Standard in regard to cutting for creaminess, it is noted that the cuts do not apply where "creamy white is specified." The only case under this exception is the Pekin Duck.

In examining birds for creaminess particular attention should be paid to the quill. And in distinguishing between brassiness and creaminess it should be observed that brassiness is most pronounced where the feathers are exposed, but creaminess is most marked in the undercolor. A bird that has a good white quill—especially in the large quill feathers—will generally, after the feathers are full grown, appear perfectly white until compared with a washed and bleached bird.

Color Defects in Shanks and Beaks—In white varieties there are three different combinations of plumage with shank color. The most common is white plumage with yellow shanks, beak and skin. This is required in all varieties but the White Dorking and White Minorca, which have what appears as the natural combination—white or pinkish white; and the White Polish and White Hamburg, which have dark shanks and beaks. In all cases but the Minorca, the color of shank and beak for the breed has been determined for the colored variety or varieties, and the effort has been made to make the color of the shank of the white variety conform to that for the breed. The usual result is that the shanks and beaks of white fowls are pale compared with those of the colored varieties of the same breed. At the same time, and especially in the more numerously bred white birds with yellow legs, many specimens are found with pure white plumage and good, rich yellow legs. The occurrence of this combination, and the method of its production can be best understood when it is noted that the case parallels that of the combination of white plumage with black or slate legs, and when it is also considered that the scales on the shanks and feet are modified feathers (or vice versa, if one chooses to take it that way), and that the distribution of scales on the leg, like the distribution of feathers elsewhere on the bird, is not uniform over the entire surface but in places the skin is bare.

In a white fowl with dark shanks we have a phenomenon similar but opposite in form or arrangement, to a black fowl with a white crest, or a dark fowl with a white hackle. The shank takes its color partly from the scales and partly from the skin, which is in some places bare and shows through the scales, and in some does not show through them at all. In just the same way the scales on the shanks of a white fowl may be yellow, and—in fact—they must be yellow if the leg is to be naturally a rich yellow. If they are not they tone down the yellow of the skin, for it is seen through a white scale, not through a yellow scale. And if the skin itself is pale the result is a very pale leg. When we breed rich yellow legs on white fowls we are making the same combination as if we breed a white fowl with a yellow hackle. When breeders of Standard poultry generally, appreciate this fact and select accordingly in breeding there will be a great many more birds with good color in the legs in breeds and varieties where now it is commonly held that the combination is "unnatural" and cannot be made with high quality in the unlike colored sections.

In the white varieties of nearly all the breeds color in the shanks and beaks is usually inferior to color in the shanks and beaks of the colored varieties of the same breeds. Consequently, in judging white varieties allowance is usually made for weakness of color of shank as correlated with white plumage. Not only so, but good strong color in yellow legs of white fowls is so easily made by artificial methods that judges, as a rule, do not attach anything like as much value to leg color as the novice usually does. Probably nine-tenths of the beginners with Standard poultry regard good color of legs as a special merit, and the lack of it as a very serious fault. Judges and old breeders like good leg color when they can get it with other good points, but extra-good leg color does not count much to the credit of a specimen generally inferior.

COLOR FAULTS—BLACK VARIETIES

There are black varieties in the following breeds: Wyandotte, Java, Cochin, Leghorn, Minorca, Spanish, Orpington, Hamburg, La Fleche, Crevecoeur, Exhibition Game and Game Bantam, Rose Comb Bantam, Cochin Bantam, Japanese. The Standard (modern) Black Java is substantially a Black Plymouth Rock, the differences in type being largely due to difference in character of plumage.

The common Standard formula used in describing black color is "lustrous greenish black." In a few cases this is varied to "lustrous, or rich black with greenish sheen." The undercolor of black fowls is variously described as black, "dull black," or "dark slate." There is quite as wide range of shades in black as in white.

Disqualifications—There is not the general uniformity of statement in regard to disqualifications in black color in the different varieties as was noted in color disqualifications for white varieties. Yet it will hardly be questioned that the intent is to make the rules for color disqualifications as uniform as is consistent with existing differences in the improvement in color in the several varieties. The best formula, and least open to dispute, is "feathers other than black in any part of the plumage,"

COMMON FAULTS IN BLACK HACKLES, WINGS AND TAILS

this being qualified in the case of certain varieties by the statements of the precise amount of white which is tolerated, without disqualification, in certain places in some black varieties. These exceptions are: In Black Wyandottes, Black Langshans and Black Orpingtons, white extending more than half an inch; in Black Cochins and Cochin Bantams, white foot and toe feathering; in Black Minorcas, La Fleche, Crevecoeurs and Rose Comb Black Bantams, white extending more than one-half inch—two or more feathers tipped or edged with positive white.

The white in black plumage appears mostly either at the top of the flight feathers, at the base of flight and tail feathers and at the base of hackle feathers near the junction of neck and back. Near the tips of the flight white often appears in only one or two feathers in each wing, and is not always a positive white, but is a shade of gray. Most judges do not disqualify for this unless the spot is quite white and also comes under the provision as to extent of white calling for disqualification. White here is often temporary—due to lack of prime condition. White at the base of the feathers is usually permanent and steadily increases with each annual molt. Either a single feather, a few feathers or a large group of feathers, may show white at the base, and it may vary from a very little white next the skin to several inches. Where as extensive as that it soon shows on the surface, but a great deal of white may exist without appearing on the surface. The defective feathers, if in the soft plumage and not too numerous, can be removed. Small patches of white may be stained or dyed black. Occasionally considerable faults are so treated, but while that may escape the notice of a judge, it is apt to be discovered if the occasion arises to compare the bird with others section by section, in a good light.

Most Common Defects—White or gray when not a disqualification; purple barring; reddish or "rusty" tinge; lack of green sheen.

COMMON FAULTS IN BLACK BODY PLUMAGE
White in undercolor and on edges, making the "frosting" which often mars the appearance of a black section, especially on body and breast of males having the black-white combination.

White or gray when not a disqualification is cut in each section where it appears at least ½ point and up to the limits of points alloted to color in each section. This gives a judge a great deal of latitude, and most judges are rather severe on this defect.

Purple barring is an almost universal fault in black color. The commonly accepted theory is that it results from an excess of pigmentation, but some breeders maintain that it is itself a character—that the barring is not dependent upon the amount of pigment, but may occur in one bird with much less intense pigmentation than another bird that is free from it. The number of black fowls that upon close examination are found absolutely free from purple barring is very small. The usual presumption is that a bird has more or less of it. The question is whether it is bad enough and found in enough sections to put the bird hopelessly out of consideration in competition. Faint purple barring is sometimes seen only when the bird is held in the right light to show it up. The barring is most prevalent on the neck, back and tail, but may be found nearly all over the bird if every part is turned to a light that will bring it out. It is usually more pronounced in males than in females. It can be largely, if not entirely, reduced for show purposes by the methods which are described in the next chapter.

A reddish or rusty tinge in black color is usually found in birds of different breeding—birds of stock that has been bred with no attention to the tendency of red to increase in it until the plumage is in places, and perhaps nearly all over, a mixture of dull black and dark brownish red. It is useless to attempt to show such specimens in competition with black birds for they look very dingy by comparison.

The green sheen on black which is so much desired, and so seldom found in perfection, appears to be in part produced by the structure of the feathers and in part by the character or condition of the pigment in them. Its exact nature has not been determined. It appears to be found only in association with good black color, and to be incompatible with much purple barring, but to be always present with good black, or in the absence of purple barring. In any ordinary class of black fowls a moderate green sheen will pass as good; indeed, it will appear very good unless it has to compete with a bird with a phenomenally deep and brilliant green sheen.

Color Defects in Shanks and Beaks—In the black varieties of yellow-skinned breeds—Wyandotte, Leghorn and Cochin—clean yellow legs are rare. The shank is usually quite dark—a blackish horn color on the upper part of the front where the scales are largest and heaviest, shading through a greenish willow to a dusky yellow, on the lower sides of the shank and the underside of feet and toes. The yellower the legs the better, but in the scarcity of good yellow legs, weakness in this point rarely counts against an otherwise good bird, especially if where the foot is yellow it is a good yellow. In these breeds the bill should be yellow or yellow shaded with black. In the Black Java the legs are nearly black and the bill black. In all the other varieties the shanks are either black or very dark. No attempt is made to make the color of the shank and beak conform to that of the varieties of the same breed which have white or pinkish-white skin farther than that where the skin is bare on the feet it must be of the color of the skin of the body. Hence, while in the few black varieties of yellow-legged breeds the preference is given (other things being equal) to the yellowest legs, in the other varieties the black leg is preferred.

COLOR FAULTS—BUFF VARIETIES

There are buff varieties in the following breeds: Plymouth Rock, Wyandotte, Cochin, Leghorn, Minorca and Buff Cochin Bantam; and buff laced varieties in the Polish and Polish Bantams.

In the solid buff varieties what is required is a "rich golden buff" uniform all over the bird. In the buff laced varieties the plumage is a rich buff laced or edged with a paler buff. This color pattern has the rather unique distinction of being the only common and popular one in which no color fault has ever been made a disqualification. The defects in it have always been treated in the Standard as serious defects—but not as disqualifications. I have known instances, however, where judges applied general formulas for color disqualifications to buff birds, and I imagine there have been many such, and also that many exhibitors in selecting have supposed that the common rules about foreign color in plumage applied.

Most Common Defects—Lack of uniformity in shade of buff; presence of black or white, or of both.

There is little to choose between these faults. Black and white, if quite strongly marked, greatly mar the appearance of a buff bird even in the eyes of a novice not critical about uniformity in shade of buff. The presence of these foreign colors seems to him a much more serious fault than any unevenness of shade which he observes.

But to the critical breeder and judge of buff color general evenness and soundness of color with a little black or white, or black and white, in the sections where these faults are most apt to appear is much more to be desired than absolute freedom from foreign color; for lack of uniformity of buff color spoils the appearance of the whole bird; while at the minimum at which it is comparatively easy to keep them black and white appear as mealiness in the flight feathers, where they are more or less concealed. Even a slight peppering of black, or a little mealiness on the coverts or in other sections may be tolerated rather than the presence of several shades of buff in the different sections of the plumage, and sometimes on the same feather.

Lack of uniformity of buff color in different sections has been in a marked degree eliminated from the best strains of nearly all buff varieties. A class of buffs at any good show now usually presents a very pleasing uniformity of shade, when seen from a little distance. But comparatively few of the most even looking birds appear sound in color when closely inspected, and it seems obvious to anyone familiar with the behavior of buff color in reproduction that the most common cause of such unevenness of color is the mating of specimens too far apart in shade of color—not what is commonly considered an extreme mating, but still with more difference between the colors of the parents than is desirable if soundness of color is to be obtained. In general, there is a tendency for the light and dark shades of parents that were themselves quite uniform in general shade of color to take the form of lacing —a light lacing on a darker buff surface, and very often this darker center is divided by "shafting" as light as the edging, making a very uneven surface. In some cases the lacing appears very regular. In others it is irregular and more or less mixed with irregular blotches of light color. The difference between the two shades of buff appears to be much exaggerated by exposing the birds to sun and weather. The plausible explanation of this is that the parents differed in respect to fading, and that the birds may inherit the capacity to hold one shade better than the other.

There is frequently a dark, instead of a light edge on a buff feather. This is most conspicuous on the backs of females. It is quite invariably a glossy edging, and appears to occur oftenest in lines in which the males have good luster in the male plumage. It is less objectionable than the white edging because it does not give the slightly washed-out appearance that goes with the other fault. Undesirable as these faults are, birds that have them only in a minor degree—that appear sound colored at a little distance or in poor lights, and are generally even in color— are much better birds to show than those of extra-good sound color but with different shades in one or more sections. In fact, judges generally will favor birds that to those not familiar with their practice, appear to lack something of "rich golden buff" rather than those showing any difference of color between sections. Most judges are also especially severe on birds in which the hackle is of a pronouncedly different shade from the back. Some difference in the shade of the tail will usually be tolerated, especially if the tail is a little dark. Many males have tails that are of a chestnut shade. If this is fairly sound in color, and the colors of saddle, lesser coverts and the remainder of the tail are graded and blend without showing too abrupt a break from back to tail color, a bird with a reddish tail may be placed well, when a bird with a like fault in the hackle, or across the back and wings, would not get looked at a second time.

Left—Good buff color, one even shade throughout. Right—Poor buff color, varying shades in different sections and mottling and shafting all over

Absolutely good sound buff color all over the bird is extremely rare. Many of the best bred birds have it on the outer surface of the body (all parts) but fail in wings and tail. Many specimens that one not trained in the examination of buff color would say were even in shade "from head to tail" are found to be uneven when the head of the bird is bent back so that a direct comparison of the color of the hackle with that of the middle of the back and the saddle can be made. Where the difference is so slight it is really imperceptible when the bird is in a natural position, the perfect blending of shades between the back and neck sections deceives the eye and gives the impression of absolute uniformity. The judge does not trust to his eye alone but tests the uniformity of color by comparison of the extremes.

In general, it may be said that a bird that appears a good uniform buff as it runs in the yard, and that upon examination is found to have no serious faults in the concealed parts of the plumage has a chance of being placed in any but the very strongest competition in buff varieties. At the best shows, and where club shows occur, it usually takes a bird of good color all through, only one or two sections at all off, to get a place. Probably more quality in color is required to win in strong competition in Buff Orpingtons than in any other buff variety, for the practice of shading birds while their show plumage is growing is general among the leading Buff Orpington breeders. They either have uncommonly good natural shade, or provide what is required.

The Buff Cochin, being far the oldest of the buff varieties and highly improved in color before any of the rest appeared, and being bred now almost exclusively by fanciers of long experience with it, has uniformity of color better established than any other buff variety, and in the stocks of fanciers few birds are found that are not of pretty good exhibition color, the exceptions being the darker birds used to keep color from running too light. Buff Cochin Bantams, being much more popular and in the

DEFINITE SELECTION OF BIRDS FOR EXHIBITION

hands of many novices, do not present the same uniform good color. Buff Minorcas are new, with good buff color very rare. Buff Plymouth Rocks, Wyandottes and Leghorns are extremely variable, but with few birds shown anywhere that appear to have had the care taken with the plumage while growing that would preserve the shade of color in its original value. Most buff birds will show better color in the coops in the showroom than in a strong outside light, because the slight differences in color are

Upper—Good red color, the same shade in all sections. Middle and lower—Poor buff color varying in the different sections, showing patchiness in many sections, and full of mottling and shafting

not perceptible in modified light. For this reason a novice in showing buff birds often underrates the value of birds he sees in a bright light in the yards, as compared with some he has seen in a subdued light at a show, and I have known men to buy a bird at a show to improve the color of their stock, only to find when they took it home that in a clear light it was inferior to their own.

While the description in the Standard has remained unchanged since 1898 the shade of buff in favor varies more or less from time to time. Different judges also have their preferences for particular shades, and in general, rather light birds get the preference at eastern shows, while in the West medium birds are favored. In some show seasons buff birds of all varieties appear unusually sound and even in color, while at other times it is hard to find a specimen that shows prime finish and quality in color. The difference is probably attributable to weather conditions. To establish the cause with any degree of certainty would require a more thorough inquiry into the conditions affecting different flocks than it is practical for anyone to make. The point of interest to an exhibitor is that when color generally is better or worse in his Buffs than he expected it to be, the probability is that most of his competitors are in the same position.

The buff laced varieties are rare. Little interest is taken in them, and anything that is passably representative of the type usually is given an award when shown.

Undercolor—The Standard requires undercolor a lighter shade of buff than the surface. This means lighter than the Standard color for the surface, not lighter in any case than the surface of the particular bird under consideration. The distinction is of more importance from the breeder's than from the exhibitor's standpoint, but in some cases may be of consequence to the exhibitor—especially when showing under a judge who has a strong bias in favor of good undercolor. It is often stated that the undercolor is invariably lighter than the surface. That is true as a general rule, but it has exceptions, and these exceptions are most likely to be found in birds with light lacing on the surface. In such cases, and particularly in score-card shows, the strength of the undercolor may offset the weakness of surface color. As a rule, at comparison shows, superior undercolor will not place a bird ahead of one equally good in surface color unless decision comes to the undercolor after the judge has been unable to decide between the birds on any external point.

The most common defect in buff undercolor is lack of buff color. In some specimens, particularly in old females, it is white—or nearly so. This is not a fatal handicap to an otherwise exceptionally good bird, for in buff varieties the penalties for white in undercolor cannot be applied as in black and black-red varieties, and it is practically impossible to avoid having a good deal of what elsewhere would be called positive white in the undercolor of buff birds as long as very light buff is favored in judging.

Slate in the undercolor of buff varieties is a fault that belongs to their early stages of development. There is no excuse for it today in any but the new Buff Minorca.

Color of Shanks and Beaks—All buff varieties, except the Orpington, Minorca and the buff laced varieties of Polish, have yellow legs and beaks, and in these faults of color are rare. In the Orpingtons and Minorcas the tendency is to yellowish shanks, so that except where they have been long carefully bred for proper color in this section many otherwise good birds have to be discarded for exhibition purposes.

COLOR FAULTS—RED VARIETIES

There are three red varieties of fowls: the Rhode Island Red (in two subvarieties, rose and single comb), the Buckeye and the Red Sussex. As these all contain some black in wings and tail, and ticking of black is required in the hackle of the Rhode Island Red female, red is not classed as a "solid" color, and in making sweepstakes awards is not handicapped as buff, black and white are. The Standard color description is different for each of these red varieties.

The Rhode Island Red is described differently in different sections as, neck: in male "rich, brilliant red"; in female "rich red"; body and fluff: in the male "rich red"; in the female "red." These descriptions indicate difference in brilliance between the male and female. But in the general description of plumage color it is specified that the less contrast there is between sections the better, and that "harmonious blending in all sections is desired." This specification being in accordance with the general principle in the making and application of standards—that self or solid colors shall be of even shade all over the bird—the differences in descriptions of color for different sections lose their significance and really become misleading to those who attempt to follow the Standard literally. The detailed color specifications in the Standard really are descriptions of the best birds of the time. The specifications for the ideal are in the general description of plumage. The birds now, in considerable numbers, have a very uniform shade of red all over.

The Standard does not describe any particular shade of red. Rhode Island Red breeders and judges favor in general two different shades of dark red. One group prefers the brightest dark red found that has no suggestion of wine color, maroon or brown. The other—more numerous—group prefers what is often described as a "rich wine color." Both groups agree in demanding one uniform shade of red all over the bird; and most judges will give a bird of uniform, even color on the surface, but of the shade they do not prefer, place over a bird of the shade they prefer but lacking in uniformity.

The Buckeye is described in the Standard as "mahogany bay" in color, with undercolor red, except on the back where a slate bar is specified. In the description of the Red Sussex the terms used are "mahogany red," "rich mahogany red," and "lustrous mahogany red"; and the requirement for undercolor is slate "shading to red at the base." It is important to note the requirement for slate undercolor throughout the Red Sussex and in the back of the Buckeye, because in the list of cuts for defects in the Standard these cases are not excepted, as they should be, from the provision for specific cuts for slate undercolor in buff and red varieties "in each section where found."

The term "mahogany" as applied to red and bay is quite as indefinite as the terms "rich" and "brilliant." Without close regard to Standard definition the tendency of judging in both Buckeyes and Sussex is toward the shades popular in Rhode Island Reds. In view of the history of standards for buff color, it is presumable that red colors will go through the same changes—they are quite evidently doing so now; and the exhibitor of Buckeyes or Sussex who has not the opportunity to see these varieties at the shows where the best classes appear, but who has opportunities to see representative classes of the more popular Reds will find it to his advantage to use his observations on Red color as a basis of determination of the proper color—in the estimation of most judges—for his variety. In the case of the Buckeye something a little lighter and duller than the lighter popular shade in Reds will meet the requirements of discrimination according to Standard descriptions. In the case of the Sussex, the requirement for slate in undercolor on a dark red fowl leads to a darker surface than can regularly be secured with red undercolor.

Disqualifications—In Rhode Island Reds and Buckeyes, entirely white feathers; in Sussex, white feathers showing on the surface. When in the soft plumage such feathers are removed.

Most Common Defects—Very light or very dark-colored birds; different shades of red in different sections; shafting; unevenness of color of red in a section or on a feather; light lacing and mottling; dark glossy lacing; black peppering or stippling in red surfaces. In the Rhode Island Red female, lack of black ticking in the hackle; and in the male, presence of red in the tail where black is specified.

It is often remarked in regard to Rhode Island Reds, that as the Standard specifications for red lack definiteness, any shade of red that might be considered a rich red, or a brilliant red is "Standard." Technically this may be correct, practically it is not, for where the language of the Standard admits of such a wide range in application, common-sense demands that custom or use shall determine what the words leave indeterminate. As has been pointed out, custom among Rhode Island Red breeders admits as meeting the requirements of the Standard only the shades of red near and between a very dark red that is still unquestionably red, and a deeper red with a wine color or maroon tint. On the light side the color runs to a bricky, yellowish red, and on the dark side to a reddish brown. A bird cannot go very far from the range of popular shades in either direction and get any consideration in good company even though uniform in its (undesired) shade of color. In a small class of ordinary quality it might be placed, but in nearly all shows of any importance Rhode Island Reds are one of the largest classes and it takes birds of the popular color to win.

Although, as has already been stated, Rhode Island Reds are well in advance of the 1915 Standard in the matter of uniformity and evenness of color, a great many birds lacking in that respect are produced and shown. There is a liberal sprinkling of them even in the largest classes and at the best shows. Absolute uniformity is in fact comparatively rare, but the birds that win in good competition mostly have a uniformity of shade which leaves little room for criticism on that point. Shafting is common in this as in all varieties. In moderation it does not set an otherwise good specimen back much, and even when it is quite pronounced a bird that has a good deal of general color quality may be placed in spite of the fault.

Unevenness of color in a section, feathers of different shades, or different shades on the same feathers, come —as in buff— from the mating of birds differing too much in shade of color. Dark glossy lacing in females results from matings to produce extreme brilliance in males. It is considered undesirable but only slightly handicaps a bird of good color and type. Peppering is most commonly found on the wing bows and tail coverts of very dark red birds, and more often on females than on males. It is often so fine and inconspicuous that it is not noticed until a bird is examined closely. Some birds have a lot of this fine peppering quite invisible except on close inspection. A little peppering is not of much consequence, but a noticeable amount of it will put a bird out of consideration.

Lack of black ticking in females is not severely punished by most judges. On the other hand, lack of intensity of black in the tails of males, and the presence of a pronounced bronze effect due to red in tail feathers, is commonly regarded as highly objectionable. A few breeders, and fewer judges, lean toward the "red" tails, but the greater number object to them.

DEFINITE SELECTION OF BIRDS FOR EXHIBITION

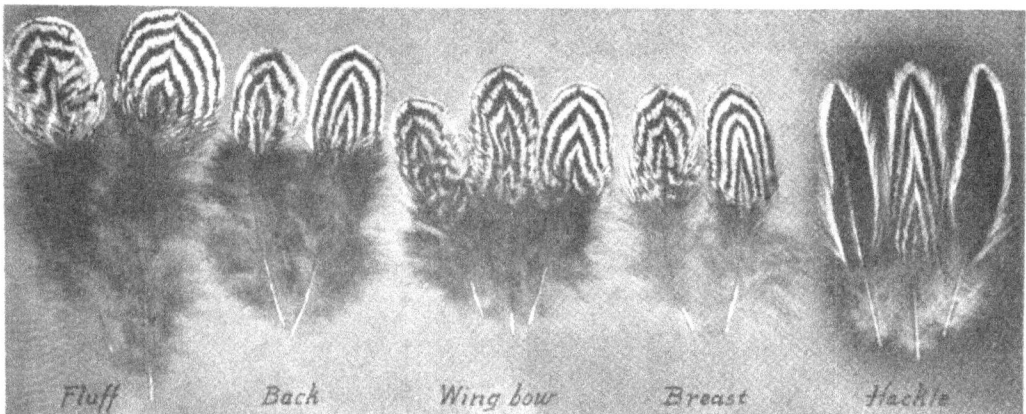

PENCILED FEMALE FEATHERS SHOWING A GOOD FEATHER AND ONE OR MORE FAULTY FEATHERS IN EACH GROUP

Undercolor—In Rhode Island Reds good undercolor counts nearly if not quite as much as surface color. Most breeders and judges are cranks on the subject of strong undercolor, and it gets a good deal of consideration along with the surface color when the values of a bird are being summed up. The chief fault in undercolor in this breed is slate, which is very apt to be present to some extent in birds with intense color in the black sections, or parts. Slate in one or two sections will not hurt a bird much in competition, but appearing in many sections it will.

Color of Shanks and Beaks—In the Rhode Island Red these may be either rich yellow or reddish horn. The most common faults are a pale yellow verging on a flesh color, and a pinkish-white leg and beak. These are not common in well-bred stock, but quite so in stock of ordinary quality. The darker birds usually have reddish horn legs and beaks, and with the dark red most in vogue yellow legs are not as numerous as in the early history of the breed. With the dark birds most favored, the general tendency is to accept the reddish horn leg and beak as most desirable. But on an interpretation of the Standard according to the general precedents that when alternate specifications are given the one first mentioned is the preferred form, a dark bird with yellow legs and beak should win over one equally good in every respect, but with reddish horn legs and beak. A point of interest to the exhibitor of Rhode Island Reds is that in this breed the red on the outside of the shank, which in most breeds is rated a blemish on a yellow leg is considered meritorious. It was made a requirement on the theory that it invariably accompanied vitality, and that emphasizing it would help to insure the selection of vigorous specimens only for exhibition.

The Buckeye is required to have a beak of yellow shaded with reddish horn, and yellow legs. The absence of horn color on the beak counts as a fault and calls for a slight cut. The Red Sussex has a horn beak and white shanks and toes—more properly described as pinkish white or flesh color.

COLOR FAULTS—PENCILED VARIETIES

A penciled color pattern is properly one with two or more darker stripes on a light ground, following the outline of the feather, being crescentric near the tip and approximating straight lines where they run parallel to the quill and edges of the feather. It is unfortunate that the Standard describes as penciling the fine barring on the two so-called penciled varieties of the Hamburg, for the same term ought not to apply to two such different patterns as transverse stripes and circuitous stripes. The cause of the inconsistency is in the loose application of terms in the early days. In systematic consideration of color patterns we have to make proper distinction and discuss the "penciled" Hamburgs with barred patterns.

Of the penciled varieties there are two groups having the same pattern, but in different colors: —one being

FEATHERS FROM HACKLE, WING BOW AND SADDLE OF SILVER PENCILED MALE
It is such irregularities as are plainly seen in the defective feathers here that give the top color of males of this color pattern a smoky, dirty tinge

golden penciled—that is, having a golden or red or reddish ground with dark pencilings of brown or black; the other silver penciled—that is, having a white or light silvery gray ground with pencilings of darker gray approaching black.

The penciled varieties take that description from the markings of the females which are penciled all over while the males have no penciling at all but have their light and dark color massed in different sections, and are in color just like the males of several varieties in which the females are not penciled. The general pattern of these males is a black breast, body and tail, with the top color—hackle, back and saddle—in one group red, in the other

MISMARKED FEATHERS FROM BACK AND SHOULDERS OF SILVER PENCILED MALE

white. These patterns are often spoken of as black-red and black-white. In females of the black-red penciled pattern the ground color is a buff, bay or red, with brown or black penciling. In females of the black-red pattern the ground color is white or a light silvery gray and the penciling is a darker gray sometimes approaching black. Except that the color terms are different the characteristics of the patterns in these two groups are identical; whatever applies to one applies to the other by substituting the appropriate color term.

The typical penciled pattern is that in which each feather has three or more distinct pencilings. The varieties having this pattern are:

Black-Red (also called Partridge)—Partridge Cochin, Partridge Plymouth Rock, Partridge Wyandotte.

Black-White—Dark Brahma, Silver Penciled Plymouth Rock, Silver Penciled Wyandotte.

Disqualifications in Partridge Varieties — White in main tail or sickle feathers or in flights.

In silver penciled varieties there are no special color disqualifications.

MOST COMMON DEFECTS

In Partridge Varieties—Gray in tail or wings; red in the black sections of males; unevenness in shade of red top color; poor striping in the hackles, backs and saddles of males; smutty edging on the hackles in females; lack of regularity and distinctness of penciling; wrong shade of color and lack of uniformity in the lighter ground color; lack of intensity in the color of the dark pencilings.

In Silver Penciled Varieties—White or gray in the black sections of males; poor striping in the hackle, back and saddle; brown or red, or smutty gray in the white of top color. In females: lack of regularity and distinctness of penciling; reddish surface tinge in the dark penciling.

In Males of Both Types—In selecting a male of either Partridge or Silver Penciled pattern for exhibition the first thing to consider is the black breast and body which are the most striking features of the bird as he stands in the exhibition coop. The blacker these sections, the better; but a little red or white should not condemn a bird without further consideration, for the proportion of birds that are absolutely sound and black and clean in color and markings of the top sections are small, especially in the silver penciled varieties where a little dark color in the wrong place is more striking than on the red ground.

The next point to consider is the unevenness of appearance of the whole top color—from head to tail. In the Partridge males the red should be as rich and dark as possible and still have the black stripes show with some distinctness. Most of the males that win prizes in the best competition in Partridge Rocks and Wyandottes are rather darker red than the winners in Partridge Cochins. In the two first-named varieties a novice will usually find the judge favoring a bird rather darker in red than he thinks meets the specification for some distinctness in contrast between the red ground and black striping. In the silver penciled males comparatively few birds are found that have clean silvery ground color. In fact, a large proportion of winners are rather brassy and smutty looking, and some very much so. It should be observed however, that in silver penciled males that win with these faults the striping is generally strong all through. Most judges prefer such birds to one with a clean looking top surface but narrow and weak striping.

In Females of Both Types—The first color considerations are evenness of general shades of surface color all over the bird, with distinctness of penciling also as nearly uniform as possible in all sections. The best exhibition female is the one that is noticeable at first sight for these qualities. To assure himself that his impression that a bird has these qualities is correct one must examine it critically in each and every section. On a snap judgment he may easily be mistaken because he is taking his impression from some one or more sections that are conspicuously good, and fails to note that one or more sections may be poor. In Partridge varieties a rich, mahogany-bay ground is preferred, but uniformity of ground color in all sections is more important than any particular shade of ground color. In the silver penciled varieties the ground color is usually uniform unless some sections have more flecks of black in it than otherwise; and the most conspicuous common faults are weakness and irregularity of penciling and a reddish tinge on the surface. In both color types the general tendency is for the breast to be lighter color than the back because the dark penciling is a little narrower and both colors a little lighter in shade.

The correction of general faults of color is a matter of breeding. When it comes to the selection of specimens for exhibition nothing can be done to remedy them. The question simply is how far can a bird fail in such respects and still have a chance of winning. It is hardly possible to indicate that degree of fault in general appearance in words. It varies according to the keenness of the competition and the ideals of the judges. Only a small portion of the stocks of these varieties—a few of the best strains, and not all the birds in those strains—is of really high quality by critical standards, for the pattern is a difficult one and it is only in recent years that most of those working with it have begun to get fair control of the production of it. Except as he may have stock direct from one of the best strains the novice is not likely to have anything wonderful. If he supposes he has, the probability is that he cannot see the common faults. If he is able to see these faults and to distinguish differences in value in different sections, I think it is safe to advise him that except in the best classes birds that seem to him pretty good—yet a little disappointing in quality of color in most particulars—will have a fair chance of winning.

Nearly all birds of these varieties (and females especially), however good the markings in general, have many mismarked feathers in the soft plumage the removal of which will greatly improve the appearance of the bird. Caution should be used in plucking these feathers for the removal of too many of them—particularly the removal of a group that are alike defective—may expose too much of the undercolor, thus spoiling the appearance of the surface more than the original defects.

Color of Shanks and Beaks—In all these varieties the beak is horn color shading to yellow in the males, and yellow or dusky yellow in the females. In scoring a bird a dusky yellow shank could not be cut, but in case of a tie a yellow shank would have the preference, as it does—other things being equal—in comparison judging.

The Dark Cornish—The color of this variety is a modification of the Partridge or golden pattern. The breed is short feathered and prized especially for its type. The males are mostly so dark that the contrast in black and red in top color is lacking and so details of their combination are not given much attention. Richness of color and high luster are the points most esteemed. In the females the color is the same as in the Partridge varieties but with two pencilings on each feather. A bird with triple penciling is not disqualified, but is subject to a discount for fault in every section where this appears.

COLOR FAULTS—STIPPLED VARIETIES

The males in this pattern have the same arrangement of colors as the males in the penciled breeds. It is the females that are stippled and have the combination of colors from which the general shade results that gives to the black-red type the designation "brown," and to the black-white the designation "silver."

Of the black-red color there are three varieties: the Brown Leghorn—now in two color subvarieties, light and dark (these being shades of the same color and

WING FEATHERS OF SILVER PENCILED FEMALE CONTAINING AN EXCESSIVE AMOUNT OF WHITE
A bird with as poor feathers as these in the wings gets little consideration when competition is at all good

pattern, and not different colors and patterns as in the Light and Dark Brahma); and the Black-Breasted Red Exhibition Game and Game Bantam.

In the black-white pattern there are: the Silver Leghorn, Silver Gray Dorking, Golden and Silver Duckwing Exhibition Games and Game Bantams.

Disqualifications — In Brown Leghorn males; lack of evenness and richness of top color; lack of stippling; gray in wings and tail. In females: lack of evenness of stippling; reddish or ashy tinge to plumage; light shafting; gray in wings and tail; in some stocks, bricky patches in the dark parts of flight feathers.

As the Brown Leghorn is now divided into light and dark subvarieties for exhibition as well as in breeding, and the breeders are about evenly divided on the wisdom of that act, while the breeders of other varieties that are regularly double mated appear to be generally opposed to making two color subvarieties, it is not possible to give exhibitors advice in regard to these varieties with the same confidence as where the precedents of the showroom are well established. Neither the New York nor Boston Show have at this writing made classifications in Brown Leg-

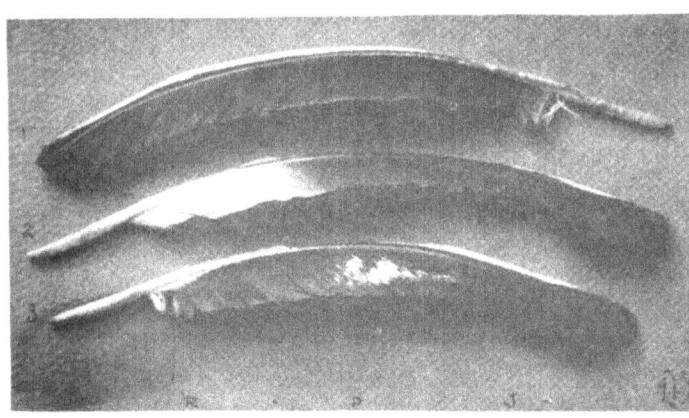

FLIGHTS OF PENCILED, LACED OR ERMINE MALE
1—average-good flight feather of black-white male. 2—flight feather with white at base. 3—white at base of quill and in wide upper web

horns to meet the changes in the Standard, and the eastern breeders of Brown Leghorns reinforced by others opposed to the policy of splitting color varieties are disposed to urge the repeal of the legislation that divided this variety. In this situation it seems to me advisable to discuss the variety first on the old basis, and then indicate the features that require further consideration where classification of Light and Dark Brown Leghorns is provided.

brown effect particularly noticeable on the back of the birds of prime color. The tone of color should be emphatically brown—neither a reddish brown nor an ashy medium, and as some judges lean to the light ashy brown, and some abhor ashiness and will favor a reddish brown bird rather than one they regard as having a "washed-out" appearance, it is quite important for an exhibitor to know the ideas of the judges on this point.

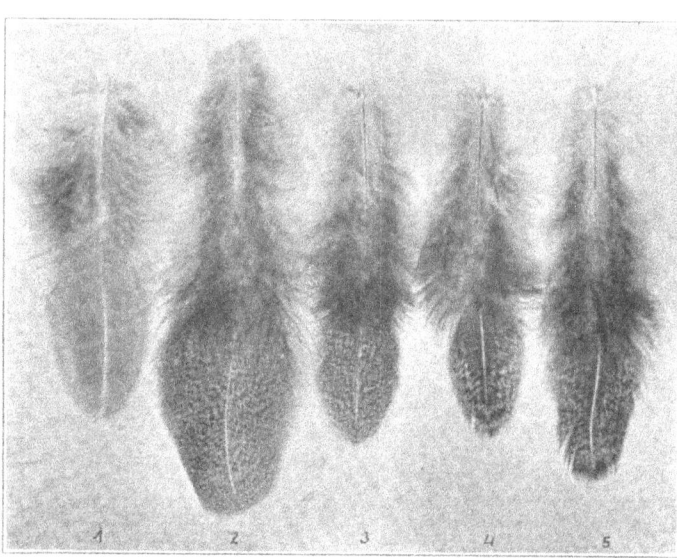

LIGHT STIPPLED FEATHERS FROM SILVER GRAY DORKING HEN
1—breast; 2—cushion; 3 and 4—back; 5—cushion

The average good exhibition Brown Leghorn male meets the full Standard specifications better than the average good exhibition female. He rarely fails much anywhere in quantity of black color, but the breast may have some purple barring not noticeable until observed in the right light. Then, if he has uniform rich and rather deep red top color, he may fail a little in striping, and may even lack striping in the saddle, yet get a place in pretty good competition—provided he has good type and style and a good head. It is only relatively gross faults of detail in color that count strongly against him, if the general effect in color is pleasing.

In the female color details are all important. To win in good competition a Brown Leghorn female must have attractive shape and carriage, and a good head, but practically all the birds against which she has to compete will be good to very good in these respects and so decisions will turn on color in a pattern where it is the little details that give finish that count, because they are everywhere conspicuous. The stippling on a stippled feather has a natural tendency to form patterns, particularly an irregular, finely penciled marking in which the pencilings are not continuous lines but broken lines formed by the arrangement of separate points.

In all Standards up to 1905 the marking of Brown Leghorn female plumage was described as "finely penciled." The velvety soft surface shade that is desired is obtained when the stippling is pure stippling with no suggestion of lines in its arrangement, but a uniform distribution of fine dark dots on a lighter surface. The first thing to look for in color of an exhibition female is the soft

The body of a hen is usually a little lighter in color than the back. The gradation of color should be hardly perceptible and the brown shade should be in evidence all the way down. Many birds that are good color on top run to an ashy, faded looking brown on the sides and underneath. The Standard description of the breast as a "rich salmon" gives most people the idea of a stronger, richer red than is desired. Prior to 1898 the Standard description of breast color was "a rich salmon brown." That description would apply to some birds, but no one ever saw on the feathers of a fowl the color of rich red salmon meat. The reddish breast of the Brown Leghorn is rich in color compared with the breast of the gray females in this pattern, which is described as "salmon" but it has a distinct brown tone altogether lacking in the color of the meat of the salmon and it also has the appearance of a semi-transparent color put on over a pale ground and brushed on very lightly where it extends from the breast to the sides and body.

A correct interpretation of the color of breast is of a good deal of importance, for it appears quite plain from comparison that a preference for strength in this section is largely if not wholly responsible for the distinct, and to many breeders and judges objectionable, reddish cast all over the bird that is seen in some otherwise very high-quality birds. Without losing sight of the requirements for a comparatively distinct red tone in the breast, the exhibitor should select first for the brown surface color as seen on the back and sides, and get as good breasts as he can to go with this. The shade of the breast is of less importance than the shade of the brown sections, and practically all exhibitors and judges of this variety agree that a good deal of latitude is allowable in that—provided the color on a bird is even and an exhibition pen of females is properly matched in color.

Absolute freedom from shafting is by no means common. Many birds of remarkable quality in other respects are more or less disfigured (to a highly critical judgment) by the presence of shafting in one or more sections. Such a bird may be well placed notwithstanding the fault, for if the quality in other respects is there the judge does not ignore it, though I think I have observed that the tendency of judges is to consider carefully every specimen quite free from shafting before placing a specimen with distinct shafting above it.

Where separate classes are made for Light and Dark Brown Leghorns, the males in the first are much lighter red, and generally not so uniform a red, and not as well striped in hackle and saddle as the exhibition male in the old Standard variety. There is also more of a tendency to red in the black of breast and body. When males of the pullet line were exhibited only in the occasional case where one was found that approached the color and quality of males of the other line, it was only the bird of outstanding quality that received consideration for exhibition. As a rule, a breeder would not even look at the males of the pullet line with any idea of exhibiting them, but judged them entirely on their relationship to winning females and ability to produce winning females. Now it is necessary in exhibiting Light Brown Leghorns to select males on appearance without regard for anything else and the Light male must have all the qualities of the other in lighter red colors. On general principles and precedents in judging the great majority of judges will pick the best looking birds—according to established standards—even though the new Standard tolerates what was previously regarded as a fault.

In the Dark Brown Leghorn the females are of a type not previously regarded as desirable — sometimes nearly black—often more or less penciled, occasionally beautifully stippled, though very dark as compared with the female of the Light variety. The best exhibition birds will naturally be the most uniformly and distinctly stippled dark females. In view of the persistence of the taste for distinctness in markings wherever markings are required in a pattern, it appears quite plain to a student of poultry history that the best exhibition Dark Brown Leghorn females will not come from the same lines that produce the richest colored Dark males. Inasmuch as the main argument for the separation of the Brown Leghorn into two color subvarieties was to do away with the necessity for double mating—in effect to make "Standard" not a bird of a certain description, but the bird of one sex used with a certain type of the other sex to produce that type—the natural supposition would be that the bird to select to show would be the same bird that would be selected to breed. The exhibitor who proceeds on that assumption will in nearly every case find himself beaten by the exhibitor who selects for finish in color details.

Black and Red Exhibition Games and Game Bantams— In these the color requirements are in general the same as for Brown Leghorns, but less rigid in the description and still less rigid in the judging. This leniency comes from the fact that "type" is the feature that is most prized in Exhibition Games, and also from the fact that the interest and competition in them is very much more limited. Exhibition Game Bantams are more popular than the large Games, but there are only a few shows where the competition in either is strong enough to bring decisions down to fine points.

Silver and Silver Gray Color Defects— In none of these varieties has color been brought to the high finish reached in the Brown Leghorn (old Standard). The Dorking is bred primarily for shape and type, and so are the Games, and both are comparatively rare in this country. Silver Leghorns are also rare, and as many of the best birds come from England where a more heavily striped male than our Standard calls for is in favor, the best classes of this variety usually appear rather in advance of the Standard in color. An exhibitor of the variety is likely to find that his males with more striping than the Standard calls for will win over those that seem to him to better meet the specifications.

On Silver Gray Dorking males a little striping at the base of the hackle is admissible but not desired. The most common defect is brassiness in the hackle, back and saddle. Silver Dorking females seen at the best shows here are usually very attractive in color. Many of them

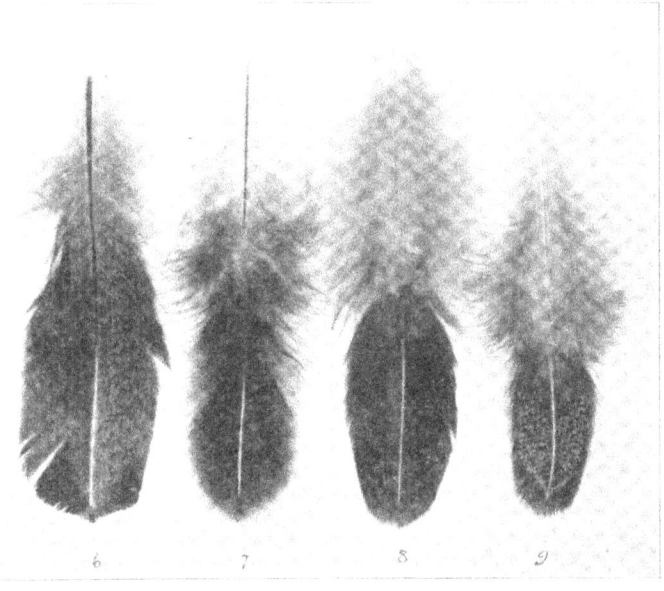

DARK STIPPLED FEATHERS FROM COLORED DORKING HEN
6—wing bow; 7, 8 and 9—back

have considerable shafting, but unless it is quite coarse it is not as conspicuous on them as on brown females.

Golden and Silver Duckwing Game males differ only in that the former is a creamy or straw color in the "white" sections, while the latter is pure white. The females are identical in description, except that the flights of the wings of the Golden female are brown where the others are black. There is so little interest in these varieties in this country that there seems no occasion to go into details of faults. Almost anywhere any bird that generally conforms to Standard description will win. Occasionally good small competitive classes appear at good shows, but this is so rare that a breeder of passably good birds is reasonably sure of winning wherever he exhibits.

COLOR FAULTS—LACED VARIETIES

In some varieties of this pattern the males have the same characteristics of top color as the penciled and stippled varieties with only the breast and body laced, while the females are laced all over. That is the case in the Laced Wyandottes of which there are two varieties —the golden and the silver.

The Golden and Silver Polish, and the Golden and Silver Sebright Bantams—all of which have dark lacings on a light ground—are laced all over in both sexes.

The Red Laced Cornish, and Buff Laced Polish and Polish Bantams are laced all over with a light-colored lacing except that lacing may be absent or deficient in tail feathers.

In Brown-Red Exhibition Games and Game Bantams the lacing is confined to the breast, both male and female having the feathers in this section black laced with a narrow edging of lemon color. The top color of the male is lemon, each feather having a fine black stripe through the center; other sections black.

In all laced varieties, and sections, the general object sought is to secure clean-cut lacings, with the space inside free from the lacing color, and the same character of lacing, and the appearance of uniformity of surface shade in all laced sections. In the Wyandottes, Golden and Silver Polish and Sebright Bantams, the lacing has been brought to a quite high degree of excellence. In the Buff Laced and White Laced Red types it is not nearly so good, and the White Laced Red Cornish is the only one of the class in which there is interest and competition enough to bring progressive improvement.

Most Common Defects in Laced Wyandottes—In the golden variety, white or gray in plumage. In both varieties irregular lacing and mossy centers; lacing too heavy, with small centers; lacing too light and fine, giving the bird or section a washed-out appearance. In Goldens, ground color so light that richness of coloring is lacking; or so dark that while the general color tone is rich the separate colors lack distinctness.

The Laced Wyandottes have this peculiarity in character of markings; while the general plumage has dark lacing on a light ground, the markings of the neck hackles in both sexes, and of the saddle hackle in the male, reverse this arrangement of colors so that the appearance is of a dark stripe on a light ground. Also in these varieties the tail is black and black predominates in the wings, so that there is much more dark pigment in the fowl, and cleanness of ground color is more difficult to secure than in the Polish and Sebrights, which are everywhere laced and have the dark pigment greatly reduced.

This the exhibitor of Laced Wyandottes must consider when he is making his selections, and in general he must be satisfied with a bird that lacks something of the snappy contrasts seen in both Golden and Silver Sebrights—and in a less, but still highly pleasing degree in the Silver Polish. Both the merits and faults in Laced Wyandottes are more conspicuous in the Silver Laced than in the Golden Laced Wyandotte. The greatest difficulty in Silver males is to get birds that are at the same time well laced on the breast and free from smuttiness on the top sections. Few specimens meet both requirements as satisfactorily as the exhibitor would like. In a Golden male smuttiness is not conspicuous if the top color is

FEATHERS FROM SILVER LACED FEMALES
Showing gradations from the irregular line of white shafting at the tip of a nearly black feather to the large open center with clean black lacing required by the Standard. Even the best birds usually have many feathers like the first three at the left.

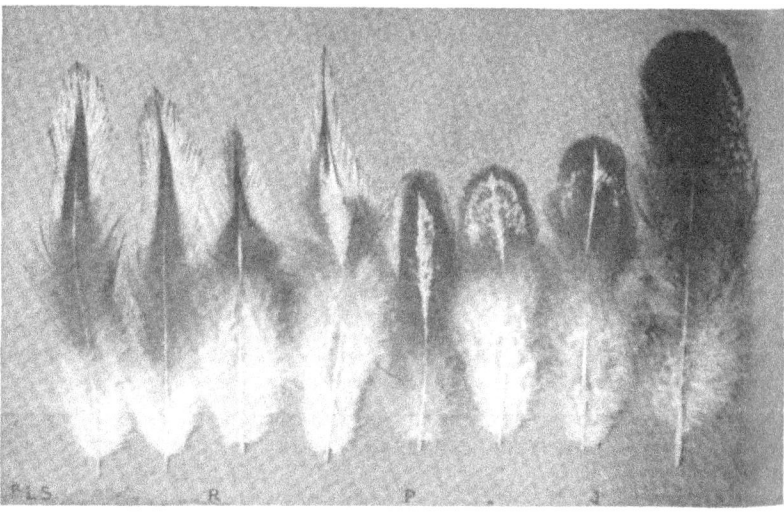

MOSSY MALE AND FEMALE SILVER LACED WYANDOTTE FEATHERS
A large proportion of males winning in good competition are decidedly "smutty" because of feathers like the four at the left prevalent in hackle, back and saddle. Feathers of the character of the four female feathers at the right are most prevalent in the cushion, tail coverts and on the thighs, but should be looked for in all sections.

rich and uniform. But nearly all good Silver males are more or less conspicuously smutty, and this fault is generally tolerated rather than sacrifice the laced sections.

In both sexes of Goldens the common tendency is for breeders to favor a ground color rather deeper and richer than the "golden-bay" specified by the Standard. This does not make any particular difference as far as awards are concerned, but the lighter birds, if well marked, make a much more attractive display in the showroom than the others. Few Laced Wyandottes are shown that are not of at least fair exhibition quality. While birds that with general good quality have the common faults are sure of consideration, a bird free from those faults and lacking strong points will get no attention in good competition.

Buff Laced Varieties—These have been considered in connection with buff varieties.

Laced Polish—Interest in these is almost entirely limited to a few veteran fanciers whose birds are exceedingly well bred for color in general but because of lack of competition do not always show on close inspection the quality of color that from a little distance they appear to have. Many of the Goldens are so dark as to quite lack the color contrasts needed to show them off to advantage. Except in the rare cases when several leading breeders come together, an exhibitor of Laced Polish generally wins wherever he shows. Buff Laced Polish have been considered in connection with buff color on page 81.

Common Defects in Red Laced White—Lack of depth and uniformity of red color; lack of distinctness and evenness of white lacing; lack of evenness of apparent color in all sections.

These faults are found in a large proportion of the birds of this variety reared, for the pattern is too new to have become highly finished. The best stocks however, produce a fair proportion of specimens good enough to show, and as the variety has had an abundance of effective publicity any important show is likely to have a class of them in which it takes a pretty good bird to get a place. Also, the variety is improving fast enough to make it always necessary for an exhibitor who expects to win in all classes to go equally strong in all classes. He cannot take a few good specimens—fill out his string with inferior birds—and expect something on every entry. In many of the varieties not more extensively bred than this an exhibitor can go almost anywhere outside of a few leading shows and make good winnings with a string of mostly very ordinary birds of their kind. In this, if there is competition at all, the awards are certain to be divided, for all stocks are near akin and not far enough removed yet from the original stock for everyone to have given his the strain character which tends to bring the poorest of a string in line with the better when prizes are awarded.

Sebright Bantams—In these it takes remarkably nice specimens with very narrow lacing and clean ground color to win at any show where large classes of Bantams are the rule. At many minor shows ordinary birds failing in nearly every point of color as well as in type may win.

Most Common Defects—In finely laced birds lack of intensity of color in the lacing. In more heavily laced birds lack of cleanness of ground color. In the Goldens a ground color lighter than the "golden-bay" specified in the Standard is much favored, and with this the lacings tend to be brown rather than black.

COLOR FAULTS—BLUE OR BLUE LACED PATTERN

The Standard blue varieties are the Blue Andalusian and the Blue Orpington. Non-Standard varieties in this color are the Blue Wyandotte and Blue Leghorn.

The Standard blue is described as a slaty blue laced with darker blue except in the top color of the male which is a solid, dark lustrous blue.

Disqualifications—Red or white in any part of the plumage.

Most Common Defects—Lack of uniformity of blue shade and of distinctness of markings where lacing is specified; drab shades in blue—particularly in the light

COMMON FAULTS IN BLUE LACED PLUMAGE
These feathers are poorly laced and also uneven in ground color

ground color; lack of blue luster in top color of the male—and the presence in it of dead black or greenish tints, or of reddish color where not positive enough to constitute a disqualification.

In general, except that the top color of the male is usually relatively dark and the hackle of the female tends to run dark, the most common fault in the blue pattern is an apparent lack of blue pigment, and a lack of contrast between the dark and the light shades. The lack of contrast and of a finished appearance seems to be due

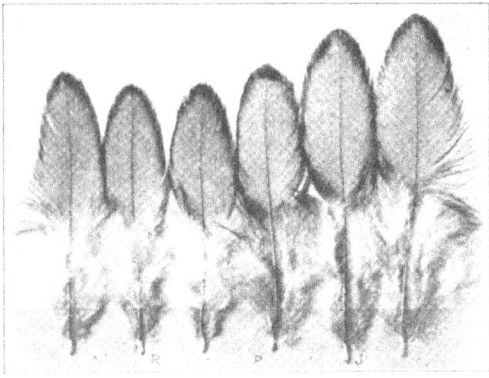

AVERAGE-GOOD DARK BLUE LACED FEATHERS
The centers here are a little cloudy but the lacing is quite strong and a specimen with plumage of this quality is very attractive

to the general vagueness of the Standard description of color, and to the fact that any "darker blue" for the lacing would meet Standard specifications. The English Standard specifies the light color as a "silver blue," and the lacings on it and the top color of the male as "black." The darkest color seen on Andalusians here lacks quite a little of being black. If in selecting, an exhibitor of blue fowls will give the preference to birds which seem to him to meet the specification of silver blue for the ground

COMMON FAULTS IN MOTTLED BLACK AND WHITE PLUMAGE

color and dark blue for the darker color he will come as near as possible to a shade and combination of colors favored in the showroom.

Uniformity of general blue shade of color is not found in competitive classes. Each competitor may have a general uniformity in the birds he shows, but except as it happens that different persons are showng closely related stock, or in an occasional bird, the difference of shade in different exhibits is usually apparent to any critical observer, and analysis of the awards will nearly always show that the particular shade of color did not influence the judging.

COLOR FAULTS—MOTTLED, SPANGLED AND SPECKLED VARIETIES

The mottled color pattern was originally no pattern at all, but simply the irregular mixture of different colors or different shades of the same color. Such mottling is now considered undesirable except in one almost obsolete variety—the Mottled Java. In the other varieties described as mottled—the Ancona and the Houdan—the plumage is black with a part of the feathers evenly tipped with white.

Common Defects—The common defects of the black (except as to disqualifications for the presence of white); excess of white tips; deficiency of white tips; irregular distribution of white tips; irregular and undesirable shape of white tips.

Theoretically, the breeder of Mottled Anconas and Houdans has all the faults of the black pattern to contend with—plus those of the white markings. As a matter of fact, attention to the requirements for white allows a little latitude in black color. It is desired to have the black as good as possible and many specimens of remarkable quality in that color are shown, but a well-marked bird that lacks a little in quality of black will usually win over one that is superior in black but inferior in mottling. The exhibitor of these varieties will also find as he meets good competition that the birds are in advance of the Standard both in regard to amount and pattern of the white tips.

The Standard calls for "about one feather in five" tipped with white, except in the back of the male, where the specification is "about one feather in ten." The birds that win at the best shows have nearer one-half the feathers tipped with white—many of them nearly every feather tipped. The Standard text does not specify any particular shape for the white tips, but the Standard illustrations show a V-shaped tip corresponding to the shape of the spangle in the Spangled Hamburgs, but much smaller. The Standard makes the same requirements as to number of tips in old and young birds, but both the number and size of the tips increase in each succeeding molt.

An excess of tips in the soft plumage can be remedied by the removal just before exhibition of as many white-tipped feathers as appears desirable. A deficiency of white tips in the plumage may be partly remedied by plucking enough feathers quite evenly distributed through the plumage several months before a bird is to be exhibited, to get the desired increase of number of white tips by the substitution of white-tipped new feathers. This process is in the nature of the case very uncertain as to results unless one knows pretty thoroughly the breeding of each specimen treated and how birds of its line usually change in the annual molt. Where the white tips on the feathers—though sufficiently numerous —are too small or somewhat smutty and lacking in distinctness their removal to allow better feathers to take their places is more regularly effective. While the V-shaped white tip is preferred, a specimen with distinct crescentric or irregular tips would probably be preferred by most judges to one with small and indistinct V-shaped

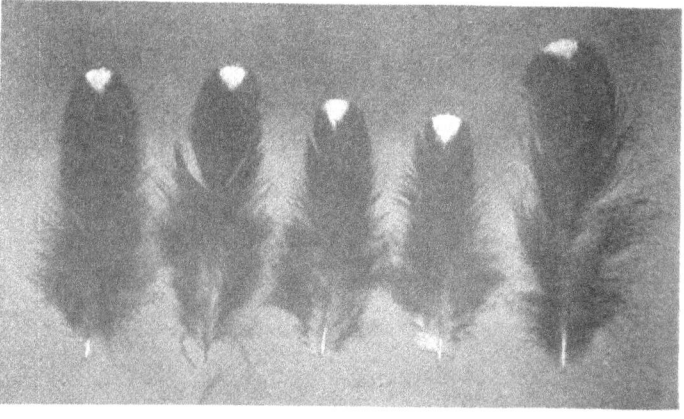

AVERAGE-GOOD MOTTLED BLACK AND WHITE FEATHERS

tips. A few judges appear to favor quality in black color over everything else. For a while there was a decided tendency to give the preference to very dark birds and some judges were partial to nearly black specimens, especially in cockerels and pullets, holding that it was both desirable and proper to make allowance for fading with age. At the present time I think it correct to say that with most judges uniform and fairly distinct mottling gets first consideration.

MARKINGS OF SPECKLED SUSSEX PLUMAGE

The varieties described in the Standard as being "spangled" are the Golden and Silver Spangled Hamburgs, and the Redcap. The Standard definition of a spangle is: "A clearly defined marking of a distinctive color located at the end of a spangled feather." This is obviously one of the definitions that do not define; for one who sought to learn from a dictionary what a spangle feather means would find that the technical usage of the term is just the reverse of the common meaning. A spangle is a bright, presumably a sparkling, ornament on a dark or dull ground. A spangled feather is one having a clearly defined marking unlike the remainder of the surface in color at the tip. In the three varieties designated as spangled the marking is a dark spot on a white or lighter ground.

Most Common Defects in Spangled Varieties—Irregularity and lack of intensity in markings; spangles so large that they overlap with the overlapping of the feathers and do not show as separate and distinct.

Interest in spangled varieties centers on the Silver Spangled Hamburg, it being the only one where the colors admit of contrasty effects that are attractive. Even in this variety the interest is almost entirely limited to breeders who began with it years ago as boys or young men. The Standard specifications for large spangles on the back, tail and wings, seem to be responsible for the spangles being all over the bird, though in some of the other sections it is specified that the spangle shall be proportionate to the size of the feather. The result is that the spangles on the breasts, bodies and backs, of females especially, are commonly so large that the sections are nearly black. According to Wright, very extensive plucking is commonly resorted to in England to remove enough feathers to show the spangles separately. He declares that a good bird will sometimes "lose half a basketful." Some plucking to improve the appearance is no doubt resorted to by Hamburg exhibitors in this country, but there is much more evidence of the disposition of judges to award prizes to birds with good spangling on the individual feather so large that the bird does not show the spangles distinctly, than of pulling of feathers by exhibitors to anything like the extent mentioned by Wright as customary in England where the Hamburg is of more importance. The plucking here is more in the way of removing mismarked feathers, and does not go beyond what is customary in all colors.

In Silver Spangled Hamburgs very good birds are required to win where the old breeders compete, but very ordinary specimens will be placed elsewhere. There is so little competition in Golden Spangled Hamburgs that almost anything shown will be placed.

Most Common Defects in Speckled Varieties—The rather indefinite term "speckled" is applied to a single large variety, the Speckled Sussex—the color is found also on the Mille Fleur Bantam.

Speckled Sussex—The pattern in this variety is really a double spangle—a black and white spangle on a red or buff feather. The Standard describes the marking as a white tip divided from the balance of the feather by a narrow black bar. As in all well-known barred varieties great emphasis is placed upon barring straight across the feather, the use of the term to describe the marking on the Sussex may be misleading unless the bar is defined as V-shaped.

The most common faults in this pattern are defective spangling and the presence of white or largely white feathers in tail and wings. Only a few breeders have so far succeeded in producing birds with good finish in color, and in general a bird that upon critical examination appears fairly good will stand well in a class of this

MILLE FLEUR BANTAM MARKINGS
Left—White spangle, black bar, red or bay ground. Right—"Porcelain" pattern, white spangle, blue bar, buff or cream ground

pattern. The removal of mismarked feathers, if not too numerous, will greatly improve the appearance of the birds.

COLOR FAULTS—BARRED VARIETIES

In Standard patterns there are three groups which have all or most of their plumage alternate light and dark transverse bars across each feather. The most important of these groups is that which includes the Barred Plymouth Rock and the American Dominique, both of which are barred in all sections, and have a bluish or bluish-gray appearance, instead of the steely or silvery gray of the gray penciled varieties. The Dominique color requirements of the Standard hardly go beyond the crudities of the mongrel barred pattern in which the highly finished pattern of the modern Plymouth Rock originated. We will therefore, consider the Barred Plymouth Rock first in full detail, and the other in a supplementary manner.

Disqualifications in Barred Rock Color—Red in any part of the plumage; two or more solid black primaries, secondaries or main tail feathers.

Red feathers are found mostly as single "foul feathers" in the soft plumage. One or more black feathers, or white feathers, will sometimes appear in tail or wings. As a general rule, where one appears there will be two, the sides of the tail being alike and the wings alike. The red feathers can be removed without detection and are commonly removed. The black stiff wing and tail feathers cannot be removed and their existence makes it necessary either to leave the bird out of consideration for exhibition, or to remove them and take the regulation cut which is from 1 to 3 points for each missing feather.

Most Common Defects in Barred Rock Color—Partly black or white feathers in main tail and flights; partly or entirely black or white feathers in other parts of the plumage—that is, the barring lacking on a feather or part of a feather; unevenness of surface shade of color, due to different shades of dark or light color in different sections, or to difference in character of barring; coarseness and unevenness of barring; irregular barring; lack of clear definition in barring.

These general statements of defects cover a great variety of variations from the ideal. It might be thought that such gross defects as red feathers or feathers to any extent lacking in barring, and therefore either black or white where the bars are missing, could all be discovered in a rapid examination of a bird. Experience shows that it is necessary to look quite as carefully for these faults as for those not so readily appreciable to the untrained eye. All these off-colored feathers are removable from the soft plumage, but not from the main tail, sickles or flights.

The best breeders and judges of Barred Plymouth Rocks agree that the most important point in an exhibition bird is an even shade of color in all sections. Possession of this is to a degree the result of the absence of some of the other defects mentioned. One even color all over the bird would not necessarily mean good color—a clean and attractive surface with pleasing contrast in barring; but as a matter of fact, a bird so well bred that it has the same general shade of color all over is quite free from bad faults in the other particulars. It may not have the finest finish in some of them, but has pretty good finish. So in selecting for exhibition color the first thing to do is to look for the most even-colored birds that appear decidedly blue. The most common variation from this in well-bred stock is in a brownish or greenish cast in the dark color, most conspicuous on the back, and usually more noticeable on the backs of males than elsewhere. Many otherwise remarkably good birds have this, and it may not be a handicap to them in competition unless they are competing with very even-colored and blue looking birds.

Evenness of color is generally associated with uniformity of barring in all sections, but not necessarily with straight barring. The latter feature is perhaps the most difficult of all to secure, and comparatively few specimens

BARRED PLYMOUTH ROCK MARKINGS
Left—Most common form of defective barring in males of general good quality. Right—Defective feathers showing the common faults and irregularities in barring. It should be noted that most of these feathers have either white tips or poorly defined dark tips.

have absolutely straight barring all over, and have at the same time the strong definition between the light and dark bars which brings out fully the beauty of barring of this character. In a great many cases the barring is straight but not clearly defined. Instead of two shades of color, each quite clean and distinct, the edges of the light and dark bars blend so that between light and dark there is a region, of varying width and intensity of color in different cases, that is of an intermediate shade. A bird like this may be even in general shade of color but will never look well in comparison with birds whose barring is of two different shades only, though slightly irregular in character.

Approximate straightness in barring may be general in the rest of the plumage while in the hackle, and especially in the hackle and saddle of the male, the bars have a wide V shape. While it is desirable to have the bars as straight here as in other sections, deficiency is so common that in ordinary competition a bird that has no worse fault in color is to be regarded as having a chance to win. Broken barring and shafting are very common faults. In the first the bars on different sides of the quill do not match. In the other the quill does not carry the color of the dark bars but cuts each with a light line.

Because of the difference in the structure of the parts of a feather that is exposed to the weather and that which is covered by other feathers, the barring in undercolor is never as clear-cut as in the surface but still should be conspicuous and distinct right down to the skin. There is a great deal of controversy about the merit of insisting upon strong underbarring, and one not versed in conditions at shows might get the impression from this that under some judges, at least, a bird very weak in that section would not be set back much on that account, and perhaps would not be set back at all. If one thinks that he has a bird that is about right in surface color but is plainly weak in undercolor, the probability is that his ideas of what constitutes good surface color need revising.

Birds with the strength of color and narrowness of barring that have been favored in Barred Rocks for some years now almost invariably have pronounced barring in the undercolor. I would not say that such birds could not have weak undercolor, but I have never seen one that did have what could be called poor undercolor. Nor have I ever known a judge or expert breeder of Barred Plymouth Rocks who, however he might generalize about looking for surface color and letting undercolor come as it would, was really indifferent to good undercolor. I think the true explanation of the attitude of breeders and judges on this point is that their attitude on the importance of undercolor has been determined by their ideas of surface color, and as a rule, those who would tolerate what they could not regard as anything but poor undercolor wanted a much lighter surface than the others. Of late years there has been pretty general agreement on the relatively dark surface shade.

In the main tail and flight feathers the barring was for a long time in the earlier history of this variety decidedly poor, and one may still find here and there exhibits and flocks of old-style Barred Rocks with poor barring in the main tail feathers, and the flights "marbled" instead of barred. In the absence of better birds such sometimes win, and I have seen ribbons in some of the smaller New England shows within a few years on birds with a great deal of white at base of the tail and extending up conspicuously on the main tail and sickles. The exhibitor who has any stock of this kind should get out of it. He is too far behind of the times. In a variety of scant popularity, in which no improvement is being made, one who likes it may get some satisfaction from winning with such birds as he has, but in a popular variety it is poor policy to do so.

The precise shade of color desired in a Barred Rock is and probably always will remain a matter of difference of opinion. As popular taste changes from light to darker shades, all tastes follow the general trend, yet never do all agree on any particular shade. The Standard gives a definition of the color which if literally accepted by all would fix one shade of color, but the insistence of judges and breeders upon evenness

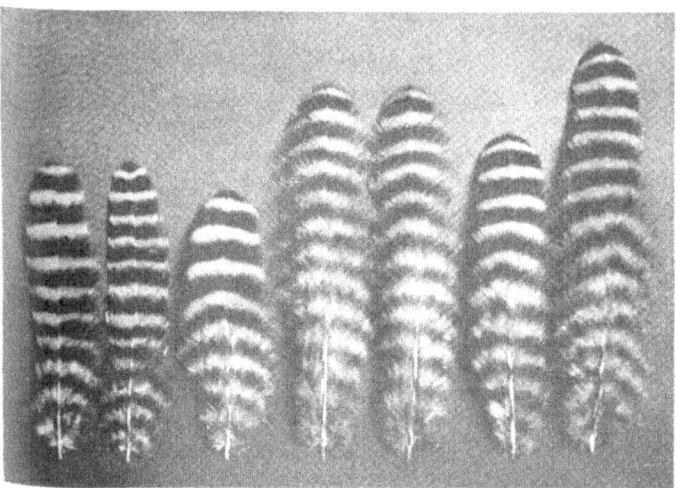
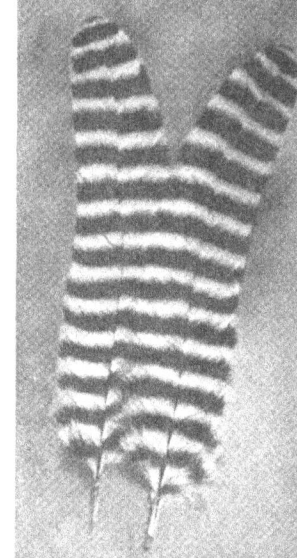

BARRED PLYMOUTH ROCK MARKINGS
Left—Well-marked feathers corresponding to the well-marked group on the opposite page. Note that these all have distinct and clean-cut dark tips. Left—Nicely marked tail feathers.

of color as the most important thing tends to maintain all the variations in shade that can be found between the darkest and lightest birds used in breeding. In the selection of exhibition pens particularly, the necessity for matching male and females in color often makes it better to take a lighter shade than is considered Standard because one happens to be able to match a better pen in it. While there has always been and still continues to be much controversy about whether a Barred Rock is really blue in color or the blue is in the eye and imagination of the person looking at it, the fact remains that the birds that look blue, that are uniform in shade of color and have straight clear, snappy barring, are the reliable winners.

The apparent surface color of a bird and the whole general appearance of the pattern depend much upon the order of the bars on the feathers, whether the first bar at the tip of the feather is light or dark; and also upon the uniformity of growth and the proper relative length of feathers in adjacent parts. A bird with good dark bars at the tips of the feathers will show the character on the surface which appears in the barring of the separate feathers. A bird with the same character of barring and strength of color, but with the light bar at the tip, will show a much lighter and more uneven surface shade. When only a part of the feathers have light tips their removal may greatly improve the appearance of the bird. The much-prized "ringy" effect in barring produced by the bars on adjacent feathers matching, can often be greatly increased by a systematic examination of the whole plumage and the removal of feathers whose bars do not align regularly with those touching them. This work to be effective

PENCILED HAMBURG BARRING
Upper—Defective feathers from a second-prize hen.
Lower—Well-marked feathers of the first-prize hen in the same class. Besides showing good and poor quality these two groups illustrate the possible range of quality in birds that win.

DOMINIQUE AND CAMPINE BARRING
Above—Dominique barring, ranging from coarse, irregular barring to a pattern resembling Barred Rock, but inferior in quality.
Below—Campine barring. Left—Well-marked feather; center—white bar too wide; right—white bar too narrow, sometimes missing.

requires careful and patient work with absolute certainty that the result of each removal of a feather will contribute to the result sought. The ringy character is primarily a matter of breeding and of full development of the plumage. Where it does not exist to a quite marked degree efforts to make it by plucking feathers are fruitless. Where the white bars at the tips of feathers are quite narrow and faint, they are sometimes removed by trimming with scissors, or by singeing with a lighted cigar. A little of this work, carefully and well done, may be effective and pass unnoticed; but it can rarely be done to any considerable number of feathers without the attention of the expert being caught by it when looking at the bird.

Dominique Color—In the American Dominique color we have the only case where the Standard specifications for color continue to call for a very inferior style of birds in one case, while in another case the same pattern has been brought to such perfection of finish as is seen in the Barred Plymouth Rock. Failure to improve Dominique color has been due to the erroneous idea of some breeders and their advisers that crude barring was a distinctive breed character of the Dominique, and to the further ill-advised attempt to repopularize the breed as a barred variety in which double mating was not necessary to produce high quality in color. The reader who admires the Dominique for its breed merits and type, and wants to exhibit and promote interest in it, will find that permanent progress is made only by developing Dominiques with the shape shown in the Standard but with color characteristics of the Barred Plymouth Rocks.

WELL-MARKED SILVER CAMPINE MALE FEATHERS

Color of Shanks and Toes—The males in this pattern

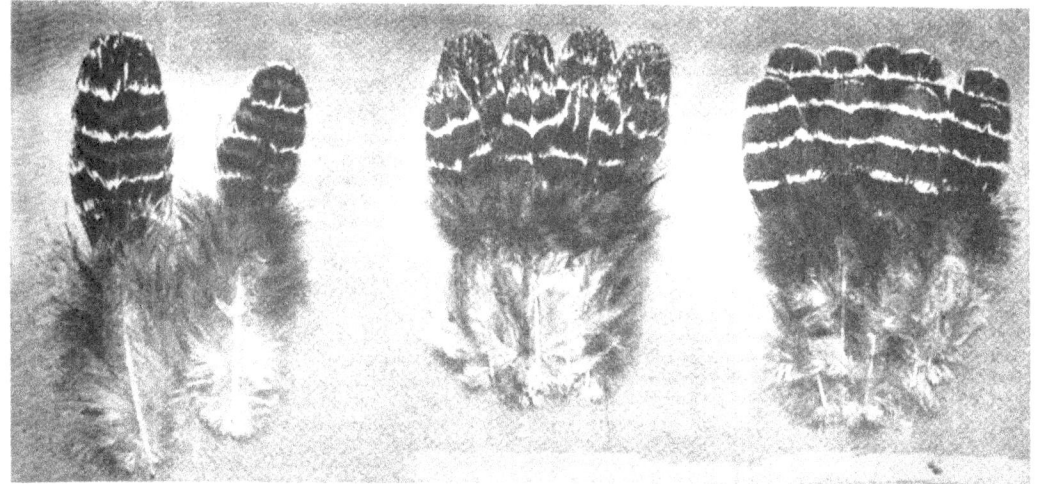

SILVER CAMPINE FEMALE FEATHERS
Left—Poorly marked feathers with secondary barring. Center—Poor barring and mossy tips. Right—Good barring

generally have good shank color. Most females have some dark color on the shanks. Where strongest it gives the front and upper part of the shank a willow or greenish-yellow shade. Pullets which finally have pretty good leg color often have very dark, cloudy-looking shanks until about the time they begin to lay. As the color clears up a few dark spots usually remain. These do not disqualify, but count as faults. They can be concealed by coloring. Judges generally are much more lenient to dark spots on the shanks of barred females than those not familiar with their work suppose they would be when judging according to the Standard, and many exhibitors make no effort to conceal dark spots.

(Penciled) Barred Hamburgs—In this pattern the female is barred on the surface everywhere but in the hackle, while the male has only a little indistinct barring on the body and fluff and wing coverts, with more distinct barring on the upper webs of the secondaries. There are two varieties, Golden and Silver, the former having bay ground color, the latter white. The main tails of the males are black, the sickles edged with the ground color, and the flights are partly black. The black in both varieties is a lustrous, greenish black, and the undercolor in all sections in both sexes is slate.

The interest in this type is limited to a few fanciers of great skill in breeding it, and while certain of these may be said to have superior lines, there is very little Penciled Hamburg stock of which it can be said that it is constitutionally poor in color quality. The faults in these varieties are principally those resulting from keeping and growing the stock in too close quarters. To show them in perfect condition requires the same attention to development of plumage and to the removal of defective feathers as in other patterns; but the lack of competition in them enables an exhibitor to win without this meticulous care at any show except where one of the breeders who always shows birds in prime condition is exhibiting.

Campines—Silver and Golden—In this breed the birds of both sexes are barred everywhere but on the hackle, but instead of the bars being of uniform width the black bar is specified as (according to the section) three or four times the width of the light bar. The black is, as in the Hamburgs, a lustrous, greenish black.

Most Common Defects—Barring at the base of the hackle; light bars in the rest of the plumage so narrow that in some places they are broken by the joining of the black bars which they should sharply divide; light bars too wide, giving a coarse appearance; dullness and lack of luster in black color; smudginess in ground color.

The beauty of this pattern is in brilliance of black color, cleanness and distinctness of light barring, and clean hackles. Dullness of color is in many cases due to the lack of vitality that characterizes much stock of this breed. Vigorous specimens usually have rich color and a high sheen.

The silver variety is much better finished than the golden. This is in part due to the fact that the color is much more contrasty and the variety therefore more popular and more extensively bred than the other, but a comparison of the most pleasing specimens of the Golden that have been shown with other Goldens and with the best Silvers shows quite plainly that the most attractive Goldens are those that have the light bar a little wider than the Standard specifies so that it stands out as clearly as the narrower bar in the Silver.

Hackles free from an objectionable degree of barring are quite rare. Indeed, it is only within a few years and in a few of the best finished stocks that really nice hackles have been obtained with strong color in adjoining sections. In general, where the hackle is free from barring, the front of the neck is lacking in barring and much too light in color. Judges will usually be easier on the fault in the hackle which the Standard treats leniently than upon the correlated fault for which it makes no exceptions.

Many specimens vary greatly in the character of barring in different sections. In general, the tendency is to very heavy and relatively dense black bars on the back, more open barring on the breast, and still more open barring with failure of the black bars to preserve a sound color on the under and rear parts of the body. The black bars are sometimes more intense through the middle, giving lighter or duller supplementary bars at each side. Or they may be broken by flecks of gray, or have bronze or purple barring in them. Except in the strongest competition in Silvers a creditable showing can be made with birds that are lacking in many of the fine points of col-

WHITE FEATHERS FROM ERMINE PATTERN
Left—Slate bar in undercolor. Right—White undercolor

or finish, but have no very conspicuous fault in any one section. A Golden Campine that shows as much character in barring as the average good exhibition Silver and has uniformity of surface and a fair quality of black color—even without high sheen—is a likely first or second in any competition.

The Ermine Color Pattern—White with Black Points

The varieties having a form of this general color pattern are: Light Brahma and Light Brahma Bantam; Columbian Plymouth Rock and Wyandotte; Black-Tailed White Japanese Bantam; and the non-Standard Light Sussex and Lakenvelder. The Brahmas, Rocks and Wyandottes of this pattern have the same Standard specifications for color. Broadly, the color specifications are: a white bird with black tail and black flights, the hackle having a black stripe, and tail coverts being black edged with white; and the undercolor a bluish slate.

Disqualifications—Black or false-colored feather or feathers in the backs of females; black spots prevalent in the web of the feathers of the back where it should be white.

The second of these provisions has been the subject of endless controversy, particularly among Light Brahma exhibitors, turning on the construction of the term prevalent. In the most explicit common usage of the term, prevalent means predominant—that is, more black spots than white surface. But it was obviously never the intent of the Standard to admit black in a white back to any such extent or even approaching it. The term is commonly construed by fanciers to mean black spots or flecks appearing in the back to such an extent that the white surface is considerably marred in appearance. When it is taken in this sense, and the specimen given the prescribed benefit of the doubt as to whether the amount of black demands disqualification, there is a little difficulty in a fair application of the Standard.

Most Common Defects—Black markings or feathers, or false-colored feathers where not a disqualification; white in black parts of tail and wings; poor striping in hackle, and coarse, ragged lacing on tail coverts; purple barring in black sections; brassiness and creaminess in white sections—especially in the white of the top color of males.

The most common and persistent forms of the appearance of undesired white in black sections is white at the base of the tail and in the wing flights. Absolute freedom from white in the tail extending part way up on the main tail feathers is rare in old cocks, but cockerels of well-bred lines will show little if any white in new plumage. Entire white or gray feathers are much more likely to appear in the wings than in the tail. Good sound black color in flights is much more common in males than in females, which have a pronounced tendency to long patches of white in black webs. In old-style Light Brahmas black in the wing of females was not insisted upon to the extent that it is now in all these varieties. The insistence on black wings leads to the tolerance of more black or gray where white is required, and the modern style in this pattern is not as white and clean in the white sections as the Light Brahma of a score of years ago.

In the old-style bird purity of the white surface sections was sought before anything else, and while good black and clean striping and lacing were desired the black was not allowed to encroach upon white surfaces. In the modern style of this color pattern strength and richness of surface color, resulting from heavier striping in the hackle, better lacing on tail coverts and lesser coverts, and the substitution of strong striping in the saddles of males where before only light ticking was tolerated, are the things most sought, and with these the modern taste calls for extra-good black in wings, though the strength of pigmentation ordinarily needed to secure this commonly brings more black in white surfaces than is at all desirable.

White or gray in black feathers may be concealed by painting or staining, but it does not appear that this is at all extensively practiced. The disposition to be more tolerant of black in the white areas leads breeders to produce birds that have no extensive faults due to lack of pigment in black parts, thus reducing the occasions for temptation to color feathers, and bringing in more of the class of faults removable with the removal of soft feathers. The present taste also favors a wide stripe in the hackle, and is more tolerant of smutty edging on the hackles than the old style; and demands more laced feathers in the coverts of the tail as well as striping in the saddle of the male. Corresponding with the wide stripe in hackle, the tail coverts are designed to be as finely laced as possible and still have the white edging distinct, and the stripe in the saddle of the males splits or breaks into a V shape at the end toward the base of the feather. The stripe that widens here and keeps the black center is regarded as doubly objectionable because it appears coarse, and because males with that characteristic produce females with an excessive amount of black in the back.

A corresponding tendency of the stripe in the hackle

FEATHERS FROM SADDLE AND TAIL OF ERMINE MALE
Left—Well-laced coverts and lesser coverts. Right—1—Sickle, all black; 2—Large covert, partly edged with white; 3, 4—Lesser coverts, intermediate both in position and color between the tail and saddle feathers; 5, 6—Long saddle feathers; 7, 8—Short saddle feathers. Note that 3-8 show the slate bar in undercolor.

HACKLE FEATHERS OF ERMINE MALE
Left—Correct markings, good stripe with clean white edge. Center—A poorer stripe with a "break" of white across it close to the surface

HACKLE FEATHERS OF ERMINE FEMALE
Left—Good stripe, but with the white edge too wide; a hackle of this character appears to be heavily spangled rather than striped and is not at all attractive. Center—Of much the same character as the feather at the left, but not clearly marked, and having the black cut nearly through the white at the tip; a neck of this character will be black and smutty looking. Right—Poor striping divided by white on quill extending into the web. When such feathers are prevalent in a hackle it appears a dirty gray. The figures above these feathers give a judge's estimate of the proper cut on hackle for each of these three faults prevalent in it

to break in the same manner is a common fault. The seriousness of the fault is largely determined by the location of the break. If it comes at or near the surface it makes a washed-out, faded-looking neck. If it is well down in the undercolor it will not materially reduce a birds chances of a place.

Purple barring is very general in birds of this color pattern, and is most conspicuous in the tails of males. Breeders and judges do not like it, but unless it is very bad it does not set a good bird back much. Brassiness and creaminess are prevalent in the Plymouth Rocks and Wyandottes, less so — but still considerably in evidence—in the Light Brahmas. Light Brahma Bantams of good color are quite rare.

Black-Tailed White Japanese Bantams are a modification of this color pattern, in which only the black tail and a part of the black in the wings remains. While a few nicely finished specimens are sometimes shown, in general it takes only very mediocre quality to win.

Color of Shanks and Toes—Good color in this section is common in all varieties of this pattern. An occasional specimen has flesh-colored or pale yellow legs, but it is seldom that these are good enough in other characters to make it necessary to consider the possibility of changing leg color.

Miscellaneous Color Patterns

There are numerous variations or combinations of patterns that have been considered which do not fit readily into any of the groups where a number of varieties of different breeds can be considered together. As a rule,

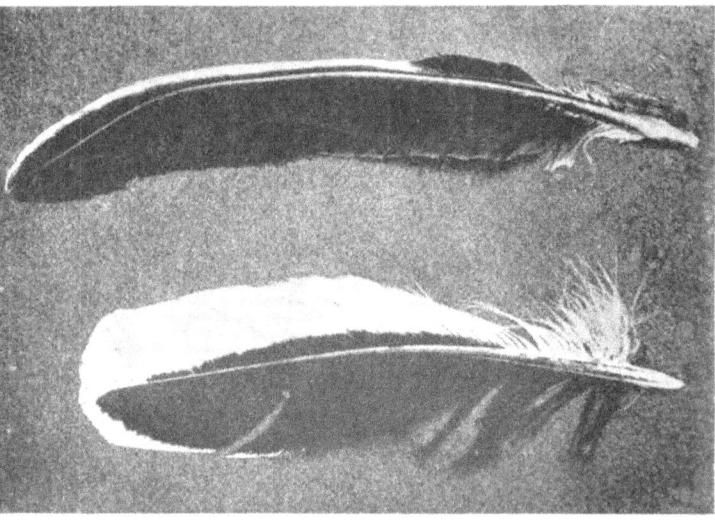

FLIGHT FEATHERS OF THE ERMINE COLOR PATTERN
Above—A fair primary. Below—A fair secondary. In both feathers there is a little more white than is desirable

these are of limited popularity and interest, and refinements of color do not often figure in judging them. In the cases where this is not so, the color points are points that have been treated in some of the foregoing statements. Hence, all that is necessary here is to mention the varieties and call attention to the few points in connection with each where knowledge of what experts do with it is helpful to a novice.

White-Crested Black Polish—The black color on this fowl is in all respects the same as on all-black fowls. The white crest is rarely naturally all white, but is more or less mixed with black at the base. The black feathers are plucked. A bird with so many of them that their removal would leave plain gaps in the plumage and spoil the shape of the crest is not suitable for exhibition.

Colored Dorking — The Colored Dorking as described in the Standard is a dark brownish-gray fowl, of rather variable shades and markings. In judging this variety awards are made almost entirely on size and type. There is not enough competition in it to bring questions of color into consideration.

Salmon Faverolles—The male in this variety has black breast and body with uneven straw top color more or less mixed with brown and black. The female is a brownish drab of varying depth and intensity, not at all uniform in either shade or markings. The pattern is very crude and irregular, and no refinements of color enter into consideration in judging.

Birchen Games and Game Bantams and Gray Japanese Bantams—The female in this pattern is black with hackle white with a black stripe. The male has the black breast laced with a narrow edging of white, and the hackle and back white with narrow black stripe. The few specimens seen are usually gray-black in color with lacing and striping not especially good. Occasionally a very nice specimen appears. There is rarely a competitive exhibit.

Red Pyle Leghorns, Exhibition Games and Game Bantams—In this pattern the male has an orange-red hackle, red back and orange saddle, red wing bows, red lower webs of primaries and secondaries. The female has a salmon breast and a "gold" edging on the hackle. The rest of the plumage is white. To one not familiar with the birds as ordinarily seen the Standard description conveys ideas of quality in color that are seldom realized. The colors are too uneven and the specifications too suggestive of faults in other color patterns to be attractive. As a novelty a few birds of these varieties add to the variety of a show. Once in a while a really attractive specimen appears. In general, an exhibitor meets no competition and anything that passes as fairly typical will win. The Games and Game Bantams are usually excellent in type. The few Leghorns seen here are mostly quite lacking in Leghorn type and character.

Buttercup—Yellow fowls with cup combs have been known in this country under various names, most frequently as "Buttercups" from a very early day. No particular interest was developed in them until about 1910 when a few breeders began to exploit them, and as the number of interested persons increased an effort was made to establish them with a unique color pattern and distinctive markings. Considerable differences of opinion and of pattern were developed. The style which appears likely to find common acceptance is a pattern which may be regarded as produced by increasing the light shafting

BUTTERCUP FEATHERS
Above—Left to right: Male breast, lesser tail covert, wing bow, saddle and thigh
Below—Left: Poorly marked female feathers. Right: Well-marked female feathers

and edging on a Penciled Hamburg type of barred plumage until the bar on each side of the quill becomes an oval spot. Considering this as the accepted standard for the female (the male being marked as in the Penciled Hamburgs) the color being buff or bay with dark markings:

The Most Common Defects are: wrong character of of markings—sometimes general indistinctness of marking, in other cases the form taken being more that of broken heavy penciling that follows the contour of the feather. In the present state of breeding of this pattern uniformity of the spots and some distinctness in contrast between them and the ground are the principal things to consider in selecting for exhibition. The next most serious fault is unevenness of color and generally lack of richness of tone in the ground color. The greater number of females have a washed-out appearance that is not attractive.

PEKIN DUCKS

DUCKS

Pekin Ducks

Disqualifications—Black marks on the bill of the drake.

This is one of the disqualifications over which there has been much controversy. At times ducks as well as drakes have been disqualified for it. Breeders are generally agreed that the largest and most vigorous birds are likely to have some dark color in the bill—if only in the bean at the tip—and all but the smallest and weakest females will, as a rule, have some dark color in the bill after the first annual molt. Insistence upon the disqualification appears to be largely due to the tendency of American Poultry Association members not concerned in matters like this to resist anything that looks like a "letting down of the bars" against faults of this character. Judges who are themselves breeders are apt to be very liberal in applying this disqualification, construing the Standard to mean that only positive black disqualifies—that a greenish or grayish mark is not a disqualification. The marks are in some cases removable by scraping and sandpapering, and some exhibitors will remove them even when they know the judge who is to pass on the class is lenient, simply because they like the clean bill and wish to make the bills of the birds they show look as well as possible.

Common Defects—Lack of size and type; in large birds lack of typical carriage. Large, heavy birds tend to carry the body nearly horizontal instead of somewhat elevated in front, as the Standard requires. This requirement is of considerable practical importance, for the strength and vitality that enable the Pekin to maintain the slightly elevated carriage of body are what gives it preeminence for commercial duck growing.

Aylesbury Ducks

Disqualifications—Bill of drake marked with black.

Most Common Defects—Yellowish bills; lack of size and type.

There is practically no interest in this breed in the United States, the Pekin being in every way preferred. Because it is in the Standard most of the leading shows regularly provide classification for it, but there are rarely more than one or two pairs seen, even at our largest shows. Anything that will pass as an Aylesbury can win almost anywhere. Once in many years an exhibitor with poor birds of this breed

AYLESBURY DUCKS

might run into a class where he would get nothing, but there has not been a class of more than four or five birds at a leading winter show in this country since 1891.

Rouen Ducks

Disqualifications—Bill of clean color—yellow, dark green, blue or lead color.

The bill of the drake is a greenish yellow, that of the duck a brownish orange with a blue blotch near the head. The bean in both sexes may be black. The pigmentation of the bills corresponds in part with that of the plumage, and as that is of a character that gives occasion frequently for double mating for Standard color there is a tendency toward yellow bills in the lighter-colored birds and green, blue or lead-colored bills in the darker ones. It also frequently happens that bills will come of one plain color without any apparent close correlation with the color of the plumage.

Most Common Defects—Lack of size and type.

The Standard has the same weights for the

ROUEN DUCKS

Pekin, Aylesbury and Rouen, but outside of the stocks of the few breeders who show birds of high quality in color, and who have kept up size, the ducks shown as Rouens are commonly much underweight and with the style of a common duck.

Cayuga Ducks

Disqualifications—None except as given under general disqualifications.

Most Common Defects—Lack of size. The Standard Cayuga should be almost as large as the three foregoing breeds. Nearly all ducks shown as Cayugas are much under Standard weight. There is little interest in the breed and, as a rule, any medium-sized black duck can be shown as a Cayuga and win.

Call Ducks

Disqualifications—None except as general disqualifications may apply.

Most Common Defects—There are no faults of much consequence. These "bantam ducks" are bred and exhibited mostly by fanciers.

Black East India Ducks

Disqualifications—None except as general disqualifications apply.

Most Common Defects—None. The few persons interested in the breed carefully maintain breed character.

Muscovy Ducks

Disqualifications—Smooth heads; plumage more than half white in the colored variety; feathers other than pure white in the white variety.

Most Common Defects—Lack of size in the females. This breed is peculiar in that the females are normally much smaller than the males of the same breeding. Exhibitors seem rather prone to exaggerate the difference, and to select very small females. An exhibit is more attractive if the females are of good size.

Blue Swedish Ducks

Disqualifications—Yellow bills.

Most Common Defects—Lack of size.

Crested White Ducks

Disqualifications—None other than the general disqualifications which apply.

PENCILED INDIAN RUNNER DUCK FEATHERS

Most Common Defects—Lack of size, many of the birds shown being far below the Standard requirements for weight.

Buff Ducks

Disqualifications—None except as general disqualifications apply.

Most Common Defects—None in shape or size. The variations from Standard requirements in these points are small, partly no doubt because the interest in the

FAWN AND WHITE INDIAN RUNNER DUCKS

breed is so far mostly confined to breeders having some confidence in their own ability to breed buff color.

Indian Runner Ducks

Disqualifications—Black in bean of white drakes.

Most Common Defects—Coarseness and lack of typical carriage. The carriage should be very erect; the bird small, as compared with the breeds cultivated for table purposes; active and energetic, handling itself much better on the ground than any others except the Muscovy. These characteristics can be developed only by giving the ducks a good deal of room to forage, and letting them pick a considerable part of their living. Heavy feeding as in growing table ducks will spoil the Runner type before the bird is grown.

COLOR AND COLOR PATTERN FAULTS

White Varieties

There are white varieties in the following breeds of ducks: Pekin, Aylesbury, Muscovy, Runner and Crested.

Disqualifications—In White Pekin, Crested, Runner and Call Ducks: feathers other than white or creamy white. In White Aylesbury and Muscovy Ducks: feathers other than white.

Off-colored feathers in the soft plumage are removable.

Most Common Defects—Creaminess in varieties that should be pure white, and excessive yellow in those which should be creamy white. These faults can be reduced or removed as desired by washing or bleaching before exhibition. As a rule, competition in ducks is not keen. At times when there is special interest in a breed or variety large classes come out, competition is strong in them and rapid progress is made in the finishing of exhibition points, both by breeding and in conditioning. Such interest however, has never been long maintained in any case, and as a result exhibitors generally do not have to be as particular, and are not as particular about fitting ducks for exhibition as they are in fitting fowls. A few exhibitors make a practice of putting everything they show in good condition; most go only as far as is necessary for the competition they have to meet.

Black and Black-and-White Varieties

There are black varieties in the following breeds: Cayuga and East Indian; and in the Muscovy the colored variety is predominantly black, the rest of the plumage being white.

Disqualifications—In the two black varieties: white in any part of the plumage. In the Colored Muscovy: plumage more than half white.

Most Common Defects — Gray or red in plumage; purple barring; lack of luster.

Red and gray in the plumage of black ducks very commonly take the form of rustiness or slaty color in the black. Many birds that look pretty well when not seen in comparison with one of fine sheen, appear decidedly dingy when in competition with Standard-colored specimens. In the Colored Muscovy the amount and location of black are not specified; but judges generally appear to favor the birds that are nearly black and have the white marks symmetrical—regular in form and the same on both sides of the bird or a section.

Buff Varieties

The one variety of this color can hardly be said to be of any particular breed. It most nearly resembles the Swedish Duck in breed characteristics.

Disqualifications — Presence of any color but buff or seal brown. This obviously applies to white, gray, red or black in any distinct form.

Most Common Defects — Lack of uniformity and evenness and soundness of color; penciling.

The Standard describes buff in ducks as "a rich fawn buff." This color is specified except for the head of the drake which should be "seal brown." Few buff ducks are bred or exhibited in this country, and the exhibits sometimes described as good in color are so only by comparison with the ordinary buff duck. By the standards for buff color applied to fowls they are far from good. This is to be expected, for the color is new in ducks and is a difficult one to produce.

Blue Varieties

The only blue (so-called) variety of ducks recognized in the Standard is the Blue Swedish. The Blue Muscovy Ducks sometimes seen come from crosses of the White and Colored Muscovy. They are blue all over and often a very nice even blue.

Disqualifications—In Blue Swedish: absence of white patch on the upper part of the breast; more than one-fourth of the plumage other than blue in color.

Most Common Defects — White in other parts of the plumage than the patch on the breast and two flight feathers in each wing; presence of other shades with blue; lack of uniformity of blue shade.

The Standard calls for a "steel

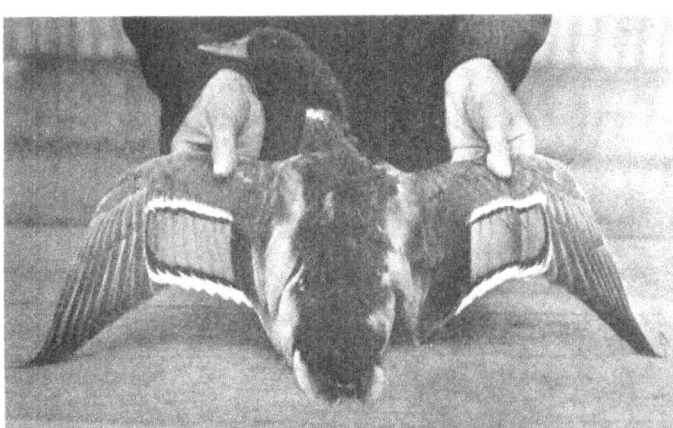

MALLARD DRAKE
Showing the beautiful color pattern which is also characteristic of the Rouen drake

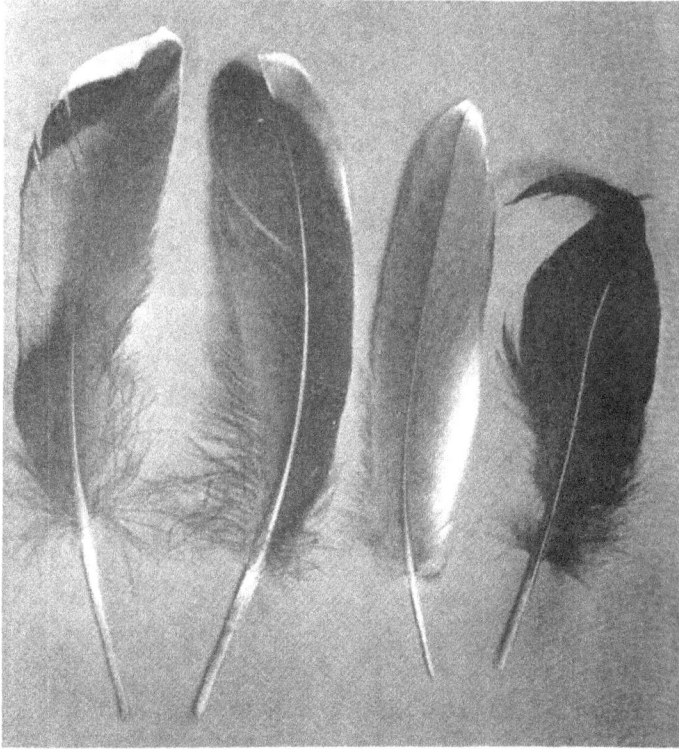

ROUEN DRAKE FEATHERS FROM WING AND TAIL
Left—Secondary flight showing the distribution of color which gives the ribbon and bar effects on the wing. Next—Plain secondary above the ribbon. Next—Tail feather. Right—Curled tail feather which distinguishes drakes from ducks when the sexes are of the same color.

FEATHERS OF A ROUEN DUCK
Above—Primary flight. Below—Left to right: Back, wing bow, breast, body, fluff, wing secondary coverts (2), tail

blue" except that the head of the drake is a darker blue approaching black. Specimens that conform to this requirement when exactly applied are not seen. The best birds are darker on the back than elsewhere and often run very light on the underside of the body. Most birds shown are decidedly patchy in appearance but birds with any considerable traces of color other than blue or white, or with extensive white where it is not called for by the Standard, are rarely shown.

Fawn and White

This combination of colors in a definite pattern is made Standard with a plain fawn color in the Fawn and White Runner, and with a penciled fawn color in the Penciled Runner. Both varieties come honestly by the faults which are relics of their common ancestry.

Disqualifications—In both varieties: claret breast; blue wing bar. In penciled variety: bright green head.

Most Common Defects—Irregular distribution of the fawn and white colors—not as required in the Standard and shown in ideal illustrations; penciling or mealiness in plain fawn color; absence of penciling in penciled fawn color; mealiness in penciled sections; unevenness of color except as due to the specifications which allow darker penciling on back and wings of females and stippling here in males.

To go into a detailed consideration of these defects would be unnecessary repetition of the discussion of buff and red colors and of penciled patterns in fowls. The general features are the same. There is this dif-

ROUEN DRAKE FEATHERS
Above—Large, soft, upper body feather. Below—Left to right: Breast, small lower body, back near tail, secondary wing covert

ference, however, that the separation of these varieties of ducks is so recent that the penciled pattern is rather rudimentary as compared with penciled patterns in fowls or in the Rouen Duck, and what would be poor penciling in these might be as good as is found in Runner Ducks.

Rouen (Penciled) Ducks—In this variety we have the color of the female corresponding in a general way with that of the Partridge varieties of fowls, and there is similar difficulty in finding males and females of equally high exhibition quality in the offspring of the same mating.

EMDEN AND A TOULOUSE GANDER

While long experience and skill, and continuous selection for the characteristic of producing equally good birds of both sexes from birds very near the Standard enables some exhibitors to do it, in selection from average good single-mated stock an exhibitor will usually find he is a little weak in either males or females, and perhaps in both. Until he finds out just how his stock compares with that of others he may wisely choose the shows where classes are comparatively large and entry fees low. This combination is oftenest found at some of the large fall fairs.

Disqualifications—White in flights; white ring, or suggestion of same on the neck of the duck.

Most Common Defects—Absence or serious deficiency of white ring on the neck of the drake; colors of the drake so dark that their rich contrasts do not show; or so light that it has a faded appearance. In the female: lack of distinctness of penciling and of evenness and depth of both ground and darker color.

The Standard for color for both males and females allows—apparently—more latitude for variation in different sections than is regarded as allowable in the partridge color in fowls. The males that are most likely winners are those that are intermediate in color with distinct contrasts between sections differing in color. A well-marked white ring on the neck adds much to the appearance of a bird, and in the English Standard absence of the ring may disqualify. One of the most objectionable features in color of the duck is a washy looking ground—especially on the breast and body—with coarse and indistinct penciling. Wide penciling is desired, but it should be distinct. Unless a light-colored duck happens to be uncommonly clean and strong in penciling, it is good policy to give the preference in choosing to the ducks with rich warm colors, even though not quite as distinctly penciled as one might desire.

GEESE

In geese the number of breeds and varieties is so small that we will consider each separately, and for shape and color together.

Toulouse Geese: Size and Shape

Disqualifications—None other than general and weight disqualifications.

Most Common Defects—Lack of size and type; too fine bone and build—lacking the massiveness of appearance which should characterize the breed; lack of depth of keel; absence of dewlap in old birds.

To show and win in the best competition in this breed one must have birds that are fully up to Standard weights—and as much above them as he can get. The principal exhibitors of Toulouse Geese are men who exhibit the same remarkably good specimens in old-bird classes year after year. In all classes but those for young specimens an uncommonly good bird is a fairly reliable winner for a long term of years, and the exhibitors who have such birds usually have young birds from them for the young-bird classes. Hence, the exhibitor who has not this class of stock should pick shows where his birds will not be out-classed.

Color Disqualifications — White feathers in flights.

Common Color Defects—None in well-bred stock. In some of the ordinary, somewhat mongrelized Toulouse—which are by no means typical of the breed and have no place in a good competition, but which sometimes help fill out and add to the interest of small shows where good exhibits of geese cannot be obtained—there is a decidedly drab tone to the gray color. This should be avoided as far as possible, but it is not desirable to insist on a clear gray if doing so means giving the preference in selection to birds lacking in size and shape.

Emden Geese: Size and Shape

Disqualifications—None other than general and weight disqualifications.

Most Common Defects—Lack of size. As in the case of the Toulouse, the best exhibition birds are larger than the Standard calls for. Some have the massiveness of the Toulouse, but as a rule the birds are not conspicuous for that characteristic, but are more on the lines of the common goose—though larger and more substantial.

Color Disqualifications—Any other color than white in the plumage, except a little gray in the backs and wings of young birds.

Most Common Color Defects — Gray in plumage. This is removable in soft plumage, but unless it is quite conspicuous and competition keener than usual exhibitors do not give much attention to it. Color counts for much less in exhibition geese than size and shape, and judges are rarely disposed to pay attention to small points in color.

African Geese: Size and Shape

Disqualifications—Absence of knob on the bill, or of dewlap on the throat.

Most Common Defects—Lack of size and type. The African Goose, so-called, is a product of a mixture of the

Brown China with Toulouse and common geese It can hardly be said to be a distinct breed, for most of the specimens shown under the name come from flocks of more or less mixed character. The Chinese type, which is next considered, is distinctive. Some authorities on geese have tried to make different forms of the knob peculiar to these two breeds, but the difference they denote in the position of the knob is purely an individual difference common to both breeds, and the Standard does not specify any particular form of knob for either. The Standard difference is in the insistence upon the dewlap in the African, and in size.

Color Disqualifications—White feathers in flights.

Most Common Defects in Color — Much lighter color than is indicated by the Standard specifications, and a decided brownish tint all over the bird. The Standard does not specify brown or any shade of brown in color. The writer cannot recall ever seeing an African goose that was not noticeably brown.

Chinese Geese

There are two color varieties of this breed—Brown and White.

Disqualifications for Shape—Absence of knob.

Most Common Defects in Shape—None. The specimens of this breed seen are very generally near Standard size and of good type.

Color Disqualifications — In the Brown variety: white feathers in flights. In the White variety: feathers other than pure white in any part of the plumage. This disqualification construed literally, excludes any trace of color other than white. Traces of gray usually are found, when present, in the neck and back, and can often be eliminated by the removal of feathers containing them. Judges are not uniformly strict in applying this disqualification and where there is no competition, or where those in the limited competition are not disposed to insist on rigid application of disqualifications, not much attenton is paid to a few traces of gray.

Most Common Defects in Color—In the Brown variety: white in the plumage. In many cases this is removable. Where it is not, and is not a disqualification, neither exhibitors nor judges are very fussy about it.

In the White variety: gray—regarded as a defect, not a disqualification.

Canadian, or Wild Geese

Disqualifications—None except general disqualifications.

Most Common Defects—None so conspicuous as to require mention. This is not an improved domestic race, but the wild species in captivity. Few competitive classes

EMDEN GEESE

appear. As a rule, any bird conforming in a general way to the type and to the Standard description is placed, and the precedence is given to the largest and most impressive looking birds.

Egyptian Geese

This is another wild race, in which there is practically no competition. The specimens seen at shows are mostly in the exhibits of "string men," or in collections of fancy waterfowl.

TURKEYS

All varieties of turkeys have the same shape description. The differences in weights for different varieties represent common differences in size—due to differences in popularity and to the development of large size and high vitality as a result of competition.

Shape Disqualifications—None but general and weight disqualifications.

Common Defects in Shape—Lack of size, and of the fullness and ruggedness of form that goes with good development. Big strong, vigorous-looking birds always get first consideration, and unless noticeably poor in some point

AFRICAN GEESE

WHITE CHINA GEESE

of color will win over birds lacking in size and apparent vitality, no matter how good in color the latter may be.

Bronze Turkeys

Color Disqualifications—White feathers in any part of the plumage; black or brown flight feathers; clear black, brown or gray, in back, tail-coverts, or tail; absence of barring on more than half the length of a primary wing feather.

Most Common Color Defects—Lack of brilliance and uniformity in the general bronze surface color; white or gray bars in main tail feathers in addition to the white tip; lack of black bar next the white in tail coverts; defective barring in tail and flights.

These faults are not likely to be present to a troublesome degree in highly bred stock, but a great deal of Bronze Turkey stock is really grade stock, with common or inferior Bronze blood only a few generations back. In such stock richness of color and finish in all minor details is rare. The birds may appear to be very good Bronze until compared with high-quality Standard birds. At the same time, these high-grade flocks often contain individuals quite perfect in bronze color, and usually have a large proportion that if well grown and of full size can win over ordinary, rather poorly developed Bronze specimens, if the exhibitor is careful to avoid showing birds with pronounced small faults in color. In most shows that get a good display the Bronze make the largest and strongest class. The exhibitor of this variety should either have stock fit for the best competition, or elect to show where the class is small and ordinary.

Narragansett Turkeys

This variety is nearly if not quite extinct. I have been informed that there are some of good type to be seen occasionally at some of the near-West shows. I have seen none shown under this name for years that could properly be said to be representative of the Narragansett Turkey, either as it was before the Bronze displaced it generally or as it is described in the Standard. The specimens shown as Narragansetts, as far as I have seen them, have simply been common gray turkeys.

White Holland Turkeys

These are, after the Bronze, our most popular variety of turkeys.

Color Disqualifications — Foreign color in any part of the plumage.

Most Common Color Defects — None other than disqualifications, which are to a large extent removable. Most of the White Turkeys seen at shows are not subject to much criticism for quality of color. Their principal fault is lack of size. Many magnificent big white birds are shown, but on the whole the Whites are distinctly inferior to the Bronze in development.

Black Turkeys

Color Disqualifications—Feathers other than black in any part of the plumage.

Most Common Color Defects — Lack of intensity of black color. Most of the few Black Turkeys shown here are dingy and dull in color, small, and very inferior in every way to the Bronze and White.

Slate Turkeys

Color Disqualifications—"Feathers other than slaty or ashy-blue, which may be dotted with black, in any part of the plumage."

Slate Turkeys as we have them can hardly be considered an established variety. There are a few breeders who have given some attention to breeding them, but much of the stock exhibited as Slate is simply slate-colored specimens from flocks with some variety in colors. Some of these specimens are remarkably good—often better than any to be found in Slate flocks. The possessor of a Slate Turkey that conforms reasonably closely to the Standard description of color, and is fair in size and shape, can win with it anywhere.

WING AND TAIL FEATHERS OF TOULOUSE GEESE

BODY FEATHERS OF TOULOUSE GEESE
Left to right—Light gray, dark gray, dark gray edged with light gray

Bourbon Red Turkeys

Color Disqualifications — More than one-fourth of color other than white in main tail feathers or flights.

Most Common Color Defects—Lack of depth and uniformity of red color; lack of purity of white color where required.

This is a comparatively new color type and difficult to produce in high finish. Birds that are really good in color when the Standard is critically applied are not often found, but many specimens are quite attractive for quality in some sections and, as a good proportion of the stock seen has good size and development, there is much more interest in it than in any other color after the Bronze and White. In selecting birds for exhibition the breeder will find it the best policy to choose birds that are of good size, rugged looking, and with strength of color in the red sections, even if in taking it he sacrifices something of purity in the white.

Trying out Single Birds and Matching Pens

Having gone through all the stock of presumed exhibition quality that he has, and selected the specimens which individually show enough quality to entitle them to consideration where prizes are being given, the exhibitor still has before him the problem of deciding which are his most likely winners in the competition for which he is preparing. This involves some nice discrimination and careful balancing of the actual merits and demerits of birds, as well as consideration of the possible preferences of judges, and of showroom conditions which may affect the conspicuousness of faults or detract from the appearance of attractive points.

As to the whims of judges, unless one has before exhibiting had the advantage of an opportunity to carefully look over the work of one or more judges of the variety he is exhibiting, who are likely to judge where he is to

BRONZE TURKEY MALE

show, he has to leave it out of consideration and take his chances with such birds as seem to him to best meet Standard requirements. Opportunity to study the work of judges is usually also opportunity to observe the birds under the showroom conditions which interest the new exhibitor; but if he has failed to observe these, or if the opportunity to do so has not been presented, he can reduce his disadvantage here very materially by making a point of studying all the birds he has selected as worth exhibition under as many conditions of light as possible, giving special attention to color in both bright and poor, dull lights; and to shape as seen at different angles of elevation from the floor.

Effect of Different Lights

There are great differences in the conditions of light in different showrooms, in the same showroom under different conditions of external light, and also in different parts of the showroom on the same day. On the whole, the light in showrooms is not as good as the ordinary daylight conditions under which the birds are examined at home. Very few of the buildings or halls in which poultry shows are held are well lighted by daylight. The best conditions are found in the buildings erected especially for poultry shows at the large fall fairs, in some of which the light is in general excellent. Even in these the light is none too good on a dull day, and there are always parts of the hall where the light is unsatisfactory.

As far as time, circumstances and his strength permit, a judge, as a rule, makes it a point to examine or compare all birds under the same conditions of light. But it is nevertheless good policy for the exhibitor to give the preference as far as consistent with consideration of other matters that have to be taken into account—to the birds that show to best advantage under various conditions of

BRONZE TURKEY FEMALE

light. As an exhibitor studies his birds to determine this point he will find the differences in this respect are more common and greater than he had supposed and further, that this line of criticism of color helps him greatly in determining which birds are the soundest and most even in color. Nearly all breeders have preferences for certain shades of color and in selecting tend to take the shade they like even though color is more or less unsound, rather than sounder color in a lighter or darker shade. Sound color may not look equally well in all lights, but it will look better than unsound color in most lights, and the less a bird is dependent upon certain light conditions to show its merits the better its chances of winning.

Effect of Angle of Vision

Where the coops for single birds are all on the same level—that is, where the floor space in the showroom is great enough to allow arranging all the coops in a single tier — only two elevations have to be considered in judging how a bird will look from the aisle: that of the single bird coops at about 30 to 36 inches from the floor, and that of the coops for exhibition pens at about 18 to 24 inches from the floor. When space is so limited that two or more tiers of coops must be used for the single birds, none of them are seen at the same angle as in a show that is single tiered, or as it is sometimes called "single decked," these terms referring always to the arrangement of the single-bird coops. Except at small shows with very limited room the pen coops are not often double tiered, nor are the coops for single large fowls and ducks often placed in more than two tiers, but shortage of space in large shows frequently leads to Bantam coops being in three tiers.

Ordinarily, however, the exhibitor has to consider that in sending his fowls to a show where double tiering is used the single birds will be at either about the same elevation as the pen birds, or the height of an exhibition coop higher. A bird that is approximately correct in type, and that naturally and quite constantly poses well, will show to good advantage in either an upper or lower coop, though perhaps not as well in either as at the usual height of single-tiered coops. A bird that is a little high in station, or one that is narrow bodied, or lacking in breast, or a coop-shy bird that hugs the corner of the coop when approached, will look much worse in an upper than in a lower coop. The elevation exaggerates the faults mentioned, and the open top of the upper coop and the greater amount of light in it make a shy bird act worse there than in the darker and more protected lower coop.

A bird that is low on the legs, and rather full in form, has these charcteristics apparently exaggerated in a lower coop, while in an upper coop he will look at least as well,

WHITE HOLLAND TURKEY MALE

and possibly better than at the usual height of single-tier coops.

The difference in light in upper and lower coops is also an important factor in its effect on observation of color. While, as has been said, the judge does his best to give every bird equal opportunities, it is manifestly impossible for him to do so to an extent that will offset all possible disadvantages to birds from being placed in places unfavorable to them. The exhibitor must look after his own interest to the extent of doing everything in his power to put his birds where they will show to best advantage. When showing where the show is single tiered he must as far as possible enter in the single-bird classes birds that look well at the elevation of the single-bird coops, and in pens birds that look well at the elevation of the pens. When showing where the show is double tiered he must either send single birds that are so near Standard in shape that it makes little difference whether they are cooped in an upper or lower coop, or he must send for an upper coop a bird that looks well in an upper coop and for a lower coop a bird that looks well in a lower coop—and he should either see to it himself or get some dependable person who is on the spot to see to it that the birds are placed right.

It happens not infrequently that an exhibitor sends to a show where the coops are double tiered birds that—whether he understood the situation and intended to provide for it or not—should be placed, one in an upper the other in a lower coop— and in cooping them each bird is placed where it ought not to go. In such a case I have known a judge, after awarding a prize to a bird so misplaced and learning from the catalogue that the owner has another entry in the same class, to change a bird to a coop where it would show to best advantage. The judge, of course, has a personal interest in doing this. It is not to his interest to have a bird he awards a prize placed where it does not look its best.

Where the coops for single birds are in two tiers the consecutive numbers are never in the same tier, so one who enters two birds in a class is always sure that one will be placed in an upper and one in a lower coop. Making a single entry in a class, one can never know until he arrives at the show whether his bird will draw an upper or a lower coop. With three entries in a class he does not know whether he will get two uppers or two lowers. With an even number of entries in the same class he always knows that he will have equal numbers of each.

Reference has been made to the effect of the difference in light in upper and lower coops upon observation of some faults of color. There are also some characteristics of color that vary in appearance according to the elevation of the bird just as the characteristics of shape which have been mentioned do. A bird with a breast very good in color, but the back not so good, will look better in an upper coop, while a bird that is weak in color of breast will look better in a lower coop. If a bird has anything wrong with its feet the fault will stand out more plainly when it is in an upper coop which brings the feet right before the eyes of a person standing before the coop. It is not necessary to go into all the points that are made relatively more or less conspicuous by the angle at which the bird is seen. The illustrations that have been given are typical and cover the most important points. To know the quality of an exhibition specimen thoroughly one must observe it critically and carefully in many lights, from many angles, at various distances, and in comparison with every other specimen of its kind and class available.

Matching Exhibition Pens

In choosing birds to be shown in the open or single-bird classes the exhibitor looks for the best individuals. In choosing birds to be shown as a pen it is in most cases necessary to select the required four females first, and then find the male that looks best with them. Inexperienced exhibitors generally fail to appreciate the importance and value of each individual of the four females in a pen. In a score-card show the male counts for half the pen, the score of the pen being his score plus the average score of the females. This method of judging may place the awards right where only a few pens are competing and the differences in quality are considerable, but is inadequate where there is close competition, for it does not take into consideration how the birds are matched. It does not necessarily bring the birds together, and for a long time it was the practice at small shows to allow the same birds to compete both in single classes and in pens, the exhibitor qualifying for the competition in pens by simply entering for that competition on his entry blank and paying the required fee.

Then when the birds were all scored singly, his highest scoring male and the four highest scoring females were considered his pen, and awards based on their scores made according to the rule. The birds might never be brought together as an exhibition pen at all, or he might show in one large pen a male and any number of females he wished, each bird being a competitor for any single-bird prize for which it was eligible, and also for a place in the "exhibition pen." This practice has long been generally discontinued, but may obtain still in some small score-card shows. It was one of the faults of early practice in score-card judging that had much to do with the bad repute into which the system fell. The common rule in regard to pens is that the birds competing as an exhibition pen must be shown together and can compete only as a pen.

While it is desirable to have every bird in the pen of as high quality as possible, it is equally important that the birds should be well matched, and it is only in flocks of the highest quality, and of great uniformity of quality, that a hen which might be selected as the best competitor for a prize in the hen class could be matched with three others so near like her that they would make the best pen of females the exhibitor could put together. For a pen he must select the four hens of highest quality that are nearest alike in every particular.

For most exhibitors this is the most difficult thing in exhibiting, and they have nearly always to make a choice between going strong in the singles and going strong in the pens—that is, they have to decide whether to put their best birds for the single classes in those classes where each bird on its individual merits may win a prize, or to make their pen as strong as possible in the expectation of winning the one prize that gives greatest prestige. When an experienced breeder gets to a show and has a chance to size up the competition against him he may shift a bird from a class where he thinks he has no chance or very little chance of winning and put it where it will improve his prospects in another class, but in selecting birds at home he has to consider only the relative merits of his own birds.

WING OF BRONZE TURKEY

As a rule, the female of outstanding superior quality can be at once set aside for single competition, because he cannot match her. If there are two females of equally good outstanding quality, both will be reserved for the single class. But if he should find three better than the rest it would become a question whether a fourth female not quite matching them, but very nearly so, would not give him a pen of pretty reliable winners when matched with the right male. The question of buying a bird to fill out such a pen also comes up for consideration. It is entirely legitimate to do this, and it is also good policy if one can find the right bird. The higher the quality of those he has for the pen, and the better they match, the more difficult it is to go outside and get a bird that matches them, but in ordinary grades of exhibition stock it can often be done.

When the one or two best females are reserved for the single classes the question of selection of a pen is a question of finding next-best birds that are good enough and uniform enough to win as a pen. At a local show where an exhibitor may wish to show as many birds as he can to fill up classes, indifferently matched pens, or fairly well-matched pens of mediocre quality (even in the competition they are meeting), may do; but when the idea is to show only where one thinks he has a chance of winning, and in good competition, one should enter only good pens well matched. He should not make his entry for one or more pens until he is sure that he can fill the entry.

When one has a pen of well-matched, or fairly well-matched females, the next thing is to find the male that looks best with them. Where male and females have the same color pattern he must match them as well as possible in color, so that the color of the male blends with that of his mates. Where they have different color patterns there must still be a certain general harmony of color tone but the male should, if possible, be a little better in color than the females, for he stands out from the group conspicuously and his small defects in color become more noticeable than those of any hen.

The type of male to use depends on the type characteristic of the females of the pen. If they are medium to fine in type, a large male with a tendency to coarseness will make them look small and they will exaggerate his coarseness. If they are of good size and rugged looking, a male that is too small and fine will make them look coarse or they will make him look undersized and insignificant.

In selecting the males for show, whether in single classes or pens, the bird that is always ready to get in the foreground and the limelight is to be preferred—other things being anywhere near equal. The experienced exhibitor likes a male that, as he puts it, "fills the coop" when penned by himself, and always is plainly seen in the exhibition pen. Most males are more inclined to show off when with a pen of females than when cooped separately, and where consideration of other points admits, it is good policy to put a male that is always forward in the single class.

Until comparatively recent years it was the general custom to make up exhibition pens with both old and young birds, matching them principally on shade of color. As more and more attention has been given to uniformity of type and shape shows gradually turn to the practice of making separate pen classes for young and old birds, the same as for single birds, and that is the rule now in nearly all the most important shows.

Uniformity of Exhibits in Single-Bird Classes

In selecting birds for a string of single entries two different general policies are followed by exhibitors. Some select for as much uniformity as can be obtained in the birds they enter for the single-bird prizes. Others select birds with slightly different characteristics. The theory of the first plan is that if the birds are very uniform in type and quality the judge who places one high in its class will have to rate the others right after it.

In the other plan the purpose is to increase the chances of getting a first prize, or a high prize, on the theory that if the judge does not like one style of bird he may hit on the other.

Among the devices favored by some exhibitors to help the chances of a favorite bird in single-class competition, is that of taking inferior birds to the show to coop at one or both sides of them to accentuate their merits by the contrast. This may have the expected result occasionally, but it is doubtful that it does often enough to justify the practice. The judge is quite sure to discover the trick after the judging—even if he does not suspect it before—and the natural result is to make him wary and particularly careful in looking over good birds so placed. I do not think it would be of any avail where the judges are familiar with good classes shown in good condition. Inferior birds do frequently appear in such classes at our best shows, and are conspicuous for their inferiority, but I have never observed that their proximity was of any advantage to better birds cooped beside them.

A novel variation of this trick was observed in one of the largest classes at a leading show in the season of 1920-21. An exhibitor showed a fine string of birds, containing some easily recognizable as coming from another breeder. Right beside the best cock, hen, cockerel and pullet entered by this exhibitor, were cooped a cock, hen and cockerel of inferior quality and in poor condition, entered "not for competition" by the breeder who sold the other birds for his string. In the pullet class circumstances in the form of an exhibitor whose entry of a single pullet slipped in between the conspirator's numbers spoiled the arrangement, and as the breeder could not help out his customer by appearing a fourth time as exhibitor of a poor bird "not for competition," he put in a very fine pullet in the pink of condition.

Except in cases, as cited, where one enters birds to help fill up a show, it is good policy to limit the number entered at any show to the birds one feels reasonably sure have some chance of being placed. This is especially true as to comparison shows where one has nothing at all to show for unplaced specimens, no matter how good. There is no advantage to an exhibitor in showing birds that do not win. It is not big entries that count, but the winnings made. Most novices enter too many birds and spend too little time fitting and training them.

Left to right—Winner of a first prize at the London Crystal Palace, winner of a first at Madison Square Garden, winner of a first at Chicago. But a pen with such an assortment of types ought not to win in any show

CHAPTER IX

Special Fitting of Birds for Exhibition

Feeding to Force Flesh and Feather Growth—Training and Posing Birds — Fitting and Faking Combs, Ear Lobes, Crests, Legs and Feet—Washing and Bleaching White Birds—Feather Bending, Fluffing and Splicing —Removing Plumage Color Faults by Dyeing, Staining and Trimming Feathers

IN this chapter are presented all the processes, legitimate and illegitimate, used in fitting or faking poultry for exhibition that the author has been able to get together. It is not certified that the list is complete, but it is not seriously lacking at any point. The general questions of ethics involved were discussed in Chapter III and no further reference to that phase of the matter is necessary here, except to say that the purpose of the book is not education of unscrupulous exhibitors in illegitimate practices, but to help scrupulous exhibitors to discover unfair practices, and to take such measures as are practical to secure fair competition. As far as possible methods of detecting forms of faking are given.

a very few ounces of additional weight on a bird may turn a decision in its favor, and when this decision may be of great moment to the owner. In such cases it is good policy for him to do whatever is necessary to help the bird make all the weight it needs.

There are also cases where the unexpectedly slow development of plumage will, it appears, leave a bird something short of the full growth of feather which brings its characteristics out to best advantage. In these cases a little extra feed may help greatly. There are cases again where a bird has to be kept up and handled

VETERAN POULTRY FANCIERS AT THE BOSTON SHOW 1914

For many years one of the pleasant features of the Boston Show has been the "Veteran Fanciers' Banquet" given by the management to those in attendance who are past sixty years of age and have been breeding poultry for twenty years or more. The latter provision is superfluous, for a man past three-score still breeding poultry or attending shows is practically without exception one who began very early in life. Most of the veterans appearing here have over 40 years' experience as exhibitors, and some nearly sixty. The author selected this particular group for reproduction here because of the number in it whose separate photographs could not be obtained for use in this book.

Standing from left to right—Frank P. Johnson, Indianapolis, Ind., 64 years (Light Brahmas); C. W. Richardson, Apponaug, R. I.; P. H. Freeman, Fitchburg, Mass., 64 years (Silver and Golden Laced Wyandottes); W. B. Atherton, Secretary Boston Poultry Association; John Lowell, President Boston Poultry Association (Hamburgs); Geo. F. Eastman, Northampton, Mass., 66 years; John D. Jodrey, Danvers, Mass., 70 years (Silver Laced Wyandottes); Frank C. Nutter, So. Portland, Me., 61 years (Light Brahmas); Frank L. Fish, Brookline, Mass., 66 years; Dr. S. Lott, Bellona, N. Y., 71 years (Dark Brahmas). Seated, from left to right—W. H. Sylvester, Brockton, Mass., 66 years; H. B. May, Boston, 82 years (originator of the Essex Strain Barred Rocks); Chas. L. Seeley, Afton, N. Y., 75 years (White Crested Black Polish); I. K. Felch, Natick, Mass., 80 years (Light Brahmas); Henry F. Felch, Natick, 75 years (Light Brahmas); Frank B. Breed, Clinton, Mass.; C. B. Travis, Brighton, Mass., 71 years (White Leghorns). Names of breeds are given in parenthesis only in the cases of men long prominently identified with a particular breed. Mr. Freeman, Mr. Atherton, Mr. Lowell, Mr. Nutter and Dr. Lott were exhibitors at Boston in 1921. Mr. Nutter also exhibited at the Garden. Mr. Johnson was an exhibitor at Chicago, December, 1920.

Feeding for Rapid Increase in Weight and Growth of Plumage

As far as possible forced feeding is to be avoided in the preparation of poultry for exhibition. If only the best good of the bird could be considered we might say that there should be no forced feeding, but each bird in competition should stand or fall on what can be made of it by feeding ordinary good rations as used for breeding and growing stock. There are many times, however, when a rather long period before a show and as a result goes off its feed and begins to lose condition and weight. A few days of liberty and light feeding may bring it back to health and with a perfectly good appetite but a little low in flesh. It needs a little extra feeding before the show, and if it is not able to stand that it certainly is not in fit condition to show.

In regard to the ability of birds being conditioned for show to stand heavy feeding, it should be said that the fancier—even more than those who engage in the special fattening of poultry for market—needs to be expert in distinguishing between birds that can stand forced feeding and those that cannot. It is worse than useless to try to force a bird to a diet that evidently does not agree with it. If it does not respond to efforts to increase its weight or to hasten its development the owner should either take his chances with it in the best condition he can get with ordinary feeding methods, or leave it at home.

In general, the safest way to go about forced feeding is by the addition of a little more concentrated and palatable feed to the regular ration. Where there is no

TEACHING A BIRD TO POSE

danger of yellow corn having undesirable effects on the color of the plumage, the best way, by all odds, is to increase the corn and corn meal in a ration, taking care that these are of good quality—sweet and bright and from good, ripe, hard corn. When it is not deemed advisable to use corn freely, or when good corn cannot be obtained, the same results can be obtained by giving good wheat under conditions that permit the fowls to get a full meal easily and quickly, and either feeding moist mash twice a day or giving one feed of moist mash and one of wheat, heavy oats, or barley soaked in water until well swelled, or slightly sprouted.

If more rapid gain than this diet will make is desired a little suet or sugar may be given in the mash. Allow suet or tallow at the rate of a piece about the size of an English walnut to every four birds every other day or, if fed daily, at about half that rate. In feeding sugar, a tablespoonful may be given each bird daily. It will put on fat very rapidly, but the fat gained with sugar tends to disappear as rapidly when the sugar is withdrawn. A little molasses in the mash will answer the same purpose, but it should be good molasses with mash of good ground feedstuffs. Cheap molasses stock feeds, in which molasses is mixed with by-products consisting largely of fibrous stuff stock would not eat alone, will not do.

Fats add gloss and luster to plumage, and may be used as freely as the bird will take them and retain good appetite, for all dark-colored and buff birds, but in feeding them to white birds caution must be used, especially with birds that have the color of the plumage much affected by fat. One of the time-honored methods of feeding to give luster to the plumage, introduced from England, was to mix the mash with linseed or flaxseed tea instead of water. A more practical and less troublesome method is to give an ordinary good dry mash—accessible to the fowls at all times with five per cent of good beef scrap, and five per cent of linseed meal.

For feather growth and increase of weight larger proportions of beef scrap might be used, but there is always in that case the danger that the amount of meat given will cause the comb and wattles to grow more than is desired. Hence, exhibitors generally try to get all the growth they can with a moderate use of animal feed. Milk, both whole and skimmed, is given in the mash, or dry bread is soaked in milk. Fowls that are to be washed may also be given milk to drink, but for fowls that are to be shown without a thorough washing it is not advisable to use milk in a form that will soil the plumage.

Green and succulent feeds should be used very liberally, the fowls having all they want of such things as cabbage, mangels, or any succulent feed available. A suitable supply of such things enables the bird's system to use rich feeds given it to increase weight and feather growth without injury to the digestive organs.

For birds that seem to be a little indisposed and to lack vitality the moderate use of tonics is often beneficial, but it is a mistake to use these continuously except in such small amounts that the purpose they really serve is that of seasoning in the feed.

Nearly all of the tonics and tonic feeds well known to poultrymen are useful general tonics. Among tonics of common formulas that have long been used by exhibitors, the following may be mentioned:

Ginger ....4 drams
Gentian root ....4 drams
Sulphate of iron ....2 drams
Hyposulphite of sodium ....1 dram
Salycilate of sodium ....1 dram

Pulverize and mix thoroughly. Dose—three or four grains a day for a medium-sized fowl.

Fenugreek ....1 ounce
Mandrake ....1 ounce
Ginger ....1 ounce
Gentian root ....1 ounce
Bicarbonate of soda ....4 ounces

Dose—a teaspoonful to each quart of mash.

Licorice ....2 ounces
Aniseed ....½ ounce
Ginger ....2 ounces
Cayenne pepper ....1 ounce
Pimento ....2 ounces
Sulphate of iron ....1 ounce

This is used as a seasoning sprinkled in mash in such amounts as appear to be appetizing to the birds.

As a special tonic for backward plumage, Wright recommends 6 to 10 grains of citrate of potash daily; but adds that in many cases it will be found that making sure that the bird is absolutely free from lice has just as beneficial effects.

Training and Posing

Birds that have been handled right in looking them over at various times before the final fitting period have usually acquired some confidence in human beings, and even though it may have been necessary to hurt them a little sometimes (as in plucking feathers), that does not seem to impress them as much as the fact that the handling generally does them no harm and that the attentions they receive add something to the interest and comfort of life. All this earlier handling however, is usually wholly incidental to examination. Now handling should be systematically directed to making the bird perfectly docile and teaching it to pose in the attitudes which will show it to best advantage, and special attention must be given to good birds that are inclined to be a little shy and wild. One should make it a point to visit the coops several times a day—the oftener the better for this purpose—and when he is in the vicinity of the coops, or has occasion to pass where the birds will see him, to stop if just for a second before each and attract its attention Whenever one has time to do a little more than this he should make it a point to do something with or for the birds, distributing his attentions as seems most advantageous in bringing all to the point where they are easily handled. If he has to spend so much time on one occasion with one backward bird that he has little left for others, he should make it a point not to let that specimen monopolize his attention again until he has been all around to those requiring a like amount.

Some instructions for posing birds are rather more insistent upon certain formulas of movement in handling them than is necessary, and instructions by equally expert handlers sometimes appear contradictory. Thus one says: "Never begin by putting your hand on a bird's back. Always put it under, rather than over, them until they are perfectly gentle." Another will say: "Stroke the bird gently from head to the tail, bearing down a little on the tail and, closing the fingers on it, continue the movement to the end."

To some extent instructions of this kind are defective and limited. The mode of approach depends upon the attitude of the bird (upon how that varies from the desired typical attitude), and also much upon the person undertaking to handle the bird. In most cases the hand will go under the bird first because the tendency of most birds that come up to the hand from the start is to lower the body and head too much. The bird's natural impulse is to strike at a hand approaching it, or to spring forward or upward past the person before it. In either case, it relaxes the body at the hips and hocks in readiness for a spring. The object of the handler is to get the bird to take a graceful upright pose with its attention still fixed on him. To do this he cautiously approaches a hand to its throat and with the tips of the fingers gently tickles the throat between the wattles. Birds like this and will generally stand contentedly while it is done.

While tickling the throat the head is gently raised as far as desired. The bird may bring the body also up to a nice pose. If it does not, the handler gradually shifts the hand down until he has it under the body and with a little upward pressure the bird may stand in the position desired. It may go too high, in which case stroking the back will bring it down. It may turn away and try to go to the rear of the coop. If it does it must be brought back to the front center of the coop and with slight touches or pressure from the proper direction gradually worked into just the position desired. Having started to make a bird take a particular pose, one should work with it as long as necessary to get it into a position that fairly approximates what is wanted. To stop at any halfway stage is to fail entirely to give the bird an idea of what is required of it. Naturally it thinks that poor pose is what is wanted.

The best handlers generally talk to a bird while posing it, and while it does not appear that birds have any actual comprehension of what is said to them, the sound of the voice soothing or reassuring them helps the process, and fanciers are wont to say that when a bird will "talk back" they can soon do whatever they wish with it.

When a bird backs away from the hand and draws itself up, one must get a hand behind or over it to bring it in position to begin tickling the throat and stroking it. A shy bird may have to be held in the front of the coop with one hand while the other caresses it. When a male is vicious, and strikes hard with his beak at a hand introduced into his coop, the hand must be kept above him and close to his head, and his efforts to pick it met by slight boxes on the side of the head until he concludes that it is no use. Always in approaching such a bird one should make it a point not to let him get a hold but always evade or block him. It is not simply a question of preventing ugly cuts on the hand, but of breaking a bird of the vice. A bird that has it and carries it into the showroom will get the very least consideration that the judge can give him, and a judge has wide latitude in the matter of posing birds to bring out what is in them.

After a bird has been trained a little in the coop, it should be taken out and the same process gone through

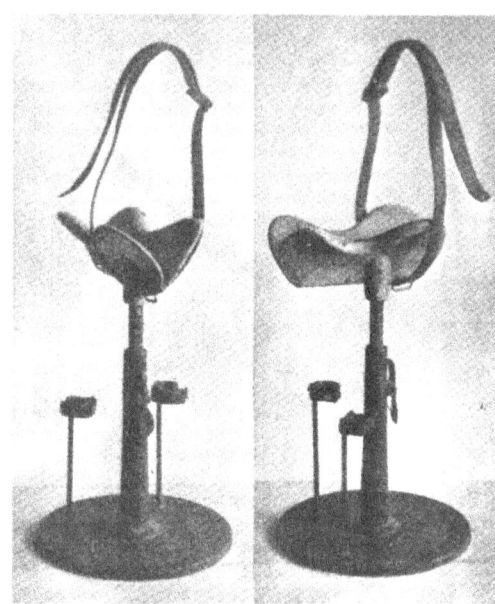

"SADDLE" USED TO HOLD BIRDS FOR CLOSE EXAMINATION AND FITTING
A strap holds the bird on the saddle, the legs being made fast in the clamps at either side of the standard which can be adjusted any height required

with the bird on a table, barrel or stand, where the opportunities to escape a handler are better. Here the handler must be more alert for the oftener a bird gets away from him the more trouble there will always be in handling it. He must, while working the bird into the desired poses, anticipate any move it may make to escape, blocking it if possible and, if the bird gets away recovering it with as little fuss and excitement as possible, continue posing it until it is in the position he wants. When the bird is in that position he should be kept in it as long as he will hold the pose, and when he changes it should be coaxed into it again, repeating this several times.

When it has learned to hold a pose for some seconds the handler should begin to back away a few steps as soon as the bird is in pose, keeping his eye on it and also pointing toward it with a finger or a hand to hold its attention. At first one should not go so far back that he cannot intercept a movement of the bird to leave the stand, but by degrees he can increase the distance until in many cases it will stay on the stand for minutes with the handler at such a distance that he could not possibly prevent it from leaving if it tried to do so.

After a bird has been got into the desired pose once, if it appears tired the training should be discontinued for the time. If it shows signs of being tired before the pose has been made it should be allowed to rest a little, but the aim should always be to have it in some way engaged with the handler until it has done what he wants it to do. He may take it from the stand and hold it, or he may let it walk about the floor always within his reach, or he may give it some tidbit to cheer it up, but he should have it under control until the lesson is ended.

There is a great difference in the aptitude of birds in learning to pose. Some can be made to take any desired pose in a few minutes, and after having been posed in a position several times will take it almost at a touch. Some it will take a half-hour, or even more, to get to pose right. The average time required for a first pose is from ten to fifteen minutes. Most birds can, with patience, be taught to pose in one position, which should be the position in which they best show the breed type and show any faults of carriage they may have least. It is not wise to try to teach a bird more than one pose. Some birds always pose well, and with these the idea is to have them learn to take the pose that shows them in profile to the best advantage whenever they are under direct inspection—to have them take it when anyone stops before the coop to look at them. Birds that habitually take poor poses are those that are not balanced right, or are lacking in vitality, and it really requires an effort on their part to hold a good pose, and they will do it only when touched up. It is doubtful whether such birds could be taught more than one pose, but there is really no object in trying it. If they will pose well in one best position for the judge, that will get them by when awards are being placed, though they will not attract the attention of observers as the natural posers do.

In posing to show breed type one must always have in mind the qualification mentioned in the preceding paragraph as to modifications of this pose to conceal any faults of shape that may be concealed by this means, and must also avoid exaggerations of breed type. It should be observed that Standard models of the breeds accentuate features of shape fully as much as is desirable and that to go noticeably beyond these in any respect is to overdo the matter.

If a bird will not move as desired when a part is touched or pressure is applied to it, the handler should take hold of it and gently force it into the position desired. Thus a bird may persist in holding its feet or one of them in a position in which it cannot pose right. The foot must be moved to the required position as often as necessary, until the bird will keep it there and relax so that it balances in the attitude desired. The handler must on no account lose his temper when a bird is stubborn, and begin to handle it roughly. No progress whatever can be made along that line, but much previous progress may be undone. After a bird has learned to pose under direction by the hand it should be taught to follow direction by a stick. The transition is easy and natural if in handling the bird one has occasionally kept a finger pointed toward it, or perhaps held a lead pencil in the hand. Judges do not all use judging sticks, but the bird should be taught to know what the stick is for and to follow its directions the same as those of a hand. Then there will be no trouble with its shying away from a judge with a stick in his hand, or dodging or fighting the stick when touched with it.

Much is said of the advantage of gaining the confidence of birds in training by feeding from the hand during or after handling. The objection to doing much of that is that the bird comes to associate the hand with feed rather than with command to show itself off, and the result is often a persistent disposition to face the handler or observer and to come close to see what is in his hand, instead of standing with the profile toward him, or in a quartering position, and coming to a nice pose in that attitude. With patience most birds can be made gentle and

FORMING THE HABIT OF CORRECT CARRIAGE OF BODY

No. 1—coop with opening in the front, represented by the dotted line, the opening being so placed that it encourages ideal carriage. No. 2—shows where the opening should be placed for a bird inclined to carry the front of body too high. Placing the opening and the feed and water cups low down, leads the bird to lower the front part of the body and get the habit of carrying it lower. No. 3—shows coop arranged for a bird that carries the front of the body too low.

SPECIAL FITTING OF BIRDS FOR EXHIBITION

as submissive as is necessary when handling. It is not desirable that birds should appear too submissive, for that gives the impression of being cowed. Nor is it desirable that the bird should be too much of a pet—inclined to be affectionate and familiar with the person handling it, instead of standing off and displaying its attractions.

There are some cases in which no amount of coaxing or persuasion has any effect, but the bird absolutely refuses to follow any direction and is either ready to dash away unless forcibly held or takes a rigid and constrained position from which it will not relax in the least. I have seen it stated that unruly birds are effectually tamed by the process of washing. I cannot vouch for the remedy as effective for all cases, but I have no doubt that in many instances it would — better than any other method —reduce the bird to a state of submissiveness.

Fitting, Fixing and Faking Head Points

A twist in a comb is due to local unsymmetrical growth of the comb. Slight lopping is due either to debility in the bird, or to the comb being too thin for its height. Lopping and twisting may occur together. A comb that shows any of these faults can rarely be made permanently free from them, but they can often be fully remedied temporarily, and permanently very much reduced.

Slight twisting in the outer sections of a single comb —over the beak, in the points or in the blade—and twisted spikes in rose combs, or spikes that do not point as they should, can be put in the desired position by massaging with the fingers and bending twisted parts in an opposite direction. Whatever improvement in appearance can be made may help the bird to a better place in the awards. One who has had no experience should practice first on some bird with faults of this kind which he does not intend to exhibit and see just how much he can accomplish, and how long a comb that is fixed in this way will keep the shape given it. Where the tendency to twist is at all pronounced it will usually return in a few days and it may do so in a few hours. Where it is slight, some exhibitors are satisfied to manipulate the comb at the latest possible moment before the class in which the bird competes is judged. Massaging commenced early and repeated too often may after a time increase the trouble.

Where a comb lops or tends to lop from general weakness of structure it may be straightened for quite a long period, and perhaps permanently, by placing on it supports which hold it in a perfectly straight position. The appliance most used for this purpose is a wire support made in the manner shown in the accompanying illustration. Cardboard supports are sometimes used, but they are rather cumbersome, are affected by moisture, and unless very carefully adjusted may make the comb sore.

When a comb lops slightly from general debility of the bird it may fully recover correct position as the bird regains good condition, but sometimes—especially in the case of a young bird whose comb is growing at the time —the comb will not recover its natural position because it has grown into the wrong position. If there is any danger of this, supports should be applied to keep the comb straight until the bird recovers its vitality.

Most of the trouble with crooked combs is in the combs of males of the large-combed breeds. The females of these breeds have corresponding faults but they do not show to the same extent because the lopped comb is required in the female. The judge will usually find them and take account of them. The most marked results from massaging the combs of females are obtained in those cases where the comb is a little too small or too stiff to lop well, or where a comb that should fall smoothly to one side tends to loop. In such cases there is little danger of overdoing the massaging, and I have known daily treatments for several weeks to put a poor comb permanently in good form. Whether it is worth the effort depends altogether on the importance to the exhibitor of showing that particular bird. It should be said that while most exhibitors do more or less massaging of combs and wattles, all seek as far as possible to avoid the necessity for it, and, as a rule, it is only the birds that are remarkably good in other respects that are considered worth treating for this class of comb faults.

As a means to prevent excessive

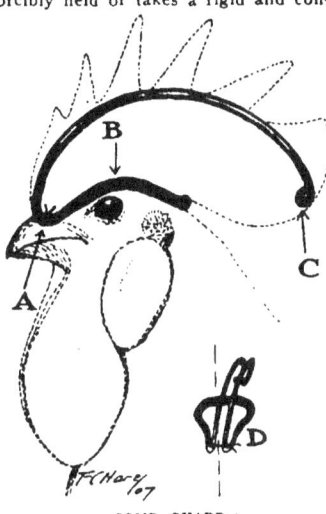

COMB GUARD

Made of soft copper or brass wire. At A a silk twist is passed with a needle through the nostril and the guard is tied in position. The section marked B is wound with waxed string to prevent its marking the head. The hook C is to keep the point of the guard from pressing into the comb. D is the front view of a guard for a comb that leans to the left (opposite way to which the guard leans).

TO TAKE A TWIST OUT OF A COMB

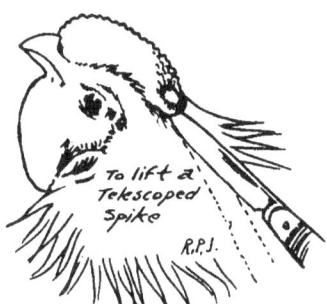

TO BRING OUT AN INGROWN SPIKE

MAKING A LOOSE CREST MORE COMPACT

growth of comb and increase of dangers of lopping and twisting, Wright mentions hazeline tincture or hazeline cream applied nightly to the comb while growing. The astringent property in these checks the circulation of blood in the comb and so retards growth. All these extraneous measures, however, are deficient in that a few hours' exposure to high temperature may offset the care of weeks, and so American breeders generally try to keep the normal comb so moderate in size that it will grow smoothly, and that the usual increases in its size made by exposure to high temperatures will not make it objectionably large.

Surgical Operations on Combs—The simplest of these is the removal of side sprigs. The earlier these are removed the better—provided a clean job is made of it; but if the growth is merely pruned, of course it will increase in size making necessary a second operation. A side sprig may be removed with a knife or a pair of scissors, or bitten off with the teeth or pinched off between the thumb and finger nails. Scissors seem to be commonly preferred, and slightly curved manicure scissors are best, for by placing a finger on the opposite side of the comb and pressing the scissors against it before cutting, the wound made is a little below the surface and the skin tends to close over it and to heal with a fainter scar. The operation should be performed several weeks before the bird is to be shown to insure that the least possible trace of it will remain. Where the sprig is quite large and the blood flows too freely powdered alum or tannin is used to stop the flow.

In the days when good five-point combs were rarer than they are now a well-known method of making a comb with too many points grow the desired number was to prick or pinch certain of the superfluous points, just as the comb was beginning to grow, so that their growth would be checked and those beside them would grow over them, making them fully ingrown points. It is said that combs so treated were far less liable than others to twist. The operation is on precisely the same basis as the removal of side sprigs, except that to succeed it must be performed at a very early stage of the growth of the comb.

A much more difficult, and also probably a much more common operation, is the "lifting" of the ingrown spike of a rose comb. This operation consists in cutting the comb around the base of the spike in such a manner that the spike can be pressed out and the adjacent parts brought together to hold it in the new position when they are healed. It is hardly possible for an operation like this to be performed without leaving rather conspicuous scars, but if the bird has at the rear of the comb a protuberance that will pass for a spike it is saved from disqualification, and as the corrugations on rose combs are often either naturally quite irregular and smooth skinned, or the skin has been made smooth by injury, scars are not so conspicuous as on a single comb of fine texture.

Accounts from various sources mention instances of much more important operations on rose combs. Wright gives it as a matter of his own personal knowledge that in a certain case where a Hamburg male had too large a comb a wedge-shaped piece was cut out of the center the entire length of the comb and the sides drawn together, making—when healed—a much narrower and very good-shaped comb. The same writer mentions as a matter of hearsay, a case where the good comb of a poor Hamburg was successfully transplanted to the head of a good bird which had a poor comb. In this case it was said that both combs were cut off clean, close to the head, and the transposed comb immediately fastened so that it would remain until healed. This case is cited as one regarded as extraordinarily brutal, but it must be said that apart from the additional pain and discomfort caused a bird by the burden of the grafted comb on its head until union of the parts took place, the brutality does not exceed that of dubbing the combs of Minorcas having abnormally large combs.

Judges have sometimes found needles in the combs of birds, either placed there to hold a point or spike in position or to hold parts together while healing after a major manipulation of them. Most of these cases are reported from England—probably because the predilection for large combs there makes more occasions of temptation, and perhaps also because as the birds are handled less by the judge the chances of escaping detection are better.

To heighten the color of the comb various washes and lotions are used and recommended. Before giving them I will say that one of our most successful exhibitors declares that clean water applied with a small brush answers the purpose as well as anything used, and has the advantage of giving more lasting effects, there being no reaction in its use as where a stimulant is applied. This washing may be done either before a bird goes to the show or in the showroom.

Sweet oil and alcohol, equal parts. Apply with a small bit of sponge or soft cloth.

Two parts of alcohol and one part of glycerine, with three drops oil of sassafras to each teaspoonful of mixture. Apply as above. Care must be taken not to get this in the bird's eyes as it is extremely pungent.

Vinegar, or vinegar and water. This is said to be objectionable for use on the combs of birds that have been washed, as it causes their combs to blister.

Vaseline alone is sometimes used on combs and wattles, both to improve the color, and to make them more resistant to frost. For the latter purpose it is often smeared on quite thickly at the close of the show when the birds may have to go from a warm showroom into extreme cold. Put on at any other time it must be wiped quite dry for otherwise it catches the dust, and the bird is likely to smear its plumage with the grease.

Olive oil, sweet oil, coconut oil and turmeric, are all used to brighten combs. Any mixture used for the shanks will, as a rule, be as satisfactory for the comb and wattles.

Treatment of Ear Lobes and Faces

Permanent white in red ear lobes and faces is sometimes concealed by painting it over with oil or water color or with a red cosmetic. The success of the deception depends entirely upon the fault being made so inconspicuous, and the application to conceal it being also so inconspicuous, that the judge does not observe anything peculiar about the lobes. Vermillion reds and permanganate of potassium are the ingredients most commonly used. India ink and Venetian red are also employed. Neither will completely cover a large patch of white. A small spot may be covered well but on close inspection will usually look enough different from the rest of the lobe to make it quite apparent that the exhibitor has resorted to coloring. Where coloring is present it will come off if the lobe is wiped with a white cloth moistened with water alcohol or ammonia, as may be necessary to remove what has been used, or if dried thoroughly hard a careful examination will show it. In many cases close inspection shows the white spot through the coloring matter.

Small spots of white are sometimes cut out of red lobes, or burned out with caustic. If the job is well and neatly done the remaining scar appears to be due to injury. A crude job may still have that appearance, but leaves very strong suspicion in the mind of the judge.

The process is thus described by Hubbard: "Take a stick of caustic and moisten the end. Then go over the whole surface of the ear lobe that shows white. Do this twice a day for three or four days, and it will cause a blister which will form a scab. When the scab comes off the white comes with it. When it heals you have a red ear lobe. Never pull the scab off or touch it in any way. If you do you will spoil the whole job." It should be added that here again the success of the job is a matter of getting the bird past the judge without special attention being attracted to the ear lobe. As soon as one familiar with the natural appearance of a lobe having normal, perfect skin looks at one that has been treated this way he knows that something has been done to it, and has a pretty good idea what has been done.

The treatment of white ear lobes to secure a perfectly white surface includes both those entirely free from red and those having more or less red. Whitening of white lobes that have no red faults may be either a matter of removing a weather-stained and sunburned outer skin, so that the lobe has a new and perfectly white skin for the show, or it may be done by the application of some white preparation. The former method is commonly preferred as being more defensible. It is, in fact, a practice of the same character as the removal of freckles or tan from the human skin. It does not change the natural quality of the white skin of the lobe but removes a worn skin that a fresh one may grow.

Oxide of zinc ointment, and toilet preparation of "peroxide cream" are the ingredients commonly used for this purpose. The ointment or cream is rubbed lightly over the lobes once a day, beginning about three weeks before the show. It causes the outer cuticle to peel and the new skin comes perfectly white. The birds should not be exposed to the sun and weather during the process or the new skin will be more or less colored. The process should also be gradual—with mild applications of the preparation used—as a too rapid removal of the old skin makes the new skin rough and may make the ear lobe sore and bring red into it. The novice in the use of the process should not get his first practice in it with good birds just before they are to be shown; but should practice on a few culls in advance of the show season.

The treatment for whitening ear lobes appears to have been originated and brought to its highest perfection by English breeders of Black Spanish fowls more than half a century ago. It seems to have had its beginning in applications of dilute sulphurous acid to cure eruptions on the white face. To prevent injury to the sound skin and temper the action of the acid preparation on sores, the face after being as carefully dried as possible was powdered or dusted with violet powder or with oxide of zinc in powder form. The oxide of zinc was preferred because where it lodged in the slight damp folds or creases of the face it did not crust as the violet powder was apt to do, and because it was found that continued applications of it to the white face increased its smoothness and whiteness. The application of powdered oxide of zinc to ear lobes today follows this method, and with the difference that whereas the powder was first used to thoroughly dry the face, in the modern process, the ear lobes are first moistened and then partly dried that the powder may adhere when applied.

The misuse of sulphurous acid, using it too strong or without proper after-treatment, is probably responsible for the idea that persists to some extent that severe applications of acids are sometimes made to whiten lobes. One writer referring to "acid treatments" used observes that it is cruelty to animals to use them, and that they always leave a scar. It should be entirely plain to anyone that a treatment that leaves a scar is of no use whatever for white lobes. For eruptions on the lobes use any of the mild antiseptic preparations common for household use, and after partly drying with a soft cloth, powder with one of the preparations mentioned, or with boracic, talcum or other powders used for toilet purposes.

An "enamel red" in white lobes cannot be removed. It is sometimes concealed when at the edges and covering only a little surface by manipulation of the lobe, stretching it gently and rolling the red edge under. Cases have been known where a considerable area of red was thus turned under and the lobe sewed from the underside to prevent it going back. The method is comparatively easy to discover, for the lobe is drawn into an unnatural position and inspection of it at once reveals the threads.

The stretching and massaging process is also sometimes used to improve the form or increase the size of a white lobe not in need of treatment for any color fault.

Treatment of Beaks and Bills

A little dark color on a yellow or flesh-colored beak or bill may be removed by scraping with a knife, or by the use of a nail file, or of fine sandpaper. Before undertaking the removal of such color it is well to be sure that it is only on the surface. If it extends into the substance of the beak it cannot be removed in this way—or any other. If it is entirely under the surface, scraping there will only increase its appearance to the eye.

In England the flesh-colored bill of the Aylesbury Duck is one of the most prized exhibition points, and smoothness of the bill and delicacy of tint are secured by keeping the birds under cover where the sun will not burn the bills, by giving clean water with an abundance of sharp, white gravel to polish the bill as the birds move their bills through it, and sometimes when these measures seem insufficient and the bill remains coarse and rough it is scraped with a sharp knife all that it will stand without bleeding, then rubbed smoother with fine sandpaper, and the bird kept in a rather dark place for several weeks until a new bill is grown. According to pretty well authenticated reports, the keen competition for perfection in this prized feature occasionally leads an exhibitor to peel the bill that an absolutely new one may grow.

Treatment of Shanks and Feet

If the shanks of a bird are not perfectly smooth and healthy looking, treatment to remedy them should begin as soon as the bird is selected for showing. The simple cleaning and polishing of the legs may be deferred until just before shipping to the show. In fact, when a breeder accompanies his birds he often finds it more convenient to put such finishing touches on there. The most common troubles needing prompt attention are unmolted scales and scaly leg. The two are often associated. The failure of the scales, or a part of them, to drop off is a condition similar to that of the failure of feathers on a section of the body to molt properly. The scales if loose can be removed by pressing with the thumb nail at one side, as shown in the accompanying illustration. The whole shank and the toes should be systematically gone over to free them of all loose scales. If the scales, though rough, will not come out easily, the force necessary for their removal should not be used for that would cause bleeding, and sometimes discoloration as a result of blood lodging under the scales where it cannot be reached in cleaning. The treatment used for these cases should be to apply freely oil or paste as used for cleaning and polishing healthy shanks and feet daily, rubbing it off as much as necessary to prevent soiling of the plumage. A paste made of sweet oil and powdered pumice is very good for this purpose. The oil is absorbed by the scales and skin, and when the parts are rubbed the pumice scours and smooths the rough scales. Continued applications of any oily mixture will frequently loosen the old scales so that they are easily removed.

This same treatment will cure mild cases of scaly leg, but where the disease is bad it is necessary to use more penetrating oils. The preparation most commonly used is a mixture of kerosene and linseed oil in about equal parts for ordinary cases, and with more kerosene for severe and less for mild cases. The method of application is first to rub off as much of the dead matter as can be removed with a stiff brush (an old nail brush), then dip the legs to the hocks in oil, hold them above it for as long as necessary to let what will run off drip, and either wipe dry or put the bird on clean straw or shavings and let the leg absorb the oil. This should be repeated daily until the leg is clean and smooth. The more kerosene is used, the sooner the dead matter will all be removed, and it is possible to get rid of all the scale on very bad legs in a week or little more. Such rapid process however, leaves the skin bare, raw and sore, and it will not look as clean after the scales grow out on it as if the cure is more gradual.

Other preparations used for scaly leg are:

Vaseline, 5 parts; oil of caraway seed, 1 part.

Vaseline and zinc ointment, equal parts.

For severe cases: One ounce of sulphur, half an ounce of oxide of zinc, one dram of coal tar, two ounces of whale-oil soap. Mix together well and apply daily.

For Feather-Legged Birds—The use of greasy preparations for feather-legged birds is not practical, and many prefer not to use them on clean-legged birds to the extent necessary in treating scaly leg with them. An emulsion which can be washed from the feathers after applications are made to reach the scale on the leg and among the quills of the foot feathering can be made in the proportions of half a pound of soft soap and half a pint of kerosene to a pint of water. This may be applied with a brush, or it may be further diluted with water and the legs dipped in it and the feathers cleaned afterwards with water and a stiff brush. It is always desirable to have necessary treatment for scaly leg in feather-legged fowls come at a time when it is not necessary to be careful about keeping preparations used for the leg off the plumage as much as possible. By giving attention to this at the beginning of the molt, or better just before it, one can do the work when the foot feathers are shortest and most scanty and interfere least with applications to the skin, and when it makes little difference if they are soiled with the preparations used.

Ordinary Cleaning and Polishing of Legs—The time and amount of this depend on the condition of the legs, and this depends generally on the conditions under which the birds are kept. Those on good grass runs should require no cleaning until just before going to the show. Those that have run in dry, bare yards, especially on alkali soils, will usually have less than the normal amount of oil in the skin of the leg, and its condition and appearance will be much improved if the legs and feet are washed and oiled several times before the final treatment. All that is necessary is to wash the legs with warm water and soap, and after wiping them dry apply a little oil or vaseline. Most poultrymen use the same preparations for this as for oiling the legs before exhibition. Many use the same mixture for both legs and combs. Some use oil or vaseline alone; others mix with it a little turpentine, camphor, alcohol or other volatile matter. The oil should be wiped quite dry after each application, and the birds should be kept where little dirt can adhere to the feet.

The final cleaning of legs before exhibition has first the object of getting them perfectly clean. Until one has compared the yellow, white or flesh-colored shanks that appear perfectly clean in the yard with those of the same color as made absolutely clean and polished for exhibition, he has little idea how dirty the cleanest shank is

HAMMOCK TO HOLD BIRD WHILE BEING PREPARED FOR EXHIBITION

No. 1—shows an easy and convenient way to hold a bird while working on its feet and legs. Above—Detailed drawing showing how the hammock is made; below—how it is hung. In No. 3 the dotted lines show how it is fastened over the bird's back to hold the wings down. No. 2—shows the method of removing old scales with the thumb nail.

SPECIAL FITTING OF BIRDS FOR EXHIBITION

in the yard. Even when a novice has given the legs of his birds what he supposes is a sufficient cleaning and polishing he is usually surprised to see how far short they fall when compared with birds fitted by an expert. He may have noticed as he worked over the legs of his birds a little accumulation of dirt everywhere under the edges of the scales, but as no easy way of removing it occurs to him he concludes that as it cannot be brushed out the only thing to do is to leave it there.

The expert does not leave it there. In most cases he removes it by the tedious but effective process of insert-

HOLDING A BIRD TO DIP LEGS IN SOAP SOLUTION

ing the point of a toothpick or small piece of soft wood under the edge of the scale and taking the dirt out in much the same manner as he would clean his own finger nails. That is the common process. It takes sometimes as much as an hour or two to clean the dirt from under the scales on the shanks of a bird. Another less laborious process preferred by some exhibitors is to dip the legs, after the bird has been washed and while the scales and skin are soft and loose, into thin soft soap, or an emulsion like that mentioned above for the treatment of scaly leg, and hold them there for a few minutes until the soap loosens the dirt under the scales when a little pressure on the scales—applied by the fingers, beginning at the hock and working down—will force the dirt out with the soap and water that have worked under the scales. Some of those who use this method say they never find it necessary to use any implement to clean under the scales, but make the leg perfectly clean in this way.

Whatever process is used, after the leg is clean it should be given an application of grease or oil, and this rubbed dry with a soft cloth. Black or slate legs, or dark willow legs do not require as much attention in washing, but they should be made clean to all appearances and then oiled and rubbed. The darker the leg the higher natural polish it will take.

Coloring and Staining Legs—Very little attention is given by expert exhibitors to the artificial coloring of legs. They are generally satisfied to do what can be done with clear oils, or oil mixtures and greases, to bring out fully the natural color of the shank. Most of the coloring is done by novices who do not realize that this is one of the easiest forms of faking to detect, and who perhaps have been erroneously informed that the coloring of legs is common. Any yellow stain will give added color to yellow legs. That most often used is butter color, three or four drops in a teaspoonful of sweet oil. Iodine and saffron are both sometimes used. For black or slate legs a little finely powdered graphite is mixed with oil or rubbed on the legs after they have been oiled.

With the birds handled as much as they are in American shows, it is generally impossible to put enough coloring matter on legs to appreciably improve their color and escape detection. If a judge's hand gets a little warm and moist as he works, coloring matter on the legs of birds may come off on them. Where coloring is suspected, but is apparently too fast to come off with ordinary handling, the judge moistens a handkerchief or bit of rag with water, alcohol, or any volatile oil available, and rubs the shank until he either gets evidence or is convinced that the color is natural. While very high color is suspicious it sometimes proves to be entirely natural. In fact, I have seen legs on stock that was never exhibited or the legs treated with any preparation for any purpose that were so deeply and intensely yellow that in a showroom anyone would at first affirm without hesitation that the legs were artificially colored. But, as a rule, very high color on the legs of birds on exhibition arouses suspicion and if artificial color has been used the fact may be determined either by the appearance of the leg or by rubbing as above described. Unless the coloring matter is carefully and successfully applied it will not be the same everywhere, and in particular the skin under the scales which the coloring matter has not reached will be lighter in color than the rest.

Trimming Spurs—This was at one time listed among the specifically prohibited practices, and it appears to be still regarded by some people as a form of "faking." As far as the exhibition room is concerned the question of deceit in the trimming of spurs can apply only where a cock is shown as a cockerel, and his spurs trimmed short to bear out the claim that he is entitled to entry in the cockerel class. In selling birds the shortened spur will sometimes make a bird that is past his prime look enough younger to be more acceptable to a customer who judges the age mainly by the appearance of the spurs. Quite

Individual perch for full-feathered birds

Rod on brackets extending around side of house and yard to prevent Cochins breaking foot feathering

Bent willow work for interior of coop or hamper to prevent foot feathering being broken in transit

apart from any question of deceit, the spurs of males which grow long and sharp should be trimmed to prevent them doing any injury to other birds or to persons handling them. Heavy males with sharp spurs often cut the backs and sides of females that are with them badly.

To trim a spur, take a sharp and tolerably strong pocket knife. Hold the bird under the left arm with the left hand holding each foot as operated on in a position convenient for using the knife in the right hand. Start about a quarter of an inch, or a little more, back—that is, toward the leg—of the point decided on as marking the length of the shortened spur, and, cutting toward you and toward the point of the spur, work around and around it, always cutting a little deeper, not completely severing the shavings, but leaving them attached to the end, until the spur is cut through and the point falls away. Care should be taken to fix the position of the point of the shortened spur beyond the quick portion of it, for if this is cut into it may bleed badly. It is better to cut a little long and then trim down than to cut too short.

Trimming Toenails—When the toenails are a little twisted, or are too long to look well the appearance of the foot may be improved by trimming them, cutting away most on the side in the direction of which the nail turns, and making the toe and nail look as straight as possible.

Treatment for Stubs and Down—The removal of very small stubs and down leaves no trace visible upon ordinary inspection. When the stub is large enough to leave a hole that is not conspicuous, but still easily visible on close examination, many exhibitors simply trim away with the point of a sharp knife any roughness around the edges of the hole, with the idea of making it as inconspicuous as possible without plugging it. Where the holes are so large that they show plainly they are filled with beeswax, soap, putty or paraffin. The "filler" can nearly always be readily seen when the shank is closely examined. If the evidence of sight is not sufficient it can be pricked out with the point of a knife, or sometimes washed out in warm water.

Washing Poultry for Exhibition

Practically all the white birds that are shown at exhibitions of any importance are washed. A white bird that has not been well washed looks poor, ill-conditioned and out of place in good competition, no matter how good its actual quality may be. Many exhibitors of buff and partly white fowls wash their birds before exhibiting, and some wash dark-colored birds. The actual necessity of washing birds for exhibition is determined by the circumstances in any case. If an exhibitor in a class of birds of a color that it has not been the custom to wash puts a string of birds on exhibition that have been well washed and otherwise properly fitted he will in almost every case win more prizes on condition than he is entitled to on quality without consideration of condition. So in any color variety where washing has not been practiced, if a leading exhibitor begins to regularly wash his birds, the rest will always have to follow suit. Breeders of other than very light-colored fowls are always reluctant to begin a practice involving so much labor, and also requiring either the invasion of the home kitchen and laundry, or special arrangements for the work outside. Probably three-fourths of the birds that are washed before being shown are washed in the laundry or kitchen of the dwelling house, and in many cases the poultryman's wife assists him in such work—not infrequently being the better washer and fitter of the two.

INTERIOR OF WASH ROOM AT OWEN FARMS, VINEYARD HAVEN, MASS.

Every poultryman who washes many birds however, soon becomes ambitious to have a properly fitted room or building especially for that purpose, and a number of prominent exhibitors have complete and convenient equipment for it. The equipment required is not at all expensive. The essentials are enough exhibition coops for the number of birds to be washed at one time, three or four tubs, a long bench or table on which to set them, one or two dripping coops, and such heating apparatus as is needed to warm the room to a temperature of 80 to 90 degrees and to furnish the required amount of hot water. The washing room at Owen Farms, Vineyard Haven, Mass., an interior view of which is shown in an accompanying illustration, is partitioned off in one of the buildings originally built for a long brooder house, the remainder of the building being used as a conditioning room, and to coop birds awaiting shipment to customers. This room has capacity for drying seventy birds. It has, to furnish heat, a large heating stove and a long stove of the type commonly used with farm feed cookers, with a tank for heating the water needed. There is nothing here but what any exhibitor could provide on such scale as he required at comparatively small cost.

The great advantage of a room especially for washing fowls, is that in it one can work with entire freedom as far as the effects of water upon the surroundings is concerned and so give his undivided attention to his job. The premises do not have to be restored after the work is done to such order and neatness as a room in the dwelling, but can have the rough cleaning appropriate in outbuildings. Some laundries with cement floors and with draining arrangements that admit of flushing the floor meet the needs even better than most outside washing rooms, and some exhibitors whose wives are equally interested in the

SPECIAL FITTING OF BIRDS FOR EXHIBITION

results do not regard the inconveniences associated with the use of the kitchen for washing poultry as serious; but a great deal of the poor washing that makes a bird look worse than when unwashed is due to washing under conditions which make the one doing the work feel that the surroundings must not be mussed up too much, and the reluctance to do the work of cleaning up undoubtedly deters many people with small strings of birds from undertaking to wash them. Of course it is possible to wash a few small birds in a kitchen, but it is a great deal more of a job to do so than to wash several times as many where everything is arranged for the work in hand, and no particular consideration has to be given to restoring order for other work when this is done.

Not only is it an advantage to have a properly arranged and equipped room for washing birds, but the work is much facilitated if the workman "dresses for the part." An oilskin or rubber apron of ample proportions and rubber boots are needed. Further than this the matter of dress is a question of omitting everything not required by decency, for the work is to be done in a dogday temperature and will require some exertion.

There are at least as many ways of washing birds as there are expert washers, and as in most of the operations of the poultry yard, a considerable number of the experts think their way is the only way. The methods used at Owen Farms are more simple than most of the instructions given. A number of the things generally mentioned as essentials are disregarded or omitted there. To begin with it is taken for granted that the feathers are washable stuff, and that fowls—though not amphibious—can be thoroughly soaked without being injured in the least if they are kept at the right temperature afterward until the plumage is completely dried. Proceeding on these assumptions, their method is, comprehensively, to wash the bird and let it dry in a temperature of 85 to 90 degrees. I will give the details of their process as staged for the purpose of illustrating it in this book. Following I will give methods as used or described by a number of others. It should be said in regard to the simplicity of Owen Farms methods that the birds here are relatively clean to begin with. There are no railroads or factories on the Island; there is none of the smoke and soot that in many places soils fowls badly; the conditions under which the birds live are ideal in every respect; and there is about the same difference between washing a bird here and washing one that has grimy, sooty, stained and fouled plumage, as there is between giving a bath to a child that has recently had one and giving a bath to a child that has not had one for months.

Four tubs of water are placed in line on a table as shown in the picture. In the first the water is as hot as the man can comfortably bear his hand in. He uses a large sponge and, before taking a bird, with the sponge and a bar of soap he makes a suds that overflows the tub. Then taking the bird between his hands, he puts it without ceremony into the tub of suds and water, submerging it at first, moving it about in the water a little to wet under the feathers better; then placing the bird on its feet, he controls it and moves the wings or parts the feathers as necessary with one hand, while with the sponge in the other he works and rubs the soapy suds into every part of the plumage. When the bird is thoroughly soaped he continues washing the plumage, working all over the bird, as long as his experience indicates is necessary to get the dirt loose, giving special attention to any parts he may have noticed as particularly soiled. Then he lifts the bird from the tub and holding it between his hands first allows what soapy water will run off freely to run to the floor, and then holding the bird with one hand passes the

WASHING A WHITE BIRD AT OWEN FARMS
Left—Putting the bird in the first tub. Right—Just after taking from first tub

other several times over the body—from the head to the rear—to take away as much more of the soap as will run out of the feathers under this slight pressure. The object of letting the soapy water drain to the floor instead of back into the tub from which the bird has just been taken is to keep the water in the first tub clean enough for use as long as possible. There is more dirt in the water that drips and is squeezed from the plumage than in the water in the tub. The bird is not transferred directly to the second tub because to do so would be to add to the water in it both dirt and soap that are not wanted.

INTERIOR OF WASH ROOM ON PLANT OF JOHN S. MARTIN
Drying coops are put on benches before the windows, supplementing the stationary coops along rear wall

The water in the second tub is not as hot as in the first, but is comfortably warm. That in the third tub is cold, as also is the water in the fourth and last tub, which is blued as for white clothes. From the time the bird goes into the second tub the process is a washing and rinsing process to remove the soap from the plumage, on the principle that with the removal of the soap all the dirt will also be removed. The bird is moved in the water, and the feathers are manipulated with the hands or sponged as seems most effective to remove soap. The operator's judgment as to when the feathers are free from soap is guided partly by sight, but more by the sense of touch. He keeps the bird in Tub No. 2 until it seems to him the water is as soapy as the plumage, and is therefore no longer effective to remove soap, then transfers it to the third tub, first letting it drip to reduce as much as possible the amount of soap added to the water in that tub when the bird is put in it. The rinsing in the third tub should remove all soap. Then the bird is passed through the bluing water, and from it directly into one of the dripping coops seen at the far end of the table.

These coops are closed on the sides, back and top, have a slat bottom, and the whole front is on hinges so that the birds are handled easily in and out. A bird is left in the dripping coop while two birds are being washed—perhaps half an hour—and is then transferred to one of the exhibition coops in the room where it remains until the next morning, by which time it is perfectly dry. The temperature of the room is kept up to 80 or more while the birds are drying. It is then gradually allowed to cool off, or if another lot of birds must be washed at once, the newly washed birds are moved to the coops in the conditioning and fitting room adjoining.

Except to remove soapy water from the plumage between consecutive tubbings, nothing whatever is done by hand to dry or assist the drying of the birds. All that is considered here a waste of time and labor. No towels or fans are used.

In giving other methods of washing I will begin with the instructions given by one of the early English breeders and exhibitors of White Cochins, as published in the first edition of Wright's "Illustrated Book of Poultry" in 1874. It is worth noting that the art of washing fowls was first cultivated and with remarkable success by the exhibitors of this most difficult of all varieties to wash to perfection. For many years prior to 1902-3 the White Cochin classes were among the most spectacular features of New York and Boston Shows, often numbering from fifty to sixty birds, and a large part of the class of remarkable quality and washed and fitted to perfection.

Quoting Mr. Elijah Smith, in Wright's book:

"Take a washtub of ten or twelve inches deep—oval shape is the best on account of the bird's tail; let the tub be sufficiently large to hold the bird comfortably. Then take of clean, soft, warm water and fill the tub about three parts full, so that the bird when pressed down by the hand in the water will be covered over it's back, up to the neck. Then take white soap and a sponge and rub it in the water until it is well mixed and you have a good suds; rub the bird well with soap on all the dirty parts, and keep sponging the bird until you see that it is quite clean, which you will be able to see plain when wet. Do not be afraid to rub the feathers, as it will do them no harm as long as you do not lay on so heavily as to break them. If the bird is rough in the water, as some that have never been washed before sometimes are, keep one hand across the bird's back and wings, by which means you will easily hold it quiet. Be sure and rub your hand well among the fluff and feathers about the breast.

"To wash the head, take it between both hands and rub it well backward and forward, as if you were washing something in the balls of your hands. Do not be afraid of the water going into its mouth, as the soap and water will do it no harm whatever, but the contrary, as it will tend to clear it out; in fact, I have often washed birds when I could not get anything else to cure them of disease, and it has answered remarkably well on many occasions.

"When you see the bird is quite clean, then take and rinse thoroughly with clean cold water; put plenty on it, until the soap is well out, for if you leave any soap in, the feathers will not come right in a reasonable time. When clear of soap, let them stand to drain a little, and don't be afraid of their getting cold, as the cold water prevents that by closing all the pores of the body; then press as much water off the feathers with your hand as possible, and don't be afraid of hurting the feathers as they will come all right again as they begin to dry, and will begin to web again in the course of an hour. When this is done, take the bird and put it before a nice fire—not too hot, but what we would call, a good hot fire—and keep turning them with the wet parts toward it, taking care not to have them so near as to blister their faces and combs, as they easily blister after washing. When the birds are nearly dry, you may put them in baskets that have got lining in—such as we use for sending to exhibition—and, if night, you may put three or four of them together, if the basket is large enough for them to lie down in comfortably. By this means it will create a warm steam that will pass through the whole of the body feathers and cause them to web beautifully, and the birds will be quite ready for exhibition in twenty-four hours.

"If the bird is looking very ill after rinsing, keep it in motion as much as possible by getting hold of it under the breast with one hand and lifting it up, when it will use its wings freely, and this will cause the blood to circulate; also give one or two cayenne pods, which will warm it as well. This is when you see a bird that goes black in comb, and looks as if it would die, which heavy birds sometimes do; also handle them pretty freely, as it will do them good. Sometimes a bird will faint when put in warm water to wash; in that case, I always throw cold water on it, when a bird will at once recover, and after a minute or so you may put it in again, and finish washing it without it showing any symptoms of fainting again."

Wright's further comments on the foregoing instructions add to their interest and value. Said he: "Many good washers prefer to dry the fowls after washing in a cage or box of ample size, littered with clean and well-broken straw. This box is wired in the front and top, but closed at back and sides to prevent draft, and placed with the open front at just such a distance from an ample fire that a genial warmth may fill the box, but avoid a scorching heat. We may add that it is in drying that judgment and experience are chiefly required, as too strong a heat withers up the plumage and makes it ragged, while too little causes it to hang together and appear draggled; but if the right temperature be hit upon and the soap has been thoroughly washed out, by degrees the

JOHN S. MARTIN'S DRAINING COOP
Left—Closed. Right—Open

WIRE SCREEN FLOOR AND DRIP PAN USED IN MR. MARTIN'S DRAINING COOP

plumage fills out, and in a few hours the birds assume their 'company clothes.' It is to assist this that Mr. E. Smith so strongly advises leaving the birds with a little dampness still in the plumage, the steam assisting the fresh webbing of the feathers. In summertime the cage may be put out in the sun if preferred; but the glare seems to distress the birds much, and we should prefer a fire. Some poultrymen are unusually clever in drying fowls, and by holding them near the fire, and carefully removing them for a little whenever they appear distressed with the heat, manage to avoid the scorching we have spoken of, and can dry a pen of Cochins in about two hours; but we cannot pretend to give the precise details of such management, which can only be successfully practiced after great experience has been obtained. As an example of what may be done by an adept, however, we may relate as within our own knowledge that the writer of the preceding remarks on a certain occasion received back his birds about ten o'clock in the morning, fed them, washed them, returned them to the hampers all wet as they were, and got them off by rail for another show at twelve; taking them out and drying them by the fire in a junction waiting room on his way to the exhibition, where he again carried off the first prize."

To complete the information from the same source, it is worth while to make some extracts from the instructions for washing a fowl given by Wright in the revised edition of "The Book of Poultry" in 1902:

"Before washing birds of his own, a novice will do well, if possible, to get a practical lesson; otherwise he will be slow to grasp the very thorough character of the process. This thoroughness is the secret of success, and most people fail in their early efforts because too nervous or squeamish about damaging the feathers. It is little or no use to sponge down the outside plumage. At least one large oval tub, not much short of a foot in depth for large fowls, must be provided, and unless there be facility for rapidly emptying and renewing the water twice, it is better to have three at once. Anyhow, plenty of hot water must be at command. Also provide a basinful of soap solution, such as washerwomen use, made by cutting up some good soap into thin slices and dissolving in hot water into almost a thin melted jelly. There is also wanted a good compact sponge, rather soft, and just as large as the hand can squeeze easily, and some soft, dry towels. In commencing operations the feet and legs should be washed first and separately*. Then the tub is filled about two-thirds with water about the heat of an ordinary hot bath and the bird is stood in this; it should be at least deep enough to come well up about the body, and if when the fowl is pushed down it covers the back, all the better. The first thing of all is to be sure that the bird is thoroughly drenched to the skin; just dipping in does not do this. The plumage must be parted and worked about with the bare hand under the water, or with the sponge, till every feather is soaked to the root. Then we begin with the soap, taking up some with the sponge and thoroughly rubbing it into the fowl, one place at a time. It is to be a good thorough rubbing, all sort of ways, except that we should not go straight against the lie of the feathers, though we doubt if even that would do much damage. But down and across, to and fro, and energetically too, with the idea always of getting down to the skin; keeping on at one part till more dirt ceases to come off. There is really no danger at this stage, and no difficulty provided the operator is not afraid to do his work and sticks to the one point that he has got to get his bird clean. About the breast it is necessary to rub almost up and down, which is best done with the bare hand; indeed we have seen a bird well washed with hands alone, not using a sponge at all. The fluff also requires the hand well worked about. Some use a brush to scrub, but this is not free from risk; not to the feather as a whole, but proper webbing afterwards; several times we have seen birds scrubbed with a brush, which did not seem to web smoothly when dry, and believe that the bristles brush out or off some of the tiny microscopic barbules which hold the web together. One very good washer we knew used chiefly a sponge wrapped in flannel, especially for the secondaries of the wings and for the tail; the slight roughness, he said, brought the dirt off well. It is best to wash the head last in our opinion, for the simple reason that most fowls stand quietly till the head is done. **** It may be well worth remarking that if a fowl has to be left for a minute to get anything, and there be no assistant, the wet sponge laid across its back between the wings will generally keep it quite quiet, believing that it is held. Sometimes a heavy patient will appear faint in hot water, or even go dark in comb, as if about to die; in that case, a good douche of cold water should at once be given, which will bring it around, and it is curious that it never, or hardly ever, faints a second time."

The remainder of the directions in the later edition of Wright's book does not materially differ from the earlier statement or add any important detail to it.

I find more useful additional details in an article by an English exhibitor which appeared in the "Fanciers Journal" in 1890, from which I quote:

"Birds should be washed two or three days before wanted for the show, and our favorite time for 'tubbing' is when the sun sets; then the birds are more docile and likely to be quiet. We get everything ready in advance.

UPSTAIRS DRYING ROOM IN JOHN S. MARTIN'S WASHING AND CONDITIONING HOUSE

*The object in washing the feet first and separately is to avoid getting the dirt that is on them into the water used for plumage. Where the feet are quite clean it is not necessary.

RACK USED BY MR. MARTIN TO PLACE COOPS ON TO DRY BIRDS OUTDOORS IN FINE WEATHER

SPECIAL FITTING OF BIRDS FOR EXHIBITION

Let us name the articles, which must naturally be procured in such quantities as the number of birds to be washed necessitates. Two or three good-sized washing tubs, or zinc baths. If the latter are used, care must be taken that they are free from any grease. Cans of hot water, and ditto cold. Water impregnated with iron must on no account be used—rain water, spring water, or clear pond water, are all preferable. Some white curd soap. Some best quality soft soap. A wooden nail brush of fair size. A sponge. Some towelling. A wineglassful of some cordial in a phial. And then last, not least, a good open fire which can be kept for the birds exclusively for the whole time of drying. Two trestles, and a well-washed scaffolding plank. This list of articles may frighten some, but in establishments where white poultry are frequently exhibited, when once obtained and kept ready on hand, they will always be useful, and where exhibiting is rarely indulged in, then the nearest approach to the requisites named can be utilized.

"We start, then, with all our requirements around us, and our birds in baskets, and we set to work. The same waters will wash three fairly clean birds. We will put in one tub of warm water, as hot as the hands can comfortably bare it, a good lump of soft soap, and rubbing it through our fingers we will make it a tub of suds. We must melt up by constant friction and rubbing every atom of this soap, then if the water is too much chilled by doing so, we must again pour in hot water to obtain a comfortable temperature, then we place in it our cleanest bird, and we soak it from head to foot. With our hands we rub the suds under the wings, in the thigh feathering, round the back—rubbing always toward the tail—and in the neck hackles; we do it thoroughly; we make the bird look like a 'drowned rat'; we take the sponge, and with the curd soap make a lather and go over all parts carefully but thoroughly. With a nail brush we wash the legs and feet, and then, with sponge in hand, we rub over and over again the whole body from head to tail; we fear nothing with a big fowl so long as the head is above the suds; and then when we think all is clean, we extract a feather from such a part of the bird as was the dirtiest, and rinsing it in cold water we see if it is white and clean. If so we stop; if not, we repeat the washing until the plumage is pure.

"We next pop the bird into the second tub, which has been filled with lukewarm water, and wash out every particle of soap; we rinse with our hands and with a sponge every portion of the body; and then when we think all the soap has been fully removed, we pull out a feather and place it in the mouth. Should it give a soupçon of soapy flavor, we rinse again, but if it tastes quite clean the bird is placed on an unlined empty poultry basket to drip, and the second cleanest bird goes through the same process, as by the time that is completed the first one is ready to dry.

"The drying operation requires great care but is equally simple. We take the bird that has been washed firmly in the hands by the thighs, and with a sudden swing, hold it up in the air, and it will stretch out its wings, and flourishing them about will well shake itself several times. Then we place it on the basket—still upside down—to obtain a firm basis, and gently rub with the towels from head to tail, then one more good shake in the air and our bird goes to the fire. The drying room must have had the clean scaffold placed upon the trestles—not too near the fire, but well in front of it—and on it the birds must be placed, tails first fronting to the heat. We generally, if we have several birds, place first a cock, and then two hens, and then another cock, and so on. They will sit like lambs, as if pinioned to the plank, and should dry gradually.

"If many birds have to be washed it will require an attendant in the drying room as well as at the washtub, to continually turn the birds for the first two or three hours, so that no part may dry too quickly, and to see that the faces do not 'catch' from the heat. Should specimens dry too hastily, their feathers will not lie smooth. Many a bird—Light Brahmas especially—we have noticed at shows to have been spoiled by having the saddle feathers curled, sometimes almost similar to an ostrich tip. This comes from drying too quickly. The whole body must be dried gradually by turning it around just as our cooks roast joints or game, by continued revolutions before an open fire.

"When all the birds have been washed and placed upon the board, those at the ends will want moving in turn to the middle, and those in the middle to the ends, that none may dry too quickly. Should a bird's ear lobes 'catch' from the heat of the fire they will become a dark purple black, and should be immediately gently rubbed with vaseline, but as blisters are usually sure to follow, where a face has been badly 'caught' the specimen will be greatly damaged for exhibition purposes for a long period, and we cannot too strongly impress on fanciers the desirability of keeping the faces and combs of recently washed birds well from the fire and constantly turned.

"We do not approve of drying in baskets. To begin with, cocks—especially Leghorns, Minorcas, White Dorkings and those with sickle feathers—are wont to crouch in corners and their tails, drying against the sides, become out of shape and are rarely carried afterwards in the show pen as they should be. But by washing the birds at sunset and leaving them all night on the board in the warm dark drying room they will in the morning be ready to be placed in exhibition pens which have been thickly littered with coarsely cut straw chaff, or in roomy baskets, and there they will continue to dry and gradually fluff out and plume themselves.

"Smaller birds—like Silkies and Sultans—require more care. The former are apt to drop their leg feathering if put in too hot water, and the latter sometimes faint. We must in conclusion mention that birds when once washed are more liable to become soiled than are those that have never been tubbed. Their plumage becomes somehow more liable to catch the dirt."

Before leaving the heavily feathered fowls, I will quote from an American exhibitor, who contributed his methods to the Reliable Poultry Journal of December, 1904, Mr. E. Wyatt. Some of the details of his method of washing are different, and he introduces some other features in drying birds. Says he:

"The fowls that I exhibit are Cochins, the hardest variety to wash, because of the length and fluffiness of their feathers. I always wash in a room in which there is a good hot fire. Everything is in readiness before the washing commences. The birds are in their shipping coops in the room, and are washed in turn. Three tubs are placed near the fire; the first two are half filled with warm, soft water heated to blood heat. The third contains cold water with a little bluing. A fowl is first placed in the tub number one. It is held firmly, with its head toward the operator, so that it cannot flap its wings. It is advisable before commencing to wash to hold the fowl quietly in the water for a short time in order that it may understand the situation. Then with a sponge commence pouring water over the fowl's back and rubbing the feathers in the direction of their natural position. When the feathers are slightly damp, begin applying the soap. White castile soap, or shaving soap, is satisfactory. Rub the soap over the sponge, and continue working the suds into the plumage until it is thoroughly soaped, and the fowl is wet to the skin. When you believe the fowl is perfectly clean, remove it to tub number two. In this the fowl is rinsed by first rubbing the sponge over the feathers, and then with a dipper, by pouring water over the fowl's back and parts. If the dipper is held about a foot above the fowl, the force of the water will soon remove the suds. When the soap is completely removed the fowl is placed in the third tub, which contains the cold water and bluing. A minute in this tub is sufficient, the main object being to induce a quick circulation of the blood and to prevent the fowl from catching cold.

"The fowl is afterwards placed beside the fire, and as much water as possible is pressed from the feathers with the hands. Later, towels are used to dry it as completely as possible. The fowl is not rubbed with the towel, but it is wrapped about, so as not to break any of the feathers. It is then placed near the stove, but not close enough to scorch the feathers, or so that the heat will curl them up. The fowl should then be given to another person who keeps fanning it and turning the bird around so that it will dry rapidly. A brisk fanning is important for two reasons. It opens the feathers and helps them assume their natural appearance. They also dry more rapidly. In drying the fluff I find it advisable to fan against the feathers instead of with them, as you work right to the skin in this manner and are not simply drying the feathers on the surface. The wings are held up one at a time, and fanned to dry the feathers underneath, for these feathers take the longest to dry. By working in this way it is not long until the fowl is completely dry and looks well. While the latter person is fanning, the first one is going on with the washing. Sometimes it is convenient to fan a number of birds at once. While each is being operated on the remainder stand drying slowly before the fire.

"For white birds it is a good plan to dampen a sponge thoroughly with peroxide of hydrogen and sponge the feathers well before you commence drying. Care must be taken that it does not touch the legs or beak. After using the peroxide a little ammonia should also be rubbed over the feathers to remove it. Fowls other than white do not need as thorough a washing, but a surface washing without soap will produce good results, as it removes any dirt and causes the specimen to appear more attractive."

Mr. U. R. Fishel's method of washing, drying and fluffing White Plymouth Rocks as given a few years ago in the "American Poultry World," is as follows:

"If you have a small number to exhibit, prepare to wash your birds four days before the show. First, clean up your coops nicely, putting in fresh straw, and see to it that there is no dust on side of coop, on wires, or in the room. If possible have a warm place to wash your birds. Heat a boiler of soft water, and secure four washing tubs. In the first tub place four inches of lukewarm water, just enough to cut the dirt nicely. In tubs two and three place five or six inches of lukewarm water. Have tub four nearly full of water with the chill off. This tub is to be used for the bluing water. Make bluing water a little stronger in blue than if bluing white clothes. Take a cake of Ivory soap, a soft sponge, several Turkish towels, a couple of palm-leaf fans, and you are ready.

"Place the bird in tub number one, thoroughly wetting the feathers in every section of the plumage. Keep the left hand on the bird so that it cannot fly out of the tub. Always rub with the plumage, never against it. After you have the bird wet, use soap, beginning at the head and hackle, washing clean; then the back, tail, fluff, breast and body in rotation as named. After you have washed the bird clean get all the water you can out of the plumage, and then place the bird in tub number two. Thoroughly rinse the bird, taking a sponge and getting clean water through every part of the plumage, using one hand to loosen up the feathers. Take plenty of time for this, and when you have all the soap washed out, place the bird in tub number three and do the work all over again. In this way you are sure to get all the soap out of the plumage.

"Washing birds is not such a difficult job once you get the knack. After getting all water possible out of the plumage, dip the bird in bluing water, let it drain, and then get all water possible out of the plumage again. Now place the bird on a barrel covered with a clean cloth so that there is no danger of the bird getting dirty, take the sponge and get all water possible out of the plumage; with the towel dry the plumage as much as possible, then take the fans and fan the bird, all the time picking out the plumage—that is, separating the feathers. This will make the bird fluffy and fine when dried. Place the bird in a warm room or near a warm stove—not too close to a hot stove, for the heat will curl the damp feathers and ruin your work."

Mr. Arthur G. Duston, who was the earliest of the prominent White Wyandotte breeders to adopt the practice of washing his show birds, and who has ever since been recognized as one of the most skillful fitters of white show birds, in the same paper gave his advice as to the essentials in washing birds in the following brief statement, which—it will be noted—brings out a very important point in relation to the effect of heat in drying:

"Fitting a bird for the show is an art and only after years of experience can one gain all the points and phases of the work. Everything must be all ready before starting: three tubs of water—one hot, one lukewarm and one cold with a small quantity of bluing in it (about what would ordinarily be used in rinsing clothes). Ivory or any other good white soap. Have plenty of hot water if you are to wash many, to renew your tubs with, and a good hot fire to dry the birds out quickly. After thoroughly wetting the bird in hot water, scrub as you would a rag, rubbing into the feathers a quantity of soap, and when thoroughly clean rinse in warm water, taking care to get rid of all the soap, then rinse again in bluing water, hold it up so it will drain off some, and your bird is ready for the coop. Of course it is dripping wet, but by watching a bit, and adding fresh sawdust from time to time, it is ready for the heat and drying. Place it in an exhibition coop beside the fire, and have two inches of dry pine sawdust in the bottom, and see that it is thoroughly dry before removing from the heat. Don't attempt to dry too quickly; if you do you will see bunches of those crinkled, curling feathers so often seen in the showroom. Lots of heat, but not too close to it, should be the rule. Feel under the wings for dampness to ascertain when the bird is dry. There is one thing I always look out for: do not take the bird out into the cold from the warm room which has been allowed to cool off, but keep in coops, reducing the fire until the room is the same as the outside temperature. With this precaution there will never be any trouble from birds catching cold after washing."

SPECIAL FITTING OF BIRDS FOR EXHIBITION

Mr. D. W. Young of White Leghorn fame used five tubs to wash the birds that for so many years outclassed all others at Madison Square Garden. He briefly gave his method in the following statement in the same paper:

"After confining the bird in an exhibition coop for three or four days in order to get it accustomed to confinement, I take a basin of hot water and soap and scrub with a nail brush his head, comb, legs and feet thoroughly. Next, take five tubs of soft water, the first being heated to a temperature of about 110 degrees. Submerge the bird, head and all, in this tub. After the feathers are soaked through to the skin, I take a cake of Ivory soap and rub it well into the feathers until a lather is formed. Am not afraid of using too much soap. After I am sure the bird is clean, I rinse off as much of the suds as possible in this water. Next the bird is put in the tub number two—in which the water is heated to about 90 degrees—

how the Greystone White Plymouth Rocks were fitted before the stock was acquired by Owen Farms where it continued to be in his charge. It will be noted that at that period Mr. Davey followed the practice of drying partially with towels, a feature of the process he now regards with an indulgent smile. It should be noted also that in presenting the subject I am coming now to the practice of more than one washing. Mr. Davey's statement reads:

"The first wash should be about two weeks before show time. If possible, have a warm room in which to wash them. Clean your coops thoroughly, and litter well with coarse shavings or cut straw. I prefer the shavings as they are less liable to stain the plumage while it is wet. Place coops quite close to the stove where you can get a heat of about 90 degrees, even a little warmer will not hurt them until they commence to dry.

WASH ROOM AT MORRIS FARM, LEBANON, OHIO
Showing the heater, hot water tank and "Rawnsley Dryer" used in the demonstration reported on page 131. A separate illustration of the cabinet, and specifications for it are given on page 132

and rinsed well in this. Then put in tub number three —same temperature—and rinse carefully in this water. Next it is placed in tub number four, same temperature, then in tub number five which is cold and blued a little more than is ordinarily used in laundry work. After taking the bird out of the last tub, he is placed in a room heated to about ninety degrees, in a training coop about three feet square, with clean-cut straw or shavings for litter, being very careful to keep it clean so that the feathers will not become soiled. Of course it pays to look after the birds while they are drying as the feathers are liable to become twisted, especially the sickle feathers of the male birds. In about twenty-four hours after this is done the plumage will be in perfect shape and the bird, if naturally white, will be perfectly clean and as white as snow."

In the same symposium, Mr. Frank H. Davey told

"Be sure to have a good supply of soft water. Fill two washtubs about half full of as warm water as is comfortable to the hand. Fill the third tub about two-thirds full of water with the chill taken off. This tub is for the bluing. If you intend washing twice, enough bluing can be used to show quite plainly when the bird is dry. This will mostly disappear, or will wash out at the second washing and will help to remove any stains or creaminess.

"For the washing you will require: Ivory soap, a good-sized soft sponge, and two or three towels. Place the bird in the tub of warm water, keep the left hand on the back to prevent flying out of the tub. Use plenty of soap, go through every section thoroughly, always rubbing with the feathers and not against them. When you have the bird thoroughly lathered, begin at the head and clean each section thoroughly; do not leave a section

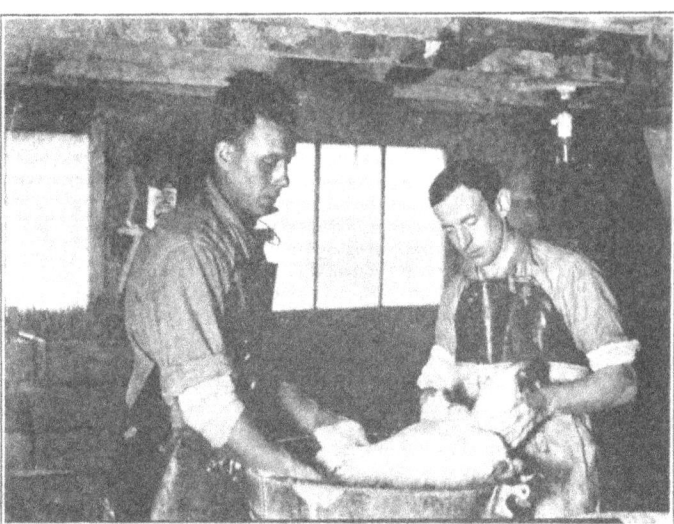

WASHING BIRDS AT MORRIS FARM
Thoroughly soaking the feathers. Left—Mr. Reynolds; right—Mr. Rawnsley

of the bird until sure every particle of dirt has been removed. If you skip from one section to another, you are sure to miss some spots which will spoil the job.

"Do not make the mistake of using too little soap; the feathers, being oily, will not take the water until they are thoroughly soaped. Legs and feet should be thoroughly scrubbed with a soft brush. When you have thoroughly removed the dirt, place the bird in the other tub of warm water and rinse thoroughly. Here you will need your sponge—go through every section just as thoroughly as in washing and remove every particle of soap. This will require nearly as much time as it did to wash the bird. If the soap is not thoroughly removed, the plumage will not take the bluing water except in the spots where it is removed and the result would be the blotchy plumage which we so often see in the showroom. It also causes the feathers to mat together, feel sticky to the touch, and catch every particle of dust.

"When the bird is thoroughly rinsed, dip two or three times in the bluing water and hold long enough to make sure the water gets thoroughly through the plumage. Place the bird on a draining board or box and get all the water out of the feathers you can with the sponge, then wipe with the towel and place in the drying coop. See that birds do not crowd in corners of the coops and bend the tail feathers while drying as it is very hard to get them back to their natural shape again. After they are thoroughly dried they should be placed in good roomy pens well littered with straw. Do not allow them any place to dust themselves; handle them all you can; place them occasionally in a training coop; pose them with your hand, also with a stick, and keep at it until they hold the pose for several seconds.

"The second wash should be the same as the first except less bluing should be used. Do not use enough so that it is visible after the bird is dry."

Mr. M. L. Chapman, now of Wilburtha Poultry Farm, long known as one of the best fitters of white birds and a successful exhibitor of both White Plymouth Rocks and White Leghorns, some years ago gave in the Reliable Poultry Journal this method of fitting birds with three washings:

"About three weeks before the show, if we are to exhibit say thirty birds, we select about forty-five and give them a good thorough wash, using Ivory soap and water about as warm as we can comfortably work in it. Use a brush on head and legs, and on any stained spots in the plumage. Rinse very thoroughly in a separate tub of a little cooler water, then dip in a third tub of tepid water that is slightly blued. Do not have water too blue, as a bird that gets streaked with blue is ruined for the time being. Do not wipe the bird with a cloth or try to squeeze out the water with your hands. Simply drain for a moment or two over the tub, and then put it in a cage with a burlap bag in the bottom to catch the water. Have the room warm, and if possible keep the unwashed birds in a cooler room. The birds should be practically dry in three hours. We put the wet birds near the stove and gradually work them back as they dry. Reduce the heat in the room gradually as the birds dry off. You can wash about five birds in a tub of water, then dump out all three tubs and replace. Of course soft water is much better than hard. It will take two or three men about all day to wash a string of this size, and it is not an easy job either.

"After your forty-five birds are all dried out, you can go over the string and pick out the birds that did not improve with a wash, so to speak. Some will prove to

WASHING THE SIDES WITH THE BIRD ON BOARD ACROSS THE TUB

be considerably whiter than others; some will probably have stained spots on the plumage that no amount of washing would remove. We select about thirty-five and take special care of them for another week and then give them another wash. This second wash improves them fully as much as did the first one. After they are dry we make our final selection of the thirty that we shall show, and in about four days give them another thorough washing. After this third wash the birds certainly look beautiful and are a sight to please the eye."

DEMONSTRATION IN WASHING WHITE ORPINGTONS BY HAROLD RAWNSLEY AND CLARE REYNOLDS AT MORRIS FARM, FEBRUARY, 1921

Note—The report of this demonstration is made from notes taken by Grant M. Curtis, Editor Reliable Poultry Journal, during an interview in which Mr. Rawnsley discussed matters relating to washing and bleaching processes, and at the demonstration made later under his observation. The author has coordinated this material, sometimes quoting Mr. Rawnsley, at other points giving only the gist of his statement as a part of the description. The washroom and the arrangements for hot and cold water as required, and for drying the birds, are shown in accompanying illustrations and described in the legends relating to them.

For the washing three ordinary metal washtubs, 24 inches in diameter at the top, 20 inches in diameter at the bottom and 10 inches deep, are used. Two of these are placed on a strongly built wooden bench, 5 feet long, 28 inches wide, and 28 inches high. The third tub is placed on a backless wooden chair at the end of the bench farthest from the suds tub.

The men prepared for work by rolling up their sleeves and putting on rubber boots and rubber aprons.

"The temperature of the room should be 65 to 70 degrees," said Mr. Rawnsley, continuing—

"We prefer cistern water in all three tubs, but if our supply of cistern water is low, we can use city water for the rinsing, but much prefer cistern water for soaking and soaping. If we have enough soft water for two tubs, but not the third, then we use city water in tub No. 2—the one used for rinsing off soap after birds have been under faucet for the same purpose."

They filled all three tubs at the start, one with hot water—as hot as they could bear their hands in; the next with water about ten degrees colder and the third with water another ten degrees colder. A good-sized hand sponge was tossed into each of the rinsing tubs, ready for use later.

In tub No. 1 nine gallons of water were poured. Into this a half pound of Lux flake soap was stirred thoroughly, then mixed and worked with the hand until the foam on the water was half a foot thick or more. "Make lather to 'beat the band,' is our rule at this

PUTTING BIRD UNDER COLD-WATER HYDRANT AFTER COMING OUT OF FIRST TUB

point," said Mr. Rawnsley. The result was six or seven inches of the water—and the rest suds. Ivory soap will answer, but birds can be washed more quickly with Lux.

"Now," said Mr. Rawnsley, suiting the action to the word. "Stand the bird in the water and thoroughly soak him, working the soapy water into all the feathers from head to shanks, taking pains not to muss the feathers unduly. Then wash the surface with an ordinary soft, but moderately stiff bristle hand brush, the kind 2x5 inches in size used for washing dishes, cleaning kitchen sinks, etc. Wash carefully the entire surface, always stroking With the Feathers, at the same time holding or controlling the bird, which you see is easy to do. The warm soapy water seems to soothe him."

Both men at the start worked on the same bird, one working the suds all over and through the feathers with his hands, while the other used the brush on the hard wing and tail feathers, brushing always with the feathers.

Next the bird was placed on one side on a board 12x30 inches in size, containing 20 to 30 one-inch holes for drainage. The head of the fowl at all times was kept out of water but no special pains were taken to keep soap out of the eyes, and the birds did not seem to mind it. One of the two men would hold the bird on the board, using the legs or a wing for the purpose, while the other would scrub with a hand brush, the other using his free hand for the same purpose. All washing and scrubbing was done from the head downward, and from the quill of the feather outward. A hand would be spread under the tail feathers and these feathers soaped and scrubbed one at a time while held in this position. Sickles also were held up against the hand and washed separately. Tail feathers were washed with head of bird held away from the man who did it. Birds stood patiently in the warm water, soaked to the skin, and did not try to get out, even when the men stepped away for a half minute or more for any purpose. Breast and body feathers were lifted up and worked into in such manner that the soap got thoroughly through them. The fingers were used to comb these feathers and rub out carefully all dirt or stain. Said Mr. Rawnsley:

"We must get the suds well up into all the feathers in every section, including breast, hackle, etc. This is what we call the soaping process and it must be done very thoroughly;

DRY SPONGING BEFORE PUTTING IN DRYING CABINET

You can see that the birds like it, because they do not make any fuss. How long does it take to wash a bird properly? Three to six minutes in the case of sales birds. In the case of birds for exhibition, we of course take more time and still more pains—simply go the limit to have them in the pink of condition."

Much of the time (in this demonstration) the brushes were used, rather than the hand and fingers—the effort being to reach and scrub every feather, especially all large feathers. While a bird is on the board one man holds the legs, as a rule—still washing the sides and under wings, etc., with the other hand, first one side and then the other. At this point a cake of Ivory soap comes into play. The brush is rubbed on the soap and then the bird is scrubbed quickly all over with this brush. They also carefully scrub the feet and wash

BIRDS IN THE RAWNSLEY DRYER—CURTAINS UP

the thighs, getting into the feathers everywhere. Wherever they go the feathers are lifted out, spread on one hand and then scrubbed carefully—thoroughly.

After the soaking, soaping and scrubbing (on board, etc.) have been painstakingly done, the suds are wiped off and stripped off by hand and are thrown back into the suds tub. Wings and tail are squeezed by hand to get out the soap, then the bird is put under the two faucets, first under the partly warmed water, next under the cold water faucet. Water is turned on full for about a minute and rapidly carries away the soapy water, while the man holding the bird strokes and squeezes the plumage to get out all the soap he can. Wings are held open, tail is quickly spread, feathers elsewhere are lifted to get out quickly all the soapy water possible. Said Mr. Rawnsley:

"This cold water treatment prevents the birds taking cold. By our method they never catch cold. Cannot recall when we ever had trouble of that kind. Yes, one person can wash birds satisfactorily, but two get on much faster—more than twice as fast.

"Head, legs and toes are thoroughly washed in the first water, using soap and brush. No soapy water is used by us after that. In washing birds for exhibition we take more pains with the feet and legs. Before washing the bird we stand it for quite a while in warm water in a tin pail placed in a narrow shipping coop that is covered over and made dark. This thoroughly soaks and loosens the dirt and grime on the legs and toes, and the scrubbing we give them in the first water cleans them thoroughly except for the dirt under and around the scales. This we remove later with a nail file, just as a manicurist does."

After the cold water shower the birds are placed in tub No. 2 for a thorough rinsing. The bird stands in the water while the man holding it goes rapidly over it in every section to finish the rinsing process. Meantime the other man has started another bird in tub No. 1.

From tub No. 2, the bird goes to tub No. 3, in which a small amount of bluing has been mixed thoroughly. Again clean water is worked quickly over and through the bird's plumage, then is carefully squeezed out of the feathers. Next he is lifted out to stand on the drain board. Here water is again pressed out of the plumage—all possible—and he is dry sponged by the use of a hand sponge which is frequently squeezed back into the tub—tub No. 3. All water that could be removed from the plumage in a minute or so was taken out, then the bird was placed on a perch in the drying cabinet and curtains buttoned down, leaving him in the dark. Said Mr. Rawnsley:

"In tub No. 3 use about as much bluing as an experienced washer woman would for white clothes. In dry sponging him, I get off all the water I can, to avoid dripping in the dryer and so the bird will dry off that much sooner. In dry sponging, take pains to straighten out the feathers nicely and in natural position."

Just before birds were placed in dryer, each was held by the feet, head downward at arm's length and allowed to flop its wings several times. Said Mr. Rawnsley:

"Hold them out this way at arm's length and let them flop their wings to throw off the water and help circulate the blood. It works like a charm.

"The aim is to dry the feathers as much as we can before placing birds in the drying cabinet. We do not use a cloth of any kind to get off the soap or water, because of danger of roughing the feathers. Be extra careful to have all feathers in their natural position when the bird is placed in the cabinet. For example, open the wings and dry sponge each section and part carefully, then feathers will not get out of shape nor will the web mat or roll up. The moisture pans in the drying cabinet take care of that, and it is remarkable how nicely the feathers, small and large, fluff out during the drying process in this cabinet.

During the soaping and rinsing process, the feathers on these birds look like long narrow ribbons. In tub No. 3, the water with bluing in it was worked well into the feathers thus to "whiten them thoroughly," as Mr. Rawnsley expressed it. No peroxide is used on sales birds shipped for breeding purposes. In cases where peroxide is employed they use "just a little" after taking the bird out of the bluing water. It is then wiped over the surface and in other sections, using absorbent cotton for the purpose. Said Mr. Rawnsley:

"Yes, I believe this takes out some of the yellowish tinge found in white plumage at certain seasons of the year and due to the oil in the feathers, or to their green condition, so to speak, but I know positively that peroxide will not remove brassiness, and I'll say further that I do not know of any process, chemical or otherwise, that will remove brassiness from the plumage to an extent that is self-evident.

"From what I've heard, one would judge that some people must stand their yellow birds in peroxide up to their eyes, in an effort to take off the brassiness, but it will not work, except to seriously damage the feathers. The one right and profitable way to get rid of brassiness in white-plumaged, Standard fowl is to breed it out of them by careful, persistent selection and elimination, both as to the males and females of a strain."

Continuing Mr. Rawnsley said: "There is one specially important thing about this washing process: TO DO IT QUICKLY! The longer a bird is wet to the skin the worse it may be for him. Clare and I can average one sales bird to every three minutes, but we should use about five minutes to each bird. We have offered to bet any two men in the United States that we can beat them at washing white fowl. For show birds we take fifteen minutes each.

"All birds while being washed are very quiet. The main thing is to soak the feathers well, as there is no danger of breaking them if they are well soaked.

THE RAWNSLEY DRYER WITH ONE CURTAIN UP AND ONE DOWN

"Have everything handy and ready before you begin washing, then while one man rinses, the other will soap and soak another bird and have him ready for the board—to be placed on his side and scrubbed that way by both men. You just saw us wash a big, vigorous cock bird and could notice how docile he was. Resting here on the drain board on one side, a man can hold him by the wing without any trouble and wash him freely with the other hand.

"In the dryer always aim to place a female between two males to prevent scrapping. The better plan is to leave the birds in the dryer all night, though we can wash sales birds in the morning between eight and nine o'clock and remove them at 5:30 that afternoon. The birds in the dryer will not fight, so long as the curtain is kept down; also as long as they remain wet they do not feel like scrapping.

"For this size cabinet twelve to fifteen males or eighteen to twenty females of the White Orpington variety will dry satisfactorily, but we must not put too many birds in the dryer or they will 'sweat.' There needs to be a quite free circulation of air among and between the birds. However, they must not dry quickly, or this will result in curled feathers; also this has a tendency to make the feathers brittle, if the same birds are washed several times the same season and dry too quickly.

"The use of this drying cabinet is a wonderful time saver. For example, if we had sixteen to eighteen birds to dry in as many single coops it would be some job. Furthermore, where birds are allowed to dry on straw, as in separate coops, they will muss and soil the breast feathers. Here in the drying cabinet they are on perches and keep clean all the time. We use fourteen quarts of warm water in each of the three large shallow pans. Sand could be used but it is no better than water, and as it gets soiled its use makes unnecessary extra work.

"We keep the temperature in the cabinet at about ninety degrees while the birds are enclosed. Care should be taken not to have the interior too hot when cabinet is filled with birds. Top frame of cabinet, which is covered with heavy felt or canvas is removable and therefore can be used for ventilation by propping it up an inch or two. The curtain buttons down both sides and across the bottom, so that inmates of cabinet can not get out. Interior of cabinet is thus kept dark so that the birds are contented and do not move about—do not crowd or fight.

"It takes the best part of the day and night to dry the birds properly. This depends partly on heat of the room in which cabinet is located. In warmer weather the birds can be dried sufficiently in one day, putting them in the cabinet before ten a. m. and removing them at six or seven o'clock that afternoon, but I prefer to dry them slowly, taking all night to it.

"Early in the morning we take them from the dryer and you will be surprised how dry they are, and how clean and bright their plumage is. Again we hold the birds heads downward and let them flap their wings, especially the male birds. They dry out last under the wings and this flapping process helps to arrange all the feathers in natural position. Next we take them upstairs to the conditioning room and place them in coops on clean straw, ready for shipment later on. After birds are put in coops upstairs they preen their feathers, straightening them out and putting them in natural shape.

"After the birds are dried, one can improve the surface somewhat, as I believe, by using a silk handkerchief, but in the case of sales birds, simply put the males down on the floor where they can flap their wings and crow a time or two, and they will be ready to ship. Another plan, when in a hurry—meaning with sales birds—is to use a cyclone hair dryer, the kind that is used commonly in barber shops. Use this cyclone dryer after birds have been in the cabinet and are nearly dry, just to hurry them along. With this cyclone dryer raise the wings and get underneath. That section is the last to dry out well in the cabinet. This cyclone hair dryer will blow up the body feathers and help dry them quickly. In five to seven minutes a bird from the dryer will be completely dry, with the feathers fluffed up and back in natural condition. If desired, finish off the job by using a piece of soft silk, like a silk handkerchief, to put on a satiny finish.

"Yes, this machine could be run by coal oil if desired. Just heat the water to the temperature necessary to give ninety degrees in the cabinet and hold it."

CONSTRUCTION OF RAWNSLEY DRYER

This dryer, or cabinet, when erected ready for use, stands 5 feet 6 inches high, is 6 feet wide and 3 inches deep.

Woodwork of cabinet is built in panels and frames, is held together with about a dozen good-sized screws and can be set up or taken down in short time.

Legs, where they extend below body of cabinet are 22 inches in length. Front legs are each made of two pieces of 1x3½ inch sound pine, one piece in each case forming part of front panel of cabinet and the other piece part of the end frame. When cabinet is erected the two pieces, where they join and lap, are screwed tightly together, making a substantial support. These front legs extend two inches above the body of the cabinet for ornamental effect. Back legs are built the same way except that they do not project above top of cabinet.

Front frame, including the two strips that help form the front legs, is made of 1x3½ inch sound pine, with a 1x3 inch strip in the middle, which divides the large open front into two squares, each 31½x31½ inches in size. Covering each of these square openings a curtain of extra-heavy canvas or ducking is used that is attached snugly to the top piece of the frame or panel and is supplied along the edges with metal-bound eyeholes for buttoning the curtain to the other three sides of the frame, including the bottom strip. Fasteners on the frame match the eye-holes for this purpose.

Attached to the bottom rail of the front frame by means of two short-butt hinges is a 2½ inch x 5½ foot one-inch strip that is used as a door for removing three shallow water pans that are located in bottom of cabinet. This door opens upward and stays in place when opened.

Back and both ends of cabinet are made of ½ inch x 3 inch tongued-and-grooved pine, built in panel form so each is handled in one piece to set up or take down.

Cabinet has removable frame top or cover, made of 1x3 inch strips with middle crosspiece covered with canvas cloth of good thickness, but not as heavy as the front curtain. Top frame rests on inside cleats, so it can be raised two or three inches in front for added circulation of air.

In both end walls of cabinet, about 2 inches from the top (or ceiling of cabinet) are ten 1½ inch round holes, located on a line and three inches apart for constant air circulation.

There is no floor to the cabinet. In place of a floor or bottom, a frame that duplicates the top or cover frame is fitted in place, on which rests the shallow tin pans, three in number, each 21x36 inches in size and 1½ inches deep.

Inside the cabinet, close above the pans and resting on 1x2 inch cleats at each end, is an ordinary coil of six 1-inch iron pipes through which hot water from a near-by boiler circulates for heating and drying purposes. Pipes are five inches apart; front pipe is six inches from face of cabinet and rear pipe six inches from back of cabinet. Flow pipe enters cabinet near front and return pipe goes back to heater from rear of cabinet at same end. Flow pipe enters at point 3 inches higher than point at which return pipe goes back to hot water boiler.

Immediately above these pipes and resting on them is another frame, practically 3x6 feet in size and removable, made of 1x3 inch material, to which is attached 1-inch poultry netting. Frame has three crosspieces and is of sufficient strength to support fowls if any are crowded off the roosts, thus preventing them getting on the hot pipes.

In cabinet, extending lengthwise are two 1x4 inch x 6 foot substantial removable perches located 12 inches above the pipes, the front perch being 13 inches from the front line of the cabinet and the other one 12 inches farther back, which brings it 13 inches from the rear wall. The 1x4 inch top perch is nailed T-fashion to the edge of a supporting 1x3 inch strip of equal length, which makes the perch rigid.

This drying cabinet can be built any size, as regards length or number of compartments. The one here described holds comfortably twelve to fifteen male birds or eighteen to twenty females of the White Orpington breed, preferably the smaller number in each case.

Bleaching White Birds

Some of the foregoing statements of methods of washing fowls have referred to the matter of bleaching. One in particular gave briefly the method of using peroxide of hydrogen. The use of "chemicals" in washing birds to increase their whiteness may be either after the washing or in the washing. Mr. E. B. Rogers, in the "American Poultry World," made this statement:

"I have experimented with everything I have ever heard of and have men in my employ who are familiar with bleaching wool, silk and cotton fabrics, who strip these goods of their original color for dyeing and who have assisted me in the experiment with no result as to bleaching. Feathers showing brass where put through numerous tests when dried out would show a shade or two lighter, in fact a great deal worse and more prominent than originally. I consider the washing of white birds an art, and one that few ever take the time and pains to learn. I have found the following method very satisfactory:

"After I have selected birds I want to condition for the showroom, I wash them thoroughly with a soap preparation I make by chipping three bars of Life Buoy soap (carbolic acid soap), two bars of Ivory soap, boiled in two gallons of soft water. Dissolve a tablespoonful of chloride of lime in a quart of boiling water, and add a tablespoonful of oxalic acid and one ounce of good ammonia. Stir the latter preparation through the soap preparation after it has been cooled and continue to stir so as to mix thoroughly, until the soap is ropey. Set in a cool place and when cool it will be a stiff jelly.

"In washing I immerse the bird in water that is as warm as the bird can stand and afterward work this preparation all through the feathers until every particle of dirt is loosened, then rinse the bird through at least three waters, and at last blue through a bluing water made of indigo bluing sufficiently strong to make the bird good and blue. After this process the birds are put back in their conditioning coops for a week or ten days, or until a day or two previous to the time they are to be sent to the show, when they are again washed in the same manner without using the bluing.

"After the birds are almost dry, I work all of the cornstarch that their feathers will hold through the fluff, tail and wing feathers. I do not think the starch has anything to do with whitening the bird, but it does aid in the drying and causes the fiber of the feathers to open up more readily."

Mr. Jos. Coleman contributed the following comprehensive statement in regard to both modes of bleaching to the same paper:

"Technically, using bluing is just as much of a sin, if sin it may be, as adopting a more complicated method of procedure—that is, whitening or bleaching. *** The

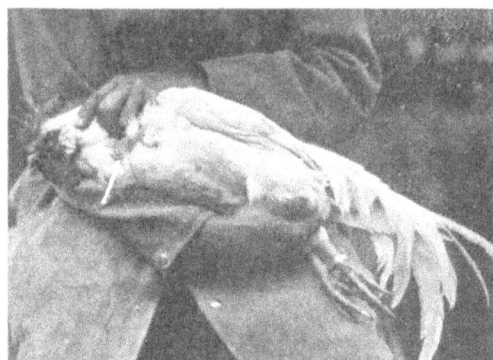

EFFECT OF EXCESS OF PEROXIDE
Arrow points to a spot where feathers are destroyed to the skin

peroxide process in the hands of some is of little effect. The ordinary method of simply sponging the bird with peroxide of hydrogen after washing, in our hands, has met with little success. A far better way is to use one part of strong ammonia and three of peroxide. After the fowl has passed through the last water, and the towels used have absorbed most of the water in the plumage (in other words, while the feathers are yet damp but not wet), quickly unite the peroxide and ammonia, rapidly sponge the plumage with this mixture, and as speedily as possible wrap the fowl in oiled silk, leaving only the head and shanks exposed. Hold the fowl in this manner for about three minutes. Practice will readily determine the length of time. Better experiment on a couple of culls in order to get your hand in. The idea of using oiled silk is to confine the gases generated by the union of the two chemicals—that is what causes the bleaching. If the washing is carried on in a hot room, and it should be, the Chinese method of plunging clothes into clear, cold water may be adopted with the fowls. This has the tendency of whitening and should be done the last thing before drying the birds. ***

"Chloride of lime. Horrors. Now chloride of lime is practically harmless and will not cause the plumage to come out if used in the proper way. Those who have met with disaster using this chemical were not next to the 'know how.' First take one tablespoonful of chloride of lime, two tablespoonfuls of oxalic acid crystals, and mix these ingredients thoroughly together, then put this mixture into a gallon of hot soft water. Stir well. Then shave up three large cakes of Ivory soap and add slowly, stirring all the while. You will now have a jelly to be used as any other soap in washing your fowls. Put in quart cans if you do not care to use all at one time. The chloride of lime is so assembled in this mixture that it can hardly be harmful in any manner whatever. If this method is followed use the bluing in the last rinse water in the usual manner. **** If a fowl is thoroughly blued by dipping in bluing water six to eight weeks before the time of exhibition, the plumage will wash out nice and white, and the fowl may be allowed to run at large during this time. If the bird is made as blue as an Andalusian the sun will have no effect on the plumage. Brassiness in fowls is only the pigment of the feather being scorched by the rays of the sun. Did you ever notice a hen that has wallowed in coal soot and is almost as black as night? Wash this same hen and she will be your whitest bird. Why? Simply because the sun had no effect on black. The plumage has not been burned."

The final clause in the foregoing statement I have quoted as made; but it is desirable to call attention to the fact obvious to one acquainted with all circumstances relating to it that Mr. Coleman's meaning was that black foreign matter on the plumage protected the pigment in it from the action of the sun. As a successful breeder and exhibitor of black fowls Mr. Coleman certainly knew that the sun affected pigment in their plumage. Many instances could be cited giving evidence of the efficacy of strong bluing and repeated washing after exposure of the blued birds to sun and weather for various periods. The essential difference between the use of bluing and the use of peroxide and chloride combinations is not in the result sought but in the different character of the undesired effects incident to the processes, and in the fact that bluing adds color to the plumage and so must always be used with the greatest caution. To put it another way, bluing spoils color before it destroys fabric, while the action of an excess of the other things is to destroy the fabric.

Bleaching with peroxide is always accompanied with some risk of damaging the feathers when enough peroxide is used to be really effective. Even those experienced in its use sometimes get too much on, or get too much in some places with the result that the web of the feather, and sometimes even the quills of small feathers, will be destroyed, though in the latter case, if not in both, it would appear that repeated vigorous rubbing of the feathers may have contributed to the breaking off of quills at the skin. The fact that it does not immediately damage the feathers to an extent preventing exhibiting the bird does not necessarily show that it has not produced injurious results. Even ordinary washing makes feathers afterwards more easily soiled and less weatherproof, and I have often seen birds that at a show displayed no bad effects of bleaching showing a great deal of damage to the web a few weeks or months later.

Detecting Bleaching—It is often stated that after washing the natural white bird cannot be distinguished from the bleached bird. The correct statement of the case is that the distinction cannot be made readily and with positive certainty for all cases. An expert might not be able to say on the inspection of a single specimen, or on comparison of a very small number of specimens, whether a particular string of birds has been bleached or simply washed; but a number of men who handle many large classes of white birds assert

WEB OF FEATHER DAMAGED BY PEROXIDE

SPECIAL FITTING OF BIRDS FOR EXHIBITION

PINCHED MAIN TAIL
Before fixing by bending the feathers

that wherever pinfeathers or partly grown feathers are found in a bird their condition is thoroughly dependable evidence of the nature of the treatment that has been applied to the plumage. Even in the whitest naturally white bird feathers in course of growth are creamy, so if it is found that immature feathers and even new pinfeathers are white it is accepted as evident that the birds have been bleached.

I have seen white birds on exhibition with pinfeathers and short new feathers absolutely devoid of oil and presenting a quite unnatural appearance. One of the breeders of White Wyandottes who discontinued showing because he was not willing to bleach at all called my attention to this some years ago, at a time when the Standard contained the provision that: "The natural white bird shall not be handicapped by the apparently bleached one, and other things being equal, the natural white bird shall win." It is a singular thing that though a very considerable number

METHOD OF BENDING TAIL FEATHERS
The bend is made close to the skin

of breeders of white varieties who were opposed to bleaching took the position that such evidence as this, and other effects on plumage, made it entirely possible to determine beyond reasonable doubt in a large proportion of cases the use of "chemicals" in bleaching; the next edition of the Standard dropped this provision and substituted for it this remarkable statement: "Bleaching by means of chemicals is such a harmful practice that where it is proven by other evidence than the condition of the specimen or specimens, such bleached specimen or specimens shall be considered faked and disqualified."

How this managed to get into the Standard I have not been able to learn definitely, but knowing something of the possibility of introducing things into it when under consideration in a committee or convention that would not be accepted if the members had them before their eyes in black and white, I surmise that someone interested in bleached birds managed to put it through. The damage done, however, is more apparent than real, for the present provision is a dead letter. No judge hesitates to disqualify a pen of birds that he finds affected by bleaches to a degree that shows beyond doubt in the mind of any one, and beyond dispute that the birds have been bleached. An exhibitor who would rely upon the letter of the Standard in such a case and, if the question came up afterward, would attempt to use it as a defense would be in a peculiar predicament.

Washing Colored Fowls

In washing colored fowls the general procedure is the same as in washing white birds but no bleaches are used, and the possible effects of soaps and also of too hot water upon colors have to be considered. The washing of colored fowls has not had attention enough yet to bring out much definite information upon any of the difficulties peculiar to it. Ordinary washing is said to have the effect in some cases of making light buffs slightly deeper in color. From the information given by Wright and cited in connection with the statement a little farther on in regard to dyeing buff and red colors, it would appear to be a question in such case whether the change of color observed was not due to the particular kind of soap used. An exhibitor of buff birds once told me that he had spoiled a lot of birds in preparation for showing by washing them as the color seemed to run, leaving the birds blotchy in appearance. In making inquiries that might throw some light upon such a case among other exhibitors at that show I happened to be talking with an exhibitor of pigeons, who in early life exhibited poultry also, but has for many years given his attention entirely to pigeons. He informed me that on one occasion when he washed some red pigeons in very warm water the color "ran" and completely spoiled the appearance of the birds.

While the practice of washing white birds was in this country limited almost entirely to Cochins, it was the general custom to wash, and strongly blue, Barred Plymouth Rocks. The advantage of the process was especially in the bluing, for in those days the color was lighter and the barring much wider than now, and the idea was to make them look as blue as possible. Sometimes it was overdone, but the taste of that time preferred an unnatural blue, especially in the smaller shows. Some reported washing with strong soap blurred the colors.

Exhibitors of some of the crested breeds wash the crest and sponge off the rest of the plumage. Very large crests are sometimes "done up" on the head to keep them out of the eyes of the birds, but the practice is attended

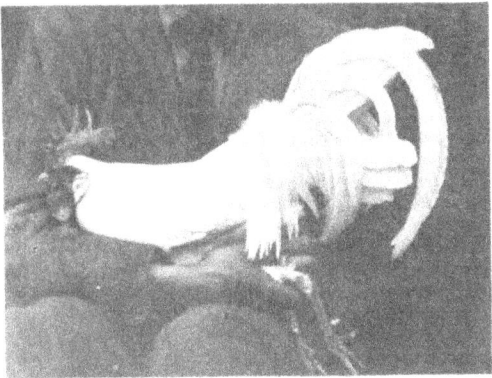

TAIL WITH HALF OF THE FEATHERS BENT
The other half in natural position

with some danger because if the feathers get wet and do not dry out readily they may rot.

Fluffing and Bending Feathers

As all processes of this nature are most readily performed on newly washed birds and are therefore most in vogue among exhibitors of white varieties this seems the most appropriate point to introduce them. A great

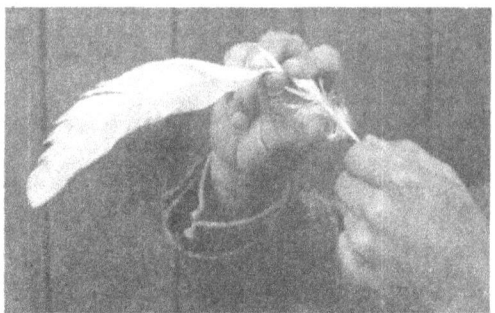

CURLING FEATHERS
Beside bending the tail feathers to change the profile outline, the expert curls them to give the desired lateral spread. Above is the position of the fingers to make the feather curve in.

difference in the appearance of the newly washed bird can be made by slight manipulation of the feathers during the final stages of the drying process. They naturally will tend to stand out a little more after the operation of working down to the skin, first to soap them thoroughly and then to remove the soap. They can be brought back to their normal position and made lie smooth and close by stroking them down as they finish drying; or by systematically increasing the fluffing the bird may be made very fluffy. Like the human hair, the smaller and more flexible quills of the feathers can have their directions somewhat changed while damp, and (drying in the new position) will hold it for some time—from a few days where the bending is produced with little force to about two weeks where the feather is bent to an extent possible only with the application of considerable force and with some damage to the quill.

Some exhibitors do ordinary fluffing with their hands, simply keeping the feathers slightly fluffed out by gently brushing the tips of the feathers the wrong way. Where it is desired to make the feathers stand out more, the work is done more systematically and with more attention to giving a decided bend to the feather, by pressing back a few feathers at a time and with a fan or small bellows assisting the drying as they are in that extreme position. The feathers are forced back farther than it is desired they should remain and in the partial recovery of their position they come about to the desired angle. This is about as far as the general fluffing process usually goes, though Wright relates an instance of a certain exhibitor in England, long ago, who was accustomed to exhibit some wonderfully fluffed birds in which investigation showed that the feathers had been bent twice, once close to the skin and again farther out, giving the fluffing a peculiar character. It was supposed that hours were spent on each bird—each feather being twice bent separately. Quite possibly that was the case, but taking the description of what was done as I understand it, I venture to say that there are a number of men in America today, so expert in the manipulation of the back and saddle feathers of males to give the bird apparent breadth and fullness in those parts, that they could accomplish remarkable results fluffing the feathers of a Cochin in handfuls in very much less than an hour.

The flexibility of the quills of feathers admits of the building up and filling out of backs and saddles that are narrow and pinched, and that show too sharp angles at the tail, and also of very extensive manipulation of the angle of the tail, and of its spread—both perpendicular and lateral. The accompanying series of illustrations shows the principal details of the method graphically, and the legend under each explains it especially. It is therefore necessary to say here in general only that in building up a saddle or cushion the feathers are bent so close to the body that the bend in the quills is not perceptible; and that in giving a tail the desired angle and spread they are bent at the root or curled slightly their entire length, or both, as the case may demand, and when the job is done with skill and care some very considerable faults of shape will be concealed and the trick is not readily detected.

It should be said further, however, that a successful job of this kind depends in the first place upon the operator having a very definite idea of the new shape he is going to give the bird, and of the amount of manipulation at each and every point that will be required to give him the result he is after. It is something like "lightning" sketching, or rather modeling. A bungling job of this kind makes a specimen look much worse than before; hence the exhibitor who wishes to fix in this way any fowls that can thereby be improved for exhibition would do well to make a careful study of the art, and to practice diligently on birds not for exhibition before trying his hand on any he intends to show.

The most conspicuous results of building up saddles and bending tails are seen in breeds of the Mediterranean class, but the practice extends to all. The development of rare skill in it is of comparatively modern date, though the practice as applied to tails goes back to the early days of the fancy. When not too crudely done this form of faking is not regarded by exhibitors generally as peculiarly heinous for it savors very much of the practice of teaching birds that are naturally of indifferent type to pose to show good type. I have to confess for myself that the good jobs in this line that are so common nowadays seem to me greatly to be preferred to what used

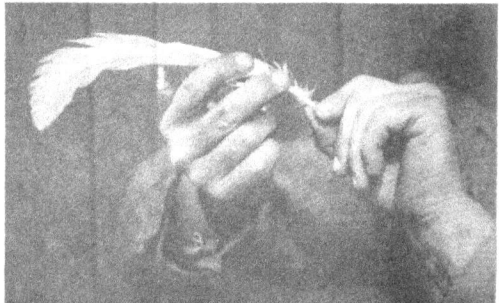

POSITION OF THE FINGERS CURLING A FEATHER TO STAND OUT MORE

to be done when Leghorn tails roughly bent dry until the quills cracked could be found on winners in leading shows. Feather bending as practiced in various forms is, as a rule, "for exhibition only." As has been already stated the feathers will come back to their natural position, if not actually broken, in about two weeks.

Splicing Feathers

The temptation to splice feathers arises from the ease with which it can be done in a manner that will pass superficial inspection, and also from the possibility in some cases of makng a half-plausible excuse for it. In fact, I have known people who never exhibited poultry, and perhaps would not be regarded as fanciers (certainly were not so in the sense of taking any interest in breeding to definite pattern and type), splice the broken sickle of a pet mongrel rooster because they did not like the un-

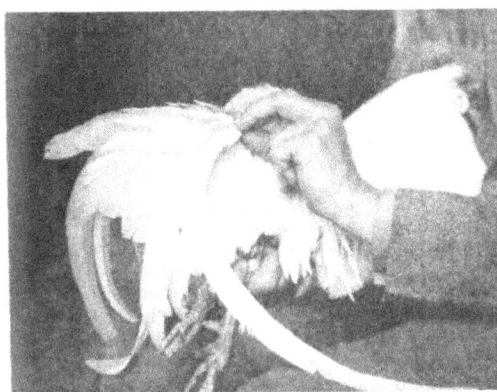

CURLING A SICKLE FEATHER

symmetrical appearance of the tail with a large part of one sickle gone. In the case of an exhibition bird there is always the possibility of the substitute feather being one of his own from an earlier coat of feathers. But in most cases appearing in the showroom the natural presumption is that a faulty sickle on an otherwise good bird has been replaced with a good one from a generally inferior specimen.

Two methods of splicing are used. One is to cut the broken feather off as close as possible to the root, and insert the new feather in the stub, using glue to make it adhere. This does not show the splice as plainly as the other, though it is easily seen when attention is directed to it. The other is to leave an inch to perhaps an inch and a half of the quill beyond the skin and the quills are trimmed to make them fit as in splice-grafting and bound tightly together with thread the color of the feather. The part put on is a little surer to stay when this is done, and in a bird with abundant tail coverts the splice would be well concealed and the feather might not be noticed by one who was not examining the bird critically. But unless it is very nicely adjusted an observant poultry breeder or judge looking at the bird would quickly note something wrong with that feather; and the same thing would happen if it did not match the rest of the tail in color.

The possibility of splicing is limited to large stiff feathers—as tail and flight feathers. Detection of splicing in the latter is so certain that it is probably never tried. Detection of tail splicing is so likely that exhibitors generally never so much as consider doing it.

Giving Luster to Feathers

Birds differ greatly in their natural luster, and their luster is also influenced by their diet. In all cases luster on the surface can be increased, and where the natural luster is good it can be made very fine, by rubbing, brushing or grooming the surface with a soft cloth or with the palm of the hand. Where the color is dull or flat looking it may be either because the feathers lack oil or because they are soiled. The treatment required to give luster and gloss will depend on the circumstances. If the feathers, though clean, lack luster, the addition of oil may give this luster, and the oil supplied by rubbing with the palm of the hand may be sufficient. If it does not appear so the palm of the hand may be made slightly oily with vaseline, coconut oil, sweet oil, or in fact any oil that is applied to the skin, and rubbing with the hand will then produce the desired gloss. The point is to add the oil to the feathers in very small amounts, so that no more will be applied anywhere than the feather will immediately absorb, and to have it evenly applied that the luster may be uniform. A cloth that is uniformly and very slightly oily may serve the same purpose, but minimum application of oil to plumage can be made better with the bare palm as the sense of touch keeps one constantly aware of the degree of oiliness.

Where the plumage is naturally lustrous but more or less soiled, the effect of rubbing is—in part at least—to show the natural luster by removing the dirt. In such cases the use of a little alcohol, benzine, gasoline, ammonia, or like material, on a cloth used in rubbing helps to clean the surface and the continued rubbing adds to the luster. Where there is any dust in the plumage of a very dark bird (especially of a black one), slapping the surface lightly with the flat of the hand as the rubbing proceeds will bring it out.

The rubbing that gives high luster, like all the operations which contribute highly finished results, is not a perfunctory process but very thorough and long continued. Mr. J. C. Punderford when exhibiting Buff Leghorns contributed an article on the conditioning of fowls to the "Reliable Poultry Journal" in which he stated that the high luster on his birds was secured by rubbing each bird well with a silk handkerchief for half an hour daily for a week or more before the show. To get the full value of such work the rubbing must of course be continued

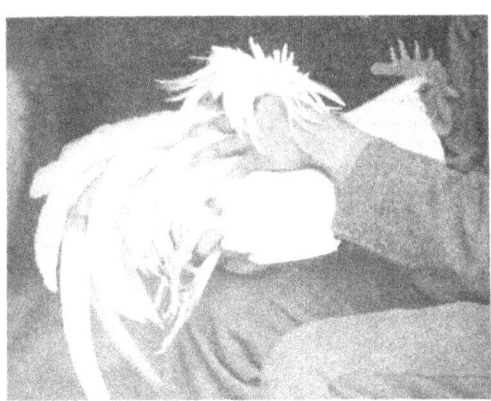

BUILDING A SADDLE BY BENDING THE FEATHERS

after the birds arrive at the show and special attention given to the final grooming just before the birds are judged. Many exhibitors postpone rubbing until just before judging. While that puts the birds in better condition than they were, it does not—except in cases where the natural sheen is uncommonly good—put them where they would have been if well groomed during conditioning.

Rubbing to make or improve luster applies only to rather hard-finished plumage. To rub where looseness and fluffiness is desired may tend to spoil those effects. That depends upon the structure of the bird—on how much its apparent type is due to the quantity and texture of feathers. Thus a Cochin or an Orpington male that has a broad back, nicely rounding to the sides, and a full and comparatively firm saddle can stand the rubbing that will bring the gloss out on the back, while a narrower bird that needs all the advantage fluffing and building out the saddle can give it cannot.

Where rubbing is inadvisable, the best way to bring out luster on black plumage is by keeping the birds in darkened pens for a week or so before showing. Wright says this was the method used by the most successful exhibitors of Black Spanish in England in the palmy days of that breed. The birds were confined in the dark in the first instance to secure purity of color of the white face. It was found that black luster was also improved. The pens were made about seven feet long by four feet wide, and boarded up four or five feet from the floor, the partition above that being of wire netting to provide thorough ventilation. Small windows high up in the outer walls admitted just enough light for the birds to see to eat and drink and fly to roost.

Purple Barring in Black Plumage—This may be in whole or in part prevented by good feeding and care, and by not allowing the bird to be exposed too much to the weather. Where it exists it may be more or less successfully concealed by one of the following methods: Applying a black stain or dye of graphite, lamp black or ink to the feathers. It is said that this method conceals the barring so that only very close inspection will reveal it. The other method, which it is claimed cannot be detected, is to rub the birds well, as in rubbing to give luster, with a mixture of equal parts of suet and lard, melted together, in which a little sulphur is mixed while hot. The method of applying is to put a little on the palms of the hands and rub the plumage until the bars are no longer visible.

Dyeing White Feathers in Black Birds

Small patches of white or gray are sometimes covered by applying dyes or inks. As a rule, it can be detected upon close examination, and the faker depends more on the chances of the dyed spot escaping notice than upon attempting to do a job that will defy detection. For large patches or feathers of white Hubbard avers that by first dampening the feathers in warm water, then applying a gallic acid preparation made by adding a quarter of an ounce of gallic acid and a half pint of alcohol to a pint of water, following this in a few minutes with an application of a preparation of half an ounce of nitrate of silver, half an ounce of gum arabic, and two ounces of ammonia in half a pint of water, and after the feathers have dried rubbing them smooth with a sponge slightly dampened in water, the feather is given a black color that will pass inspection.

Removing Tips or Edges of Different Color from Feathers

This is done either by trimming with scissors, or by burning with a hot wire, electric current used as in pyrography, or a lighted cigar (a flame must be avoided) or by rubbing and wearing off the edge or tip of the feather with pumice stone. None of these processes leave a natural edge on the feathers, and it is hardly possible to apply them at all extensively without giving the section of the plumage treated a peculiar look, which leads an expert to take careful account of its condition.

Removing Black from White Quills—A little black on the quill of a white feather—corresponding to ticking in the web—may be removed by rubbing with pumice stone. The damaged quill is then rubbed with chalk and the fingers to give it a more natural appearance. The treatment is discernible on close inspection but will often escape notice.

Dyeing Buff and Red Plumage

Reference was made on page 42 to the situation in regard to the coloring of Rhode Island Reds. Since that chapter was written I have learned that any exhibitors of the breed who had copies of the book "Secrets of Expert Exhibitors" published in 1909 and sold under an agreement (of doubtful validity) that the buyer would not divulge any information therein contained, have been in possession of a secret formula for coloring red birds with an article in common use in poultry yards for medicinal purposes—permanganate of potassium. The method of using this is the same as that of using peroxide of hydrogen to bleach white feathers. The bird is washed and the permanganate of potassium applied after the last rinsing. The plumage is dampened all over with a permanganate solution of a strength previously determined by testing feathers taken from it. The hands of the operator are stained, and to remove the stain a saturated solution of oxalic acid is used. It is to be presumed that such a solution applied to the processed bird would give positive results; but I am not in a position at this writing to say that it would. The whole matter is one for further investigation and tests. What is known of dyeing buff color may throw some light on the subject. The best source of information on the subject is Wright, from whose "Illustrated Book of Poultry," 1902, the following is taken:

"Buff dyeing has been so far chiefly found in Leghorns and Orpingtons, and has usually been done by washing in 'Maypole' soap or some other form of what is really aniline dye. Much harm was done at one time by a school of amateur detectives who professed to test buff dye by solutions of caustic potash. It is now known that potash solution dissolves out color from honest plumage and is no test for aniline dye. There is fortunately a real test, as published by Dr. Mossop in "Poultry." Test tubes should be provided, about six inches long and three quarter inch bore, and, if possible, a genuine feather tested in one tube along with the suspected one—for only plucked single feathers ought ever to be tested; no one has a right to risk disfiguring the bird itself. The feather should be pushed right to the bottom of the tube, strong hydrochloric acid poured in a half inch or more deep, and the tube a little shaken about (without spilling the strong acid) so that the feather, at first greasy, may become wetted by the acid. A feather dyed buff with Maypole soap or aniline dye will quickly turn a beautiful violet color; the genuine feather is unchanged beyond the darkening from being wetted. Dr. Mossop found one or two dyes that resisted this test, but these responded in another way, by losing all the dyed color to a solution of crystals of protochloride of tin dissolved to saturation in hydrochloric acid diluted with its own bulk of water. This solution spoils by keeping and should therefore be freshly made, dissolving a few crystals when wanted in the diluted acid. Sometimes a feather that seems to resist will bleach out the dye if the test tube containing the tin chloride is warmed over a spirit lamp. If no decided bleaching takes place then, and the feather has not been turned violet by the acid either, the plumage is probably honest, and at least no ordinary dye has been used."

CHAPTER X

Care of Exhibition Birds in Transit and at Shows

Advantages of Accompanying Exhibits—Timing Shipments to Shows—Methods of Shipping—Cooping at the Show to Display Each Bird to Best Advantage—Grooming for the Judge—General Care to Keep the Birds in Good Condition, Prevent Colds, Avoid Contagious Diseases—What to Do in Anticipation of Delay in Returning Birds

AN exhibitor who intends to "stay in the game," and make his string one which competitors will have to reckon with, must either accompany his birds to the show and look after them from start to finish, or have some competent and dependable person to do so. It frequently happens that an outstanding single bird is sent a long way to a show without other attendance than the expressman and the attendants at the show give it, and comes out on top when the awards are placed: but I do not recall an instance of a string of birds winning under such conditions where they had real competition with birds fitted to the minute.

All show associations advise prospective exhibitors that their birds will have "the best of care" while at the show, frequently stating that it is "not necessary for you petition as they affect the present, or may figure in the future, which he ought to have. It should be accepted at the outset that if one wishes to exhibit regularly and successfully anywhere there must be some personal acquaintance with the situation and contact with show managers, judges, and competitors. For that reason practically all of the regular exhibitors at important shows are breeders who can attend them for at least one day, and most of them are people who can attend daily or several times during the week.

Timing Shipments to Shows—Methods of Shipping

Shipments of exhibits to poultry shows are usually timed to arrive on the day before that announced as the opening of the show. Most shows are held from a

MANAGERS OF POULTRY DEPARTMENTS OF FOUR GREAT EXPOSITIONS
Left to right—George H. Burgott, Pan-American, Buffalo, 1901; T. E. Orr, St. Louis, 1904; John A. Murkin, Jamestown, 1907; T. E. Quisenberry, San Francisco, 1915

to accompany your birds." These advices are to be considered by the prudent exhibitor with such qualifications as a knowledge of the situation suggests. It is not "necessary" for an exhibitor to accompany his birds to insure that they will have general good care, and that all reasonable precautions will be taken for their safety and health, and that judges will consider them impartially as they see them. In many cases a bird of extra high quality, or a string that has easy competition, may make a good winning without having had any other attention at the show than the regular attendants give it. But for every case of this kind there are many more of failure to win because the birds of the absent exhibitor did not receive the special attention at the show and the grooming before judging that the exhibitors present gave their birds.

So if one expects to show regularly and win a satisfactory share of the premiums where he shows, it is quite essential that he see that his birds are as well looked after at the show as any of their competitors, and it is important that he be present at the show enough to be fully informed of the situation there, or have a representative capable of acting promptly in his interest, and on whom he can depend for information about the conditions of com-

Tuesday to the following Friday or Saturday, and as express service over Sunday is usually slow and subject to extraordinary interruptions there is more or less uncertainty about the arrival of shipments which cannot be started on Monday morning with reasonable assurance that they will arrive at the show at some time during the day or evening. Many exhibitors, rather than have their birds on the route for more than the shortest time required to make the journey, will start a shipment for a show on Monday, knowing that the chances are even that it will be delayed and not reach the show in time for the judging, but trusting to the indulgence of show managers and competitors to postpone judging their class until their birds arrive.

The bulk of the exhibits of birds at any show come from points near enough by rail to make the journey on the day before the opening, provided they are started early in the morning. An exhibitor so far away that he knows the chances are against his birds getting through in a day ought to consult the secretary of the show in advance of shipment in regard to facilities for taking care of his birds if they arrive ahead of time, and be governed largely by that official's advice. The situation as it affects this may not be the same at any particular show

FOUR OF THE MEN WHO RUN THE MADISON SQUARE GARDEN SHOW
Left to right—D. Lincoln Orr, Secretary, Harvey C. Wood, Harry M. Lamon, Chas. D. Cleveland, Directors. Pictures of T. A. Havemeyer, President, and W. A. Stanton of the Board of Directors may be found on page 10

from year to year. That depends much upon the length of the interval between the removal of the exhibits of a show preceding the poultry show, and the beginning of the work of setting up the coops for the poultry show. Sometimes the poultry show management does not get possession of the hall until the day before their opening. In that case they might prefer that exhibitors should delay shipment and not have birds arriving earlier than the day before the opening but sometimes, with an interval between the poultry show and the one preceding it, the work of setting up the coops and laying out the show begins four or five days before the date of opening, and in that case the management might prefer to have exhibits from distant points arrive early, and give them the necessary care, rather than take chances of their failure to arrive on time delaying judging in a class. The exhibitor in such cases could rely on his birds being well cared for after arrival: that is—there would be no neglect of them. They would be cooped as soon as they arrived or as coops were ready, and would be properly fed and watered.

Exhibitors living within easy trucking distance usually prefer to ship their birds to the show, over the road, by auto truck, rather than subject them to the transfers and possible exposures incident to the regular express service with its numerous transfers and conditions beyond their control. This method indeed was in favor long before the days of speedy auto trucks when a trip of twenty-five to thirty miles was a day's journey for a truck horse. An exhibitor, or perhaps two or three exhibitors in the same locality, arrange for a truck to take their birds and to return them from the show. They can then load the birds so there is no question about their safety and comfort; protect them from the weather as may be necessary; and be sure that the load will arrive at the show with all the birds in good condition, and with no delays on account of other stuff handled by the truckman. The great advantage to an exhibitor, or small group of exhibitors in handling birds in this way is that the birds can have the owners' personal supervision in loading, and again in unloading.

Where a few exhibitors of poultry cannot charter a truck, it is often possible for the exhibitors in various classes at a poultry show, pigeons, pet stock, etc., who live in the same town, to arrange so that the local express company will make a special load of their stuff and send it over the road by auto truck instead of by rail with the transfers incident to that mode of transportation. To secure this service exhibitors should notify the express company far enough in advance for them to determine whether the amount of stuff to be carried warrants it, and make arrangements accordingly. Failing this, it oftens happens that the local express company is making deliveries at a show at intervals through a day or more in taking stuff to the show, but knowing what is to be returned makes a single load and quicker delivery in getting the stuff home from the show. This it does primarily for its own advantage, but it would be equally to its advantage and very much to the advantage of exhibitors to handle exhibits to the shows in the same way. Ability to do so depends upon exhibitors notifying the express company of shipments to the show a day or two in advance, and then having them ready at an appointed time.

SOME POPULAR JUDGES ALWAYS SEEN AT BIG EASTERN SHOWS
Left to right—John Kriner, H. W. Schriver, F. W. Rogers, W. H. Card

Some well-known exhibitors who have been leaders in several varieties. Left to right—Wm. Barry Owen, Maurice F. Delano, Chas. J. Fiske, Rufus Delafield

Shipping Coops

Most exhibitors ship their birds to shows in ordinary shipping coops; but some of the larger exhibitors and a few others who show often, have for this purpose more substantial coops, painted a uniform neat color and stenciled with the name and address of the yard or farm. Sometimes these coops are built in long sections, combining several coops and requiring two men to handle. The object of this is primarily to keep a group of exhibits together, insuring that they will not be separated in transportation, and perhaps one or more coops go astray. A secondary supposed advantage is that the large coop cannot be tossed about by careless expressmen, as a light single coop might be, and the shipment is therefore more carefully handled. This is a doubtful advantage in the case of a section of coops for large fowls, for the fact that the coop is not easily handled may lead to its being kept on the floor, pushed into a corner of a car, and buried under lighter coops and other merchandise, and the chances of birds in it being smothered considerably increased. Sectional coops are more used for bantams, pigeons and small pet stock, where single coops would be so small that there is much more danger of their being damaged by rough handling than the coop for ordinary fowls.

Some exhibitors have the coops used for shipping birds to shows constructed so that they can be securely locked with padlocks. This practice, again, is most in vogue with exhibitors of small birds and stock which a thief can conceal about the person until he can dispose of them without being seen. In some cases where it is followed by exhibitors of large birds the object is to prevent the birds being taken from shipping coops and placed in the exhibition coops by the attendants at the show, the exhibitor preferring to have his birds handled only by himself or an employee in charge of his exhibit.

It would seem unnecessary to say that shipping coops should be of good size, affording the birds both ample head room, and room for the males with long tails to keep the tails in good condition; yet there is hardly a large show that does not afford instances of exhibitors who ought to know better sending birds in coops much too small for them, and from which they are taken with damaged heads and tails, generally ruffled plumage, and sometimes so cramped that they cannot stand in a natural position for hours. I have seen a Leghorn cockerel arrive at Madison Square Garden in a box about eighteen inches long and ten inches high, literally jammed into the box. This of course is an extreme case, but there are many cases where the single-bird coops are too small for the birds sent in them. All large males with full furnishings should be shipped in square coops.

The most careful exhibitors have each male in a separate coop, to insure that no possible damage can be done by hens picking the comb or feathers of a male between the time they are cooped for shipment and the uncooping at the show. The necessity for doing this will in each case depend on the circumstances. If an exhibitor has a good deal of trouble of this kind at home it is wise to take precaution. Ordinarily there is little danger of hens damaging the combs of males, or eating feathers, until after a few days' confinement at the show.

Exhibitors showing a full string of birds usually ship a few extra birds to use in case anything goes wrong with

Past masters in the arts of breeding and showing. Left to right—Frank H. Davey, M. L. Chapman, T. G. Samuels, L. C. Bonfoey

LEADING BARRED PLYMOUTH ROCK BREEDERS AND EXHIBITORS

Left to right. Top row—E. B. Thompson, A. C. Hawkins, A. C. Smith, C. H. Latham. Second row—Victor Bradley, Chas. H. Welles, M. S. Gardner, E. L. Miles. Third row—Valentine Thompson, H. P. Schwab, Newton Cosh, W. L. Russell. Fourth row—W. D. Holterman, F. W. Richardson, W. W. Henderson, D. F. Palmer

any of the birds selected for entry. This is a matter of more importance to the large exhibitor competing for prize on collection, or hoping to make a clean sweep in a class, than to others. In general, the regular entry exhausts the exhibitor's selection of birds most suitable for the competition, and any alternates taken to the show are of second quality. It does sometimes happen, however, that on arrival at a show a substitute bird is found to show better in the position secured than the original choice.

Some exhibitors make it a point to follow their birds in transportation to a show, leaving home on the same train with them, getting in touch with their shipment at transfer points, and perhaps managing to speed it along by the judicious distribution of cigars or cash tips. It should be said that the process is by no means uniformly effective. Unless one knows the route and some of the men on it pretty well he may easily lose track of his shipment at a busy transfer point, and there have been cases where exhibitors who undertook to keep an eye on their birds en route wasted time looking for them at a junction when they had gone on ahead and arrived at the show before the shipper did. While it is true that railway express service—especially of late years—has been subject to many delays, it is doubtful whether an exhibitor of poultry gains anything by trying to keep in touch with his long-distance shipment.

The Exhibitor's "Kit"

Many experienced exhibitors make it a point to take with them to a show supplies of simple remedies for the first stages of ordinary ailments, and also things used in fitting and conditioning birds which there may be occasion to use before the birds are judged: as preparations for the head and feet, small sponges and soft cloths. Some exhibitors also make a practice of taking to a show a few heads of cabbage, some turnips, or apples, to have handy to give their birds from time to time. Such things can of course be obtained in the city where the show is held, but it is often more trouble to look them up than to take what is needed from home.

Besides these things a small hammer and screw driver, tweezers, a few tacks and small nails, a supply of the exhibitor's cards, and of circulars if he has them, may come handy. A long linen duster is also a desirable part of his equipment if he is to be at the show much and give a good deal of attention to his birds. Many exhibitors take a supply of small padlocks to lock their coops to prevent their birds being handled except by judges or when they are present. Locks, it should be said, only prevent the ordinary inquisitive handling. An unprincipled competitor who wishes to injure a bird can easily do so in spite of locks, and this occasionally happens. Such instances however, are very rare, for suspicion almost immediately attaches to the guilty party, and even when no proof against him can be obtained he is apt to find that he has injured himself more than the person whose birds he damaged.

Cooping the Birds at the Show

Except where shipping coops are locked, or in cases where the management expects exhibitors to attend to the cooping of their own birds at the show, the attendants at shows—after the exhibition coops are up—put the exhibits in their proper coops as fast as they arrive. While in setting up the coops at a large show a gang of unskilled laborers as available for such work may be employed, when it comes to handling the birds that work is entirely in the hands of the men employed for the period of the show for feeding and other routine work, and these are either regular employees of the cooping company or poultrymen engaged especially for the show. On the whole there is little occasion to take exception to their handling of exhibits. It is inevitable that in the hurry and rush there should sometimes be a little rougher handling of birds than the owner would approve, but there is surprisingly little of it. Indeed, in my observation, while I have seen an attendant sometimes very rough, I have seen a great deal more manhandling of birds by exhibitors than by attendants. In general, however, though the attendants are not rough they are not as careful as the exhibitors who handle birds with strict care to avoid any

TYPICAL COLLECTION OF CUPS, MEDALS AND SILVERWARE SPECIALS OFFERED AT POULTRY SHOWS

injury to plumage or unnecessary excitement of the bird.

Except for the first numbers in the breed and variety coming first in the classification at any particular show, the exhibitor can have no definite idea until he arrives at a show what sort of position and light his birds wll have. Most shows now make their classification follow the order in which the breeds are described in the Standard, and No. 1 is a Barred Plymouth Rock cock. Boston (alone I believe of large shows) retains the old order with the Asiatics first, and here No. 1 is a Light Brahma cock. Barred Rocks at the Garden, and Light Brahmas at Boston will have the same position year after year, the differences in the numbers of entries from year to year making no material difference in the positions of these classes. And at most shows the Barred Rocks have the same position year after year. But as to the other breeds and varieties the farther down the list we go the more variations in the entries in preceding classes affect position of a class, and sometimes the result of this is to bring an entire class into different conditions of light in a certain year than they ever had before.

At New York and Boston (in the main exhibition hall) and wherever the principal daylight comes from high windows or skylights, the light is generally better near the middle of the hall than at the sides, and the difference is greatest where there are galleries and the ends of the rows of coops extend under them. Further, at nearly all large shows only a part of the exhibits can be staged on the main floor in the principal hall, and the result is that very often a part of some class is in that room and a part elsewhere under very different conditions of light.

The exhibitor knows the numbers of the coops his birds are to occupy, for they have been furnished him with his shipping tags. On arriving at the show his first care is to see whether his birds have arrived, the next to find the location of their coops, and get them from the shipping to the exhibition coops, if that has not already been done by the attendants. His next concern is to size up the competition. In case his birds are late in arriving he may have an opportunity to do so in advance.

Where a poultry show is double-tiered, particular attention should be given to placing the birds as far as possible at the level at which they will show to best advantage. The points to be considered were treated in detail in the chapter on Selection of Birds for Exhibition, and need not be repeated here. Where there is any choice of position, that is where an exhibitor happens to have an end coop, or one in better light than others, it is usually policy to put the best bird in this. On the other hand, where a bird has faults it is hoped the judge will overlook, it may help matters to give it poor light.

COOPING BIRDS AT THE GARDEN
Left—A. C. Smith with a winning S. C. Brown Leghorn male
Right—C. H. Latham with a winning Barred Rock pullet

In cooping white birds it is always well to thoroughly dust and wipe the inside of the coop that the clean plumage of the birds may not be soiled by dust or contact with smudgy metal. The management of the show is supposed to look after all matters of this kind, but much work has to be left to indifferent and unskilled laborers, and so the only way to be sure that a coop is perfectly clean inside is for the exhibitor himself to wipe it out carefully. In the cooping of colored birds this is not so important.

An experienced exhibitor, familiar with the work of the judge who is to pass on his class, and not biased in judgment by partiality for his own birds, can usually tell on short inspection of a class just about what he may reasonably expect to win in, provided the birds show no faults in handling that put them below the position they seem to be entitled to when viewed from the aisle. It is here that the exhibitor present on the ground often finds that by making some changes in his entries he can probably win something where as they stand he seems destined to get low places or nothing all along the line. Thus his best cock or cockerel may be plainly beaten in the open class but may be good enough, with the females he has entered as a pen, to give him first on pen. So he takes the male first selected out of the pen and puts him in the open class, knowing he has not the remotest chance of winning there, and with his best male in the pen wins that prize. Or if he is beaten on females in the open class, it may be that one or more of the females entered there will strengthen a pen enough to enable him to win with it.

The converse too may happen: unable to win on a pen, he may be able to take birds from it to the open class with some prospect that they will be placed there. He can do this, of course, only in so far as his entries in both classes admit of exchanges—he cannot change entries as they appear on the records, making new entries or cancelling old ones. There is a species of irregularity in this practice of substitution or exchange in entries, for the birds as entered are supposed to be identified by the bands they wear, and the absent exhibitor has no chance to take like advantages. The practice is one that show managers do not appear to take cognizance of officially. But it has always been done, and will be as long as exhibitors take charge of their own exhibits at shows and have the opportunity to "jockey" them in this manner. Exhibitors who desire to do so must always have charge of their exhibits at shows, and the opportunity to shift them to suit conditions of light or competition will exist as long as show specimens are not registered.

The uncertainties and inequalities of light in most exhibition halls furnish the best excuse for such manipulation of entries. Within a class, as cockerels, an exhibitor has little opportunity to save himself if he happens to draw a number in entering that when the cooping is arranged comes in a very bad light. He may help matters by splitting his entry—that is, entering some birds early, and others just before the entries close—but there is a possible marked advantage in this only in a large class. But as between the light where his single birds are cooped and that where his pens are cooped there may be a great difference, sometimes favorable to one, sometimes to the other, and as it is impossible to foresee such things custom sanctions the practice of shifting entries where the classification permits at any time before judging begins.

Grooming the Birds at the Show

An exhibitor cooping his own birds at the show looks each bird over as he handles it for any derangement of plumage and smooths feathers that are out of place. If his fitting of the birds before shipment has been very thorough and they have come through in good shape he is not likely to do anything more to them until the morning of the first day of the show when the judging is supposed to begin (and does), if all classes and judges and their books or cards are ready. If an exhibitor has been pressed for time in fitting his birds and has not done the work as thoroughly as he wished, he usually proceeds with it as soon as they are cooped at the show, and continues doing what he can for each bird until the judging of his class begins. In any case the morning of the day judging begins finds every exhibitor who has attended his birds to the show giving them a final examination and grooming.

The things of most importance at this time are to brighten up the head parts and legs, to see that the plumage is clean and free from dust, and that the birds are free from traces of down on the shanks, and on or between the toes. Many exhibitors do not look for down on shanks and toes until going over the birds just before judging, because it is well established that down will sometimes appear overnight, and the only way to be sure that the birds are free from it when judged, is to look them over just before. If, as frequently happens, the judging is not completed the first day, the birds to be passed on should still have attention on the morning of the next day. It is generally assumed that exhibitors will give their birds these attentions early in the morning, and not be engaged grooming birds in one class while the judge is at work on another, but as shows are arranged it will sometimes happen that an exhibitor can handle his birds very close to the time the judge begins on a class without any suggestion of his being trying in anyway to attract attention to them.

Some Matters Relating to Judging in Progress

At many of the minor shows, particularly where most of the exhibitors are novices, it is a common practice for a judge to do his work with a "gallery" of exhibitors and visitors crowding about him, while he makes audible comments on the birds as he handles them, and sometimes stops to explain points of interest. This practice is adapted to small shows, and to small classes where decisions are easily made; but is not considered practical in large classes or in close competition, where the judge cannot make his decisions with the same readiness and accuracy, and wants to concentrate all his thoughts on the one thing of placing the awards right. That the judge may do this without distraction it is the general rule—sometimes unwritten, sometimes more or less specifically set forth in the regulations of the show—that there is to be no interference of any sort with the judge in the performance of his duty, that exhibitors are to keep away from him, or if—as is often the case—the judge tolerates their presence at not too close range, they are to refrain from any remarks regarding birds in the class that might be considered suggestive and designed to influence his decisions.

American exhibitors generally accept the decisions of a judge without protest. While the regulations provide for protests in certain cases, it is only where something is grossly wrong, and so bad that it ought not to be passed over, that protests are usually made. The general sentiment is against them, and many exhibitors consider it poor policy to make protests no matter how great the apparent provocation. The whole tendency of reform measures in this country, in the poultry shows as elsewhere, is to overlook what cannot be remedied, but to take precautions that the same thing shall not happen again. There are a few old and generally successful exhibitors who are known to the fraternity as "bad losers" and are apt severely to criticise judges under whom they do not get all the awards they think they should; but on the whole the disposition of exhibitors is to accept things as they come, and if a judge's work is very unsatisfactory to them to express their disapproval by refusing to show under him again.

General Care of Birds at the Show

Until the judging is over an exhibitor's chief concern is to keep his birds in the pink of condition that nothing may mar the impression they make on the judge. After the awards are made he may relax something in the way of grooming, yet should bear in mind that it is a very decided advantage to him to keep them looking as well all through the show as when they faced the judge. To do this he must still give some attention to their appearance, and he must further give a great deal of attention to their condition and appetites and habits. Most birds are at their best at a show after a day and a night in their exhibition coop has enabled them to rest from the effects of their journey to the show, and to become accustomed to their surroundings. There are birds that will come out of the shipping coop perfectly fresh and be immediately at home in the exhibition coop, but most show

WAITING FOR THE WORD TO START JUDGING
Two old-timers. Left, Julius Frank; right, F. B. Zimmer

better after a night's rest, and when they have begun to feel at home. Thus birds arriving on the morning of the day of judging are often handicapped by their tired condition and are placed lower than they would have been if judged on the following day. Meantime, the competitor that won has lost a little of his condition, and for that day the later arrival of two birds very equal in merit may appear to stand out as very superior, especially in form and style. But in another day, he too begins to show the effects of confinement, extraordinary excitement, more or less miscellaneous handling, unaccustomed feeding, and late hours; and unless special care is given to keep birds in good condition they are likely to be considerably reduced in vitality and somewhat rough looking before the close of the show.

Birds of the heavier breeds, with a pronounced tendency to put on fat, are usually most affected. Birds with large combs and crests are more affected by showroom conditions than those without crests and with small combs. The heat of the showroom often makes a comb grow much larger and the suddenly increased size and weight of it makes a bird very uncomfortable. The heat and weight of a large crest seem to have much the same effect, though there may not be an increase in weight of the crest such as occurs in weight of comb.

Some exhibitors try to keep their birds in better condition throughout a show by attending to the feeding and

VETERAN PLYMOUTH ROCK BREEDERS
Left to right: U. R. Fishel, White Rocks; H. S. Burdick, F. C. Shepherd, Buff; H. W. Halbach, White

watering themselves. They have cards printed with the legend "Do not feed or water these birds," and place one of these on each of their coops. There are two objections to this practice. The first is that it interferes with the systematic feeding and watering of exhibits by regular attendants. They do their work much better when all they have to consider is to put practically the same amount of feed and water in every one of a long row of coops. As poultry are cooped in competitive classes at a show, an exhibitor who places the above notices on his coops might have in one place a single coop, in another place two or three coops, and so on. In many cases the attendant does not notice a sign of this kind until after he has fed a bird. Then he omits feeding others beside it. When the owner or special attendant comes around he feeds all alike. The result is that some are overfed.

The second objection is that exhibitors who request that their birds be not fed by attendants do not always get around and look after them regularly. Thus at most large shows some of the exhibitors who prohibit attendants of the show feeding and watering their birds go home, perhaps thirty or forty miles at night, expecting to be back by the middle of the morning next day, and are detained until late in the day, or perhaps are not able to get in that day at all. An exhibitor in such a case may telephone the management. The management with many other details to look after may or may not give the matter attention. The exhibitor is quite as likely to fail to notify anyone to look after his birds. The emergency might be met in any one of several ways, but it is not, and the birds go hungry and thirsty.

I recall one instance in particular where an exhibitor was taken sick and his birds went without feed or water for three days before anyone in a position to act in the matter discovered it. I have known of so many cases where from accident or forgetfulness exhibitors who objected to having their birds fed by the management of a show neglected the birds themselves, that it seems very clear to me that it is much better for exhibitors to have their birds fed grain with the rest, and counteract the possible bad effects of heavy grain feeding in close confinement by giving the birds succulent green feed in small amounts, at least daily, and several times a day if possible. Many exhibitors make it a point to always have something of this kind with them when they are about their coops during a show. They will stand before a coop, or go from coop to coop, whittling bits from a turnip, an apple, an onion, or a piece of cabbage, and feeding these to the birds, or will allow the birds to pick from the object in the hand.

Besides the effect on the diet such attentions serve to interest the birds and prevent them from seeking diversion in the vices of comb and feather eating, which are very likely to develop when they have too long periods with nothing in particular to do. Birds that get some little attention from one in charge of them at frequent intervals during the day are also more alert when others pass their coop than those that get no individual attention at all after the judge has been through the class. One of the most forlorn objects imaginable is such a neglected bird toward the end of a show. It will remain in the corner of its coop, taking no interest at all in what is going on, and showing all symptoms of utter misery, while right beside it a bird whose owner has looked after its wants carefully through the week is right up on its toes challenging the admiration of every passer. It is by no means all

WHITE PLYMOUTH ROCK BREEDERS OF THE YOUNGER GENERATION
Left to right: W. A. Halbach, A. F. Poltl, J. T. Thompson, J. G. Jones

EXHIBITORS WHO PUT AND KEEP THEIR FAVORITES "ON THE MAP"
Left to right: H. C. Sheppard, Anconas; F. C. Stier, Anconas; Geo. B. Inches, Silver Gray Dorkings; M. R. Jacobus, Campines

in the bird: What the owner does for it makes a very great difference.

Conditions in a showroom are generally much more productive of colds than the ordinary conditions of poultry life. The temperature in a large show hall is likely to be uniform only when the outside temperature is mild. The colder the weather the more unequal temperature conditions in the show become: the birds near the radiators will be much too warm, while those distant from them may be in an atmosphere more chill than the outside air. Drafts are much more plentiful than in the ordinary poultry house, and as the vitality of the birds becomes a little lower day after day they contract colds much more readily than when at home.

For these conditions only partial preventive measures are possible. Where a current of hot air rises near a coop it may be diverted and its effects largely reduced by placing a screen so that its direction will be diverted as much as possible. Where a cold current strikes a coop, the effect of the draft may be largely reduced, sometimes entirely so, by covering the back and top of the coop. If, notwithstanding these precautions, birds are still evidently uncomfortable under the conditions existing, the only thing to do is to remove them to a place where they will be safe. This may be done either by transferring them to exhibition coops in other parts of the hall, or by moving the coops in which they are, or by putting them in their shipping coops, off exhibiton, and in a comfortable place.

The measure in any case must depend upon circumstances. It is not always possible to change the position of a section of coops. In most large shows there are here and there through the show coops that are vacant because the entries assigned to them were not sent. To the extent of their numbers these are available for birds that should be moved from their places. As a last resort, the return to the shipping coop, and putting that in a comfortable place is always available. The common trouble is that exhibitors do not act promptly when they know that the conditions are wrong and are injurious to their birds, but wait until the birds are seriously affected.

Fall fairs are usually not open in the evening, but at winter shows, and especially at smaller shows, the evening attendance is large, and the long day, added to other irregularities of life, is very hard on many of the birds. The brilliant overhead lights in particular seem to distress many of them. To relieve this as much as possible exhibitors who are looking carefully after the comfort of their birds shade the tops of the coops with newspaper, or large sheets of heavy wrapping paper.

At the Close of the Show

Some shows keep all exhibits in place until the closing hour on the last day. When this is 10 o'clock of a Saturday night there is little opportunity to get the birds out before Monday. A few exhibitors whose birds can be started for home at once, if they can get the express companies to take them before midnight, make a special effort and get them away. Those who have brought their entries in by auto truck also often have it arranged to start them home as soon as they can be put in their coops after the close of the show.

Sometimes it happens that another show is coming into the hall immediately, and the dismantling of the

EXHIBITORS WHO PUT AND KEEP THEIR FAVORITES " ON THE MAP"
Left to right: John A. Gamewell, Golden Penciled Hamburgs; Dr. J. S. Wolfe, Silver Spangled Hamburgs; Watson Westfall, Silver Gray Dorkings; W. S. Templeton, Cornish

VETERAN WYANDOTTE BREEDERS AND EXHIBITORS
First row, left to right: B. M. Briggs, originator of both White and Columbian Wyandottes; F. A. Houdlette, Silver Laced; E. O. Thiem, Golden and Buff; W. A. Doolittle, originator of Partridge. Second row: Ira Kellar, Golden and Silver Laced and White; Henry Steinmesch, White; F. L. Mattison, Silver Laced and Buff; Arthur G. Duston, White. Third row: L. Whittaker, originator of Silver Laced; W. E. Samson, Silver Laced; Dr. N. W. Sanborn, Buff; Julius Bachman, Silver Laced

whole show has to begin at once and go on all night. In that case the birds are all cooped and delivered to the express companies as rapidly as possible and it rests with them to handle them as best as they can. As a rule, however, the birds can remain in the hall over Sunday, and the general removal take place Monday in a leisurely manner. Where a show begins and ends in the middle of a week the conditions relating to shipments in and out are usually more satisfactory, especially in the return, for the birds all go out promptly the morning after the show.

Birds that are not attended by the owner or his representative are usually left in their places in the exhibition coops when they have to stay in the hall over Sunday. Where the owner or his agent is present his attitude in this will be governed either by the condition of the birds, or conditions as they affect the birds, or whether it is more important to him to get back immediately to his home and poultry or other business, or to continue careful attention to his exhibition birds until they are delivered to the express company.

Some breeders coop their birds for return shipment immediately upon the close of the show even though they know they will not be taken by the express company until the second day following. They may or may not put a little grain and something succulent in the coops for them to eat. In many cases they do not do so because they consider that after the full feeding and inactive life the birds have had for the past week or more a two or three days' fast will be more beneficial than harmful.

LEADING WYANDOTTE BREEDERS AND EXHIBITORS

First row, left to right: John S. Martin, J. W. Andrews, Chas. Nixon, A. J. Fell. Second row, C. I. Fishel, W. W. Graves, L. J. Demberger, Chas. V. Keeler—all these breeders of White Wyandottes. Third row: R. G. Williams, J. F. Van Alstyne, Silver Laced; S. A. Howland, Buff; Levi A. Ayres, Columbian. Fourth row: H. B. Hark, Partridge; Fred C. Lisk, H. J. Riley, Dr. W. H. Humiston, White

LEADING RHODE ISLAND RED BREEDERS AND EXHIBITORS
First row, left to right: Lester Tompkins, Harold Tompkins, F. W. C. Almy, I. W. Bean. Second row: R. G. Buffington, G. W. Tracy, B. H. Scranton, T. L. Ricksecker. Third row: E. W. Mahood, A. A. Carver, E. H. Rucker, C. R. Baker.

The urgency of the need for the exhibitor himself to return home probably is the determining factor in most cases, and the fasting theory an excuse that relieves his mind of the feeling that he is not giving the birds proper treatment. What to do is sometimes difficult to decide. One exhibitor with a large string of valuable birds is apt to stay with them and leave them in the exhibition coops until just previous to delivery to the express company. Another may consider that his birds and interests at home need him most urgently, and therefore the show birds must now take whatever risks there may be in leaving it to the show management to see that they are properly cared for until started home.

At most fall fairs, and at many minor shows where the attendance is light on the last day and evening, it is the practice to begin to take out exhibits about the middle of the afternoon. This admits of getting the birds all started for home that day.

In general there is little trouble with birds being missent in returning from shows. An occasional error is unavoidable, and once in a long time there is what appears to be a deliberate case of thievery of a desired bird, but where birds of desirable quality which would lead one to take chances of wrongful appropriation of a coveted bird are as well known as is usually the case, the chances of detection are too great to be lightly risked. Most of the birds that are missent eventually get back to their rightful owners. Where a bird is lost the loss usually falls on the exhibitor, though in some cases the circumstances may be such that the express company or the management of the show can be held liable. It is a common practice for show managers to put in the rules of the show a disclaimer of liability for damage or loss of birds entered. The principal effect of this is to deter complaints, and to make intending complainants very careful to be sure they have a case before putting in a

LEADING ORPINGTON BREEDERS AND EXHIBITORS

First row, left to right: Ernest Kellerstrass, P. A. Cook, J. M. Williams, C. S. Byers. Second row: J. I. Lyle, Harold Rawnsley, Wm. Hobbs, G. E. Greenwood. Third row: C. H. Barnes, A. G. Goodacre, J. E. Church, R. E. Sandy. Below, Len Rawnsley.

claim for damages on account of damage, death or loss of birds. Where there is plain negligence or tolerance of conditions extraordinarily prejudicial to the health of birds and damage results, a show association cannot protect itself by making a rule to that effect, and associations do frequently pay for losses—chiefly of birds lost on the return home—when they lack evidence that they have done their part in due form.

The number of birds that as a result of being shown contract "temperature diseases" which developing at or after the show impair their usefulness for months and perhaps lead to death, is many times greater than the loss through misshipment or theft. For such damage and loss an association is in no measure liable. It is an ordinary risk of showing—and very largely dependent upon weather changes. Such effects are also much more common with birds that are shown at a number of shows and sometimes kept under very unfavorable conditions for a week or two between shows at which they are entered, than with birds that are shown but once and have proper attention. The fact that a bird which went to a show in prime condition is returned from it "roupy" does not prove that it was exposed to special contagion at the show. It indicates only that unfavorable conditions, with some lowering of vitality, made it a prey to germs usually everywhere present in small numbers, which normally it would resist. Experienced exhibitors at a show where the changes of outdoor temperature have been extreme and it has been impossible to keep the large hall always comfortable, either for birds or for people, often remark: "There will be a lot of good birds die on account of this."

The hall that has been heated more or less uniformly during a show always has the heat turned off at the close of the show. The opinion is very general among exhibitors that there is less risk of loss of birds in bad weather if they remain in the hall for some hours after it has cooled thoroughly than if they are shipped from the hall while it is warm. All this however, is largely conjectural. In most cases the birds after leaving the hall have to take the usual chances of riding next to steam pipes in express cars, of standing on trucks outdoors, or of remaining on wind-swept platforms until thoroughly chilled. That being the case it must often be doubtful whether the precautions taken to harden the birds have benefited them. There are times when with a blizzard raging as a show closes it is plainly wise to hold the birds until it moderates. Under all ordinary circumstances it is best to get the birds home as soon as possible. There is not really anything to choose between the hardships of remaining cooped at the showroom and the hardships of the journey.

Care of Birds After Return

Birds that have been shown should be quarantined upon their return long enough to insure that they have not brought with them germs of any disease which they would communicate to the flock. Nearly everyone admits that this rule ought to be followed. The fact that exhibitors on the whole totally disregard it shows quite conclusively how slight are the risks in this respect. Colds affecting the birds returned are common, but there are so few cases of epidemics brought home from shows, that the quarantine precaution is commonly disregarded with impunity. While that is the fact, and the risk is light, the exhibitor should bear in mind that the risk is always present and that every once in a while disease introduced in this way will go through a flock and cause very heavy loss. The risk to the bird shown cannot be entirely avoided. The risk to those at home can be.

After consideration of precaution against disease communicated to the flock the most imperative thing is to get the birds that have been shown in good breeding condition. Nearly all birds that have a tendency to put on fat, and that keep well and hearty through a show, come home a little overfat and soft. They should at once be given a diet and exercise that will reduce weight and restore energy. Nearly all birds have vitality and fertility more or less affected by showing, and their future usefulness as breeders depends on their being brought back to good breeding condition as promptly as possible. The owner can now disregard all the precautions which before showing he took to protect the bird from the slightest damage to plumage. It must take all chances of ordinary wear and tear both to feathers and to head parts.

On the average it is a matter of three or four weeks after birds return from a show before they are in good breeding condition. An occasional bird does not seem to have been affected at all, but many take much longer than the average to get back in form, and in some cases a bird does not show good breeding form (as evidenced by fertility and the vitality of its offspring) until after settled warm weather. No special methods of restoring breeding condition are used. All that can be done is to give regular good care and as good yard or range conditions as possible and wait for Nature to do her work. There are some cases, mostly in old birds of the heavier breeds, where the experience of showing apparently permanently spoils a bird for breeding. The risk is one that cannot be avoided in individual cases. But as a matter of general breeding policy much can be done to escape such results by not breeding at all from specimens which do not show good breeding form within a comparatively short time after being shown.

PROMINENT BLACK LANGSHAN EXHIBITORS
Left to right: M. S. Barker, L. E. Meyer, W. A. Meyer

CHAPTER XI

Analysis of Methods of Judging

"Points" in Standard Poultry—How Standard Descriptions Affect Judging Practice—The Per Cent Method of Cutting for Defects—Why It Is Impractical—Symmetry, Historically and Technically—How Misunderstanding of the Term Arose—Objections to Taking the Best Bird As the Standard—Improvement in Score-Card Judging

THE second chapter of this book gives the information about the judging of standard poultry needed by an exhibitor in selecting and fitting birds for exhibition. In this and following chapters the philosophy and evolution of methods of judging will be presented for those who wish a thorough understanding of the subject. While these chapters cover in greater detail topics treated in a general way in Chapter II the treatment of the subject for purposes of that chapter emphasizes various things which are not absolutely essential to the purposes of these later chapters, yet do help to a clearer understanding of them. The student of the subject as here presented should consider Chapter II preparatory to this more exhaustive treatment of the subject.

At the outset of this study it is desirable to get a true view of the services of the pioneers in developing methods of judging poultry and of the value of their ideas and methods. Considering a method of judging that has been long in use, or any theory regarding it that is widely accepted as having stood the test of time and use, we are quite apt to credit the originators of such ideas, theories and methods with superior sagacity and intuition. That seems quite natural and logical in the abstract, but it happens that in this matter of judging poultry some of the men who had most to do with devising and popularizing systematic judging of poultry lived a long time after it was established, and while their contemporaries duly honored them for their general services in the promotion of poultry interests they did not especially defer to their views in matters relating to judging. On the contrary, when some of them discovered faults in the very methods they had devised, and advocated improvements on them, they were unable to secure any general following. In other cases men of great influence in developing one method appeared quite unable to grasp simple principles in another.

So in a general way it may be said of the men to whom we owe the forms first given to systematic poultry judging that their ideas were as good as could be produced at that stage of the improvement of poultry, and in the circumstances of the time. Of no one who has since taken an interest in the matter can it be said that he has demonstrated an understanding of the subject which would suggest that in the same circumstances he would have done better. But the breeding of poultry has advanced a long way since the methods of judging in vogue were worked out, and—as we shall see—improvement in methods of judging has not kept pace with this progress. And the modern student of the subject who thoroughly informs himself of its history and the present conditions is unquestionably better qualified to critically analyze methods and theories of judging than were the men who originated them. He has the opportunity to see them applied to better birds, in stronger competition, by more competent judges.

FOREMOST POULTRY ARTISTS
Franklane L. Sewell, Arthur O. Schilling

What we need to do in studying judging today, is to recognize and retain the wise features of the early systems, and at the same time to recognize and reject their faults. It is neither to our credit nor to that of our predecessors for us to continue in their errors. At one of the meetings of the American Poultry Association where errors in a recent edition of the "Standard of Perfection" were under discussion, the late Mr. Fred L. Kimmey made this pithy statement: "Now, I am going back to the time the Standard started. We fools are doing what those fools did, AND WE ARE DOING IT NOW."

SUCCESSFUL BIG SHOW MANAGERS
Left to right: L. D. Howell, Robt. Seaman, Grand Central Palace Show, New York; W. B. Atherton, Boston; Theo. Hewes, Chicago, Milwaukee, Indianapolis

We shall see as this subject is presented analytically that, notwithstanding the tendency to extreme conservatism in the improvement of methods of judging, there has been consistent progress in certain definite directions, and that this indicates quite clearly the trend of future developments.

"Points" in Standard Poultry

A **point** of an object or animal is a particular character, feature or attribute. A **point** in computing scores in a competition is the unit of numerical valuation placed upon excellence or performance.

In the Standard formulated by the London Poultry Club, and reprinted in America, "with alterations and additions," in 1867, the description of each breed or closely like group is followed by a list of "points" for the breed or group of which the following is an example:

Points in Brahmas

Size	3
Color	4
Head and Comb	1
Wings. Primaries well tucked under secondaries	1
Legs, and feathering of ditto	1
Fluff	1
Symmetry	2
Condition	2
	15

It is easy to see from this how points were reckoned in judging—the method is so obvious that the early Standards had not a word in explanation of it. If a Brahma had what was regarded as good full size it was credited three points on that score. If color was considered good it had a credit of four on that score, and so on through the list. The maximum score possible was 15. The basis of judgment might be either an ideal in the mind of the judge or judges, or the best specimen—or best in the character under consideration—in the class. It is evident that with such a system of scoring the practice of considering values of the sections in the list in terms of fractions of the maximum allotted each would arise.

It is also easy to see how the idea that different valuations should be placed on the same section in different breeds and sometimes in different varieties of the same breed arose. A comparison of lists of points brings it out at once. Thus in the Dorkings there were three different lists of points, none like that for the Brahmas.

Points in Silver Gray Dorkings

Size	3
Color	3
Head and Comb	2
Legs, Feet and Toes	2
Symmetry	3
Condition	2
	15

Points in White Dorkings

Size	4
Purity of White Plumage	2
Head and Comb	2
Legs, Feet and Toes	2
Symmetry	3
Condition	2
	15

Points in Colored Dorkings

Size	5
Head and Comb	2
Legs, Feet and Toes	2
Symmetry	4
Condition	2
	15

Here there are three different maximum scores for size, and for color, and two different maximum scores for symmetry. These differences were not made according to the theory developed in later days of a careful appraisal of values in different sections, varieties and breeds in their ideal forms, but were crude adjustments of the points to the varieties as they were, to the popular ideas of value in different characters and breeds, and to the requirements of the very common practice of having different varieties, and sometimes different breeds compete in the same class. White had to be handicapped in competition with the parti-colored silver gray, therefore a point had to be taken from color on the white variety and put somewhere else. The Colored Dorking of the time was really any other colored Dorking—specimens neither silver gray nor white, and with no excellence in color that might entitle them to credit in that section. So color is omitted and extra credit given size and symmetry.

Comparing some of the other lists of points we find everywhere the same practice of giving large credits to a few conspicuous points, and of singling out certain characters as most worthy of the attention of the breeder and judge. The Black Spanish afford one of the extreme cases of this kind.

Points in Spanish Fowls

Comb	2
Face	3
Ear Lobe	3
Purity of White Face and Ear Lobe	2
Symmetry	3
Condition of Plumage	2
	15

Here 10 of the 15 points are awarded to head and adjuncts and neither size nor color of plumage are considered at all—no matter how good a bird might be in those respects it received no credit. Nearly every list of points in the book will furnish similar evidence of how the practice of placing high valuations on some points and ignoring others arose. Space will permit only the mention of a few of the worst cases:

In White-Crested Black Polish 9 of the 15 credit points had to go to head and crest. In the Houdan the fifth claw had a possible credit of 1 point in the total of 15, otherwise, legs and toes had none at all. In La Fleche the deaf ear had a whole point to itself. In the Rouen Duck the shape and color of the bill might count for 3 points, one-fifth of the total possible credits.

It was by comparison judging with such a system of credits for points that poultry were judged at American shows from the time this Standard appeared until the publication of the first edition of the American Poultry Association's Standard of Excellence in 1875.

How Standard Descriptions Affected Scoring

The tabulations of points that have been given show how in judging of birds attention was fixed on a few points, and it has been explained that this practice became established as the result of conditions developing when there were no written or printed standards, and each judge or committee of judges made standards to suit them. Seeing only the list of points from the early standards, one familiar with the "Standard of Perfection" would very naturally suppose that the descriptions were

also deficient—treating the few points mentioned in the scales in elaborate detail, and either omitting the others or saying little of them. On the contrary, the outline of the descriptions was much the same as that in use in our Standard now, the style of statement was that which is still used, and the specifications described the breeds and varieties as they were at the time quite as well as the Standard does today.

To show this, and for purposes of references in further discussion of the subject the descriptions of the Light Brahma and the White Leghorn are here reprinted from the 1867 Standard, and in parallel columns the

BRAHMAS 1867
General Shape
The Cock

Beak—Very strong, taper and well curved.
Comb—Pea, small, low in front and firm on the head without falling over to either side, distinctly divided so as to have the appearance of three small combs joined together in the lower part and back, the largest in the middle, each part slightly and evenly serrated.
Head—Small and slender.
Eyes—Prominent and bright.
Deaf Ear—Large and pendant.
Wattles—Small, well rounded on the lower edge.
Neck—Long, neatly curved, slender near the head, the juncture very distinct, hackle full and abundant, flowing well over the shoulders.
Breast—Very full, broad and round; carried well forward.
Back—Short, broad, flat betwixt the shoulders, saddle feathers very abundant.
Wings—Small, the primaries doubled well under the secondaries; the points covered by the saddle feathers.
Tail—Small; carried very upright, the higher feathers spreading out laterally. Broad.
Tail Coverts—Very abundant, soft and curved over the tail.
Thighs—Very large and strong; abundantly covered with very soft, fluffy feathers, curving inward around the hock so as to hide the joint from view. Vulture hocks, that is, those with hard, stiff feathers projecting in a straight line beyond the joint, are objectionable but not a disqualification.
Fluff—Very abundant and soft, covering the hind parts, and standing out behind the thighs, giving the bird a very broad and deep appearance behind.
Legs—Rather strong and large; standing well apart, very abundantly feathered down the outside to the end of the toes.
Toes—Straight and strong; the outer and middle toe being abundantly feathered.
Carriage—Very upright and strutting.

The Hen

Beak—Strong, curved and taper.
Comb—Pea, very small, and low, placed in front of the head, and having the appearance of three very small serrated combs pressed together, the largest in the middle.
Head—Small and slender.
Eye—Prominent and bright.
Deaf Ear—Large and pendant.
Wattles—Small, rounded on the lower edge.
Neck—Rather short, neatly curved, slender near the head, the juncture very distinct, full and broad in the lower part; the feathers reaching well onto the shoulders.
Breast—Very deep, round, broad and prominent.
Back—Broad and short; the feathers of the neck reaching to betwixt the shoulders, and abundance of soft, broad feathers rising to the tail.
Wings—Small; the bow covered by the breast feathers, the primaries doubled well under the secondaries, the points of the wings clipped well into the abundance of soft feathers and fluff.
Tail—Small; very upright, almost buried in the soft rump of feathers.
Thighs—Strong and well covered with very soft feathers, curving round the hock so as to hide the joint from view. Vulture hocks are objectionable, but not a disqualification.
Fluff—Very abundant and soft, standing out about the hind parts and thighs, giving the bird a very broad and deep appearance behind.
Legs—Short, very strong, wide apart, abundantly feathered on the outside to the toes.
Toes—Straight and strong, the outer and middle toe being well feathered.
Carriage—Low in comparison to the cock.

LIGHT BRAHMAS
Color of Cock

Comb, Face, Deaf Ear and Wattles—Rich bright red.
Head—White.
Neck—White, with a distinct black stripe down the center of each feather.
Breast, Underpart of Body and Thighs—White.
Back, and Shoulder Coverts—White.
Saddle—White striped with black.
Wing Bow and Coverts—White.
Wing Primaries—Black.
Wing Secondaries—White on outside web, black on inside web.
Tail—Black.
Tail Coverts—Glossy greenish black; lesser coverts silvered on the edge.
Legs—Scales bright yellow; feathers white, slightly mottled with black.

Color of Hen

Comb, Face, Deaf Ear and Wattles—Bright red.
Head—White.
Neck—White, distinctly striped down the middle of each feather with rich black.
Breast and Back—White.
Wing—White, the primaries above being black.
Tail—Black, the two highest or deck feathers edged with white.
Thighs and Fluff—White.
Legs—Bright rich yellow; feathers white, slightly mottled with black.

DISQUALIFICATIONS

Birds not matching in the pen, combs not uniform in the pen, or falling over to one side, crooked backs, legs not feathered to the toes, or of any other color except yellow, or dusky yellow.

AMERICAN STANDARD OF EXCELLENCE 1875
LIGHT BRAHMAS
The Cock

Head—Broad, of medium length, slightly projecting over the eyes; color of plumage, white; Eyes, large and bright; Beak, short, stout, and, in color, yellow, with a dark stripe down the upper mandible.
Comb—Pea, small, lower in front and rear than in the centre; firm on the head and distinctly divided, having the appearance of three small combs pressed together, the largest and highest in the middle, and each part slightly and evenly serrated. Color, rich, bright red.
Wattles and Ear Lobes—Of equal length, the wattles being well rounded. Color, rich, bright red.
Neck—Rather long and well arched, the hackle flowing well over the shoulders; plumage of the upper part white, the lower two-thirds being distinctly striped with black, the stripe tapering to a point at the extremity of the feather.
Back—Broad, flat between the shoulders, and as long as is consistent with the size and symmetrical proportions of the bird; saddle feathers well developed—surface color, white; undercolor, either white or bluish white.
Breast and Body—Breast full, broad and round, and carried well forward. Body, round at the sides, and deep;—color of both, white.
Wings—Small, the bows covered by the breast feathers,—color of bows white; the primaries closely folded under the secondaries;—color of primaries, black or nearly so;—color of secondaries, white on the outer web, and black on the inner web.
Tail—Full, well spread, carried tolerably upright, and well filled underneath with rich curling feathers; color of tail, black;—Sickle feathers, short and spreading laterally, and in color, black;—coverts, glossy greenish black;—lesser coverts, black, with white edge.
Fluff—Abundant and soft, giving the bird a broad appearance behind—color, white.
Legs and Toes—Thighs, strong and well covered with soft white feathers;—Shanks strong, standing well apart, of medium length, and well feathered on the outside; color of scales, yellow, inside of the legs, yellow or reddish yellow; shank feathers white, or white mottled with black;—Toes, straight and strong, the outer toes well feathered to the ends thereof; the feathering of middle toes optional with breeders.
Carriage—Bold and attractive.

The Hen

Head—Broad, of medium length, and slightly projecting over the eyes; plumage white;—Beak, short and stout—color, yellow, with or without dark stripe down the upper mandible;—Eyes, large and bright.
Comb—Pea, small and low, with delicate but distinctly defined serrations, firm and even upon the head, and in color, bright red.
Wattles and Ear Lobes—Wattles, exceedingly small;—Ear lobes well developed,—color, rich red.
Neck—Of medium length and well arched,—hackle feathers, white, with a broad black stripe down the center, the edge of the black running nearly parallel with the edge of the feather, and reaching well over the shoulders.
Back—Broad, flat between the shoulders, and as long as is consistent with the size and symmetrical beauty of the bird; feathers broad and soft, and rising to the tail; surface color, white,—under color, either white or bluish-white.
Breast and Body—Breast, full, broad and round, and carried well forward;—Body, round at the sides, and deep;—color of both, white.
Wings—Small, the bows covered by the breast feathers; the primaries smoothly folded under the secondaries;—color of primaries, black, or nearly so,—color of secondaries, white on the outer web, and black on the inner web.
Tail—Rather small and spreading,—color black;— the two highest, or main tail feathers, edged with white;—tail coverts, black edged with white.
Fluff—Abundant and soft, giving the bird a broad appearance behind—color, white.
Legs and Toes—Thighs, strong, and abundantly covered with soft, white feathers;—Shanks, strong, standing well apart, and well feathered on the outside with white feathers, or white mottled with black;—Toes, straight and strong, the outer toes being well feathered to the ends,—the feathering of middle toes optional with breeders.
Carriage—Low, in comparison with that of the cock.

DISQUALIFICATIONS*

Birds not matching in the show pen; comb falling over to either side; twisted feathers in the wings; shanks not feathered down the outer sides, and to the extremities of the outer toes, or of any other color than yellow; vulture hocks; undercolor any other than white or bluish white; crooked backs; wry tails; cocks not weighing nine pounds; hens not weighing seven and a half pounds; cockerels not weighing seven and a half pounds; pullets not weighing six pounds.

*Note—In the 1875 Standard the list of disqualifications precedes the descriptions. The position is changed here for comparison with the older Standard.

descriptions of them in the first edition of the American Standard of Excellence, 1875. The Brahma description in the American book of 1867 is taken verbatim from the original English version. The White Leghorn description originated here, Leghorns being little if at all known in England at that time. It is in order to note here that the White Leghorn was the first variety of the breed to be recognized in the making of standards, and that it was given first place in the list of varieties in the 1875 Standard. Later the Brown Leghorn was given first place—probably because it was the most popular, but possibly because it was supposed to be the original variety of the breed.

Comparison of the specifications of these Standards eight years apart, and representing the ideas of two quite distinct groups of breeders and judges, will bring up many questions in the mind of the reader, but we are more concerned here with the form and the general aspects of the different standards, and in general an adequate treatment of the evolution of standards for particular breeds can be undertaken only in a book especially devoted to one breed. There is, however, one point here where the change is so extreme that it should be explained:— that is the change from the white face to the red face in White Leghorns. No one familiar with the references to cross-bred and alleged purebred fowls in the fifties and sixties of the last century can doubt that White Leghorns of this description contained a large admixture of Spanish blood. Such fowls were sometimes called White Leghorn, or White Italian, and sometimes White Spanish. When the Leghorn grew in popularity, and the Black Spanish declined, it was natural to eliminate the characteristics of the latter, and also traces of White Hamburg crosses, as quickly as possible.

Turning again to consideration of the relation between the descriptions and the lists of points in the first Standard, the list of Brahma points in it was given on page 154. The list of White Leghorn Points is:

Points in White Leghorns

Comb	2
Face and Ear Lobe	3
Purity of Plumage	3
Size	3
Symmetry	2
Condition	2
	15

Now, comparing the number of different characters, or attributes, or groups of such, included in the description of a bird, with the number included in the list of **points** of merit or excellence, it is evident at a glance that there are several times as many of the points of description as of the points of scoring. Counting them it appears that in the shape description of the Brahma male there are 17 sections, in the shape description of the female 16 sections, in the color descriptions of the male 12 sections, and in the color description of the female 8 sections, while in the list of points there are only 8 sections. In the White Leghorns, male and female, there are 16 sections in the description, shape and color being described together in each, and only 6 sections in the list of points.

In comparing the descriptive and recording lists, it is seen at once that the discrepancy is actually very much greater than is indicated by the comparison of numbers of sections, and that there is very little correspondence between the two lists. In each list of points there are 3 sections that do not appear at all in the descriptions in this standard—size, symmetry, and condition, and in each case they take collectively 7 of the 15 points, leaving only 8 credit points, divided in the Brahmas among 5 sec-

---

WHITE LEGHORN 1867
The Cock

Beak—Rather long and stout.
Comb—Bright red, large, erect, single, straight, and free from twists or falling over to either side, deeply serrated, extending well back over the head and free from side sprigs or excrescences.
Head—Short and deep.
Eyes—Large and full.
Face—Opaque white, free from wrinkles or folds.
Ear Lobes—Pure opaque white, rather pendant, thin, fitting close to the head, smooth, and free from folds and wrinkles.
Wattles—Bright red, long, thin and pendulous.
Neck—Long and well hackled.
Breast—Full, round, and carried well forward.
Body—Rather square, but heaviest forward.
Wings—Large, and carried well up.
Tail—Large and full, carried very upright; sickle feathers large and well curved.
Thighs—Medium length, and rather slender.
Legs—Long, white or yellow, yellow preferable.
Plumage—Pure white throughout. The neck hackle and saddle may be tinged with gold or straw color.
Carriage—Upright and pleasing.

Disqualifications in White Leghorn Cocks

Comb falling over on one side, or twisted; decided red about the face; plumage any other color than pure white, with a golden tinge on neck, hackle and saddle.

The Hen

Beak—Rather long and stout.
Comb—Bright red, large, single and drooping to one side, serrated and free from side sprigs.
Head—Short and deep.
Eyes—Large and full.
Face—Opaque white, free from wrinkles or folds.
Ear Lobes—Pure opaque white, rather pendant, thin, and fitting close to the head, smooth, and free from folds or wrinkles.
Wattles—Bright red, thin, and rounded on lower edge.
Neck—Long and graceful.
Breast—Full and round.
Body—Deep, broader in front than back.
Wings—Large, and well tucked up.
Tail—Large and full, carried very upright, feathers broad.
Thighs—Rather long and slender.
Legs—Long, white or yellow, yellow preferred.
Plumage—Pure white, the more free from yellow tinge the better.
Carriage—Not as upright as the cock.

Disqualifications in White Leghorn Hens

Duplicature of comb, any red about the ear lobe or face, prick-combed, plumage the least marked, or any other color than white.

AMERICAN STANDARD OF EXCELLENCE 1875
WHITE LEGHORNS
The Cock

Head—Short and deep, color, pure white;—Beak, yellow, rather long and stout;—Eyes, full and bright;—Face, red, and free from wrinkles or folds.
Comb—Red, of medium size, erect, firmly fixed on the head, single, straight, deeply serrated (having but five or six points—five preferred), extending well over the back of the head, and free from twists, side sprigs, or excrescences.
Ear Lobes and Wattles—Ear lobes, white or creamy white, fitting close to the head, and rather pendant, smooth and thin, and free from folds or wrinkles;—Wattles, red, long and pendulous.
Neck—Long, well arched, the hackle abundant, and flowing well over the shoulders,—color, pure white.
Back—Of medium length and width—color, white, as free as possible from yellowish tinge.
Breast and Body—Breast, full, round, and carried well forward;—Body, rather broad, but heaviest forward—color, white.
Wings—Large and well folded—color white.
Tail—Large, full and somewhat expanded, and carried very upright; sickle feathers, large and well curved;—tail coverts, abundant —color, pure white.
Legs—Thighs, of medium length and rather slender; plumage, white;—Shanks, long, and in color, bright yellow.
Carriage—Upright and proud.

The Hen

Head—Of medium size; color, white;—Beak, rather long and stout, and in color, yellow;—Eyes, red, full and bright;—Face, red and free from wrinkles or folds.
Comb—Red, of medium size, single, drooping to one side, evenly serrated, and free from side sprigs.
Ear Lobes and Wattles—Ear lobes, white or creamy white, fitting close to the head, rather pendant, smooth and thin, and free from folds and wrinkles;—Wattles, bright red, thin, and well rounded.
Neck—Long and graceful, and pure white in color.
Back—Of medium length, full, and in color, pure white.
Breast and Body—Breast, full, and round, and in color, pure white. Body, deep, broader in front than in the rear. Color, white.
Wings—Long, well folded, and clear white.
Tail—Upright, full and long, and in color, pure white.
Legs—Thighs, of medium length, rather slender, and in color, white;—Shanks, long and slender, and in color, bright yellow.
Carriage—Not so upright as that of the cock.

DISQUALIFICATIONS

Comb falling over to either side, or twisted in cocks, or pricked or duplicate in hens; red ear lobes; legs other than yellow; plumage other than white, or with colored feathers in any part thereof; crooked backs; wry tails.

tions, and in the Leghorns among 3 sections. Attempts to judge by such a crude system of scoring applied to comparatively elaborate descriptive standards could not give satisfactory results. When judges made awards right in close competition, using this system, results were creditable to their own discernment, or to their skill in supplementing its deficiencies. It was inevitable that there should be a general lack of uniformity in the work of those using it. Such records as were made by it were of little value to anyone, and so far as is now known they were not made in any regular form, or used for any purpose but the guidance of the judges in making awards.

American judges and breeders soon became dissatisfied with some features of the first Standard, and very much so with the conditions of judging under it. There was also some confusion because of conflicting versions of the printed Standard. To remedy these conditions, and issue a better Standard, and establish better and more uniform judging, was the prime object of the organization of the American Poultry Association. Its revision of the original Standard published in 1875 was designed to be used with a practical system of scoring, and of making permanent records of the work of the judge in placing awards in competition. The plan devised was not perfect, but it was a wonderful improvement over the old method, and it was the first of the several factors that have been generally and highly effective in the development of the general high quality, the uniformity of type, and the perfection of finish found in America standard-bred poultry today.

In its main features, if not in its entirety, the system was devised by I. K. Felch, who was then and for long after the leading poultry judge in America, and the only one whose major occupation was judging poultry. The system he created grew out of his efforts to improve upon the first crude methods of scoring.

Briefly, the principal change and improvement consisted in making the sections for recording scores correspond with the sections of the description, making the maximum number of points large enough to meet the requirement of cutting for many different faults, and applying the system of cutting for defects directly and positively instead of by reduction of the credits given. Here again the features of the methods introduced were not the result of elaborate and ingenious calculation to make all the parts of the system fit perfectly and to secure consistency throughout, but were simply the easily made improvements on the old practice which experience had suggested, and the adaptations of common practices in marking and scoring for other purposes—specifically in the grading of the work of pupils at school and in the marking of contestants in academic contests. The suggestions to judges in this Standard and the lists of points for the two breeds and varieties that have been used as examples are given herewith:

SUGGESTIONS TO JUDGES
To Be Considered in Applying the Standard

The American Poultry Association, in placing the American Standard of Excellence before the Breeders and Fanciers of the United States and Canada, recommend that, in its application, Judges shall determine the merits of competing specimens by a careful examination of all the points named, commencing with "Symmetry," and following the schedule through in the order named in the table of values, and deducting such per centum for defects, as may be apparent, from the full value of a perfect bird.

In all competitions where "Size and Weight" are considered points of merit, the largest and heaviest specimen in its class shall be deemed a perfect bird; and other contestants shall be rated comparatively, losing as many points to the pound as are designated in the scale for such variety.

In adjudicating the merits of Light Brahmas, no value shall be placed on middle toe feathering, or any preference given to either white or bluish undercolor of the plumage.

Combs turning slightly over to one side of the head, but firm in position, while considered objectionable, shall not be a disqualification, under the specification in the Standard, "combs falling over to either side."

It is respectfully recommended that State and County Societies appoint but one Judge, and one alternate to each class, or variety in each class, and that their names be published in the regular premium list.

It is also recommended that eighty-five points shall be the minimum value of a bird to which a first prize shall be awarded.

Points in Light Brahmas

Symmetry	10
Size and Weight	13
Condition	8
Head	5
Comb	8
Wattles and Ear Lobes	5
Neck	10
Back	7
Breast and Body	7
Wings	8
Tail	7
Fluff	5
Legs and Toes	7
	100

Comparisons in Size and Weight shall be in the ration of 2 points to the pound.

Points in White Leghorns

Symmetry	10
Size	10
Condition	10
Head	7
Comb	15
Ear Lobes and Wattles	15
Neck	5
Back	5
Breast and Body	8
Wings	5
Tail	5
Legs	5
	100

I have quoted the suggestions to judges in full, although there are several paragraphs not relating to matters to be considered here, because they are the beginnings of explanations of the use of the Standard, and because, brief as they are, they constituted all the general guidance the judge or breeder had in interpreting the Standard for thirteen years after this one was issued. At the time this 1875 Standard was made and what has since been called the "official" score card was devised and the methods of using it worked out, the common method of marking pupils in schools in this country, in both recitations and examinations, was on the basis of a mark of 100 for a perfect recitation or paper, with deductions for errors.

In examinations of various kinds it was a common practice to give a list of ten topics or questions, each of which counted 10 in the mark for the work. It was also a common practice to excuse from examinatons at the end of a month or term pupils who had a grade of 85 in their daily recitations or in examinations on their work for short periods. In contests in essay writing, oratory, etc., it was a common practice to mark the best—in the opinion of the judge—100, and to grade other performances accordingly. We can see in these things the origin of the idea that a bird scoring less than 85 points ought not to receive a first prize, the origin of the idea of making the largest bird in the class the standard for size and weight in judging it, and the origin of the idea of dividing the characters or parts of a bird into 10 sections for consideration in judging it.

The list of points for Brahmas contains 13 items. But the first three of these relate to general characters or conditions. In consideration of the specimen in detail it is divided into 10 sections. Table No. 1, on page 172, giving a list of points for other breeds and varieties in this Standard, shows how the other breeds came to have more or less than 10 sections of detail. It was simply a matter of a prominent character more or less. In the allotment of points to the different detail sections shape and color were lumped together. Theoretically, at least, this was a backward step. It was later corrected by making the division regarded as appropriate to each.

The principle and procedure followed in the division

158   POULTRY FOR EXHIBITION

of the maximum of 100 points among the sections to be marked in making the score were the same as in the making of the first Standard. Conspicuous and favorite features were given high values, so were some general characters, and what points were left were divided with some approach to uniformity among the other sections. Thus the neck of the Brahma, which was the feature to which the breeders of that time were devoting most attention, was given 10 points while the tail was given only 7. By referring to Table No. 2, page 173, the reader will see that eventually neck, back, wings, tail and breast were given the same valuation of 10 points. Similarly in the White Leghorn, the comb was given 15 points, the ear lobes and wattles another 15 points, while most of the other detail sections were given a valuation of only 5 points. Table No. 2 shows in this case also how the valuations have since been equalized.

The Per Cent Method of Cutting for Defects

The instructions as to the method of scoring in the 1875 Standard were not clear. They do not plainly show that the per centum cut was to be computed upon the valuation of a section in the list of values of points, and that the deduction "from the full value of a perfect bird" referred to the method of computing the score after the cuts had been made. As the mode of practice—the theory of it—was frequently explained both verbally and in print by judges, there was no difficulty in understanding it.

But one has only to look at the lists of points to appreciate the practical difficulties of the actual application of the per cent theory to the cuts for defects. An estimate that a section or character lacked ten per cent of perfection applied to a list of sections of different numerical values gives a lot of cuts in dissimilar fractions, which cannot

LEADING LEGHORN BREEDERS AND EXHIBITORS
These are all breeders of White Leghorns except as otherwise stated. First row, left to right: D. W. Young, Edward Knapp, C. H. Wyckoff, Irving F. Rice. Second row: R. J Thompson, Bion H. Naldrett, C. S. Phelps, P. Sciarra. Third row: Wm. M. T. Sherwood, Geo. B. Ferris, Frank Gloeckl, Howard L. Goss, Black Leghorn

be quickly reduced to a common denominator. It is not necessary to give examples here in detail. Any child who understands the simplest operations in fractions and percentage can see by a glance at one of these lists of points that only a mathematical prodigy could calculate values of defects in this way. No one in practice under the ordinary conditions of judging could apply the rule regularly and consistently. Judges, however, did undertake to compute the values of the most common faults and degrees of fault by applying the per cent rule to the valuations given different sections, and the result was that each gradually established his own scales of specific cuts—some cutting much harder for certain defects, or harder all along the line than others. The fact that each judge at this time divided points to shape and color in each section to suit himself tended to make still greater variations. There was also great difference of opinion as to the proper way of judging for symmetry.

"Symmetry" Historically and Technically

"Symmetry" in the 1867 Standard plainly means either the whole shape of a bird, or such points of shape as were not especially mentioned in the list of points. In the descriptions of breeds and varieties the shape of many characters is specified, as is the "carriage," but in the lists of points there may or may not be reference to shape of detailed sections, usually there is not, though often the form of the statement would seem to imply that the credit was for good shape.

Examining the list of points in Brahmas in this connection, it is seen that there is hardly an item where the application is not doubtful. Thus if we consider that "Color" means all color on the specimen, the next four points will appear to refer only to shape in the characters mentioned. But if we assume that such is the case we get the absurd result of head and comb, wings, legs and fluff, given each 1 point of special valuation, while all other sections combined have only 2 points. On the other hand, if we assume that the points given the four sections mentioned apply in part to color, we get into a similar difficulty there.

When we examine the White Leghorn schedule to find what "symmetry" applies to in it, we find that it appears to apply to shape of everything but the head. In the Dorking it appears to apply to shape of everything but head and feet. So we might go through many varieties without finding a definite meaning for "Symmetry"; but it happens that in the list of points in Black Hamburgs "Shape" is used instead of "symmetry"; that in the list for Crevecoeurs "Shape and symmetry" are coupled together and given the number of points usually accorded "Symmetry"; and that in a number of other varieties the two terms are used interchangeably. A critical consideration of the use of the term in the first Standards leaves no doubt that "symmetry" there meant shape both as applying to descriptive shape points generally, and in the broad sense of typical shape.

In the 1867 Standard of Excellence "Symmetry" was placed at the head of every list of points, and given the uniform valuation of 10, while shape in every section described in detail had such proportion of the points given that section as each judge saw fit to give it. Under the general instructions as to the practice of scoring, the judge was to cut for faults in "Symmetry," and then cut for shape faults in each and every section as he reached it. The Glossary of Technical Terms defined it thus:

"Symmetry—perfection of proportion—often confounded with carriage, but quite distinct, as a bird may be

PROMINENT LEGHORN BREEDERS AND EXHIBITORS
First row, left to right: Wm. F. Brace, S. C. Brown; Aug. D. Arnold, S. C. Buff; E. E. Emerson, S. C. White; W. W. Kulp, R. C. Brown. Second row: E. E. Carter, S. C. Brown; W. H. Wiebke, H. V. Tormohlen, W. G. Warnock, S. C. Brown

nearly perfect in his proportions, and yet 'carry' himself awkwardly."

The distinction which it is here sought to explain is one that the ordinary mind does not grasp, and that those who think that they apprehend have never succeeded in conveying to others. Few things have been more exhaustively explained with as little result. To most minds it appears that due consideration of all points of shape, section by section, necessarily includes or results in due consideration of the sections collectively, and that if the results of the detailed consideration of shape do not agree, one or both must be wrong. To most of those who had to consider the matter it appeared that to cut for "symmetry" and also cut according to instructions for shape faults in the detailed sections was to "punish" shape faults twice.

From what has been said of the use of the term in the first Standards, and of the methods by which the systems were evolved, it seems quite plain that the inclusion of "Symmetry" in the list of points in the 1875 Standard, which otherwise fully provided for **Shape** in every section, was an error. Referring to Table No. 2, page 173, the reader may note that in all but a few neglected breeds the valuation of "Symmetry" in the list of points has been reduced from 10 to 4. As will be fully explained farther on, this has come about because in the system of scoring symmetry is not an essential section and so can be robbed of its points of valuation as the need of an additional point elsewhere arises without creating either theoretical or actual difficulties.

In common judging practice the custom early arose of giving all birds scored a light cut on "symmetry" and making this heavier when a specimen was conspicuously off type for its breed. In most cases, however, the cut for symmetry was uniform and perfunctory. I have seen many judges mark a cut of ½ for symmetry on every card before they had looked at the class at all. They justified this on the ground that they had never seen a specimen so perfect that it did not merit that much of a cut. I have also seen judges in scoring a specimen where their estimates of the values of faults in several sections happened to come about halfway between two customary specific cuts—as ½ and 1—give the lower cut in each section and then increase their cut on symmetry to make the final score what it should be. This procedure too has a certain amount of reasonableness in it, though it cannot be said to be in the line of accuracy.

The "Best Bird" As a Standard

The 1875 Standard provided that in the matter of size and weight the largest bird in a class should be considered perfect, and the other should be cut at the rate of 2 points to the pound for the differences between its weight and theirs. Such a provision was so directly contrary to the purposes of a Standard, and its bad effects could be so easily foreseen, that its acceptance can only be explained on the ground that it had only hasty consideration—as we know has often been the case with errors in later days. As it applied to weights it was corrected in the 1883 edition of the Standard, but its other possible applications continued long after and caused a great deal of unsatisfactory work in score-card judging.

Many good judges are unconsciously influenced in their work by the impression made upon their minds by the birds they see before them, or have recently seen. Their mental standards for a breed or variety tend to go up and down with the quality they have seen latest. As the illustration of standard poultry has been developed in the last quarter of a century pictures of good models and of sections superior in finish help to correct this tendency. Also breeders and judges everywhere are much more familiar with the best, and near-best models than they were in the earlier years of the Standard. It was not until 1905 that the Standard itself showed ideal profiles. A few crude outlines of profiles were put in the first edition of the 1888 Standard, but they were so unsatisfactory that the edition was declared obsolete.

For fully twenty years after the introduction of score-card judging illustrations of ideal birds were unknown, and both breeders and judges were more influenced by the peculiarities of the most attractive specimens they saw than has been generally the case since ideal illustrations appeared in the Standard. This is one of the facts that needs especially to be considered in viewing the faults of score-card judging in the period when it was used exclusively in America.

Improvement in Score-Card Judging

Aside from the giving of definite standard weights to a number of breeds, and the abolition of the rule that the largest bird should fix the standard of size and weight for his class, no material change affecting judging was made in the Standard until 1888. Then the points for shape and color in the detailed sections were divided, each class of characters being assigned its definite number. The general tendency was to divide them equally, but in operation this was modified slightly by the number of possible faults under each in any particular section. As in the earlier Standards many sections had small, odd numbers representing their total valuation, it became necessary to find extra points for them. So these were drawn systematically from the three general sections—Symmetry, Weight and Condition.

By this time the authority of the Standard and the general popularity of the score-card method of judging—notwithstanding the obvious faults of both—were sufficiently established to warrant the American Poultry Associaton taking a little different attitude toward judging. Instead of the short list of "Suggestions to Judges," this book contained several pages of "Instructions to Judges." The points of most general importance in these were: the regulations in regard to applying standard weights; the provision that where no weights were required size should govern to some extent in the application of awards, but not to the prejudice of any other quality; the requirement that judges should mark cuts for shape and color separately; and the raising of the minimum scores entitling birds to prizes to 88 points for a first prize, and 85 points as the lowest score that should be given any prize. These changes were in the right direction, but fell far short of what was really necessary in order to make standard judging keep pace with standard breeding.

CHAPTER XII

Philosophy of Judging and Relation of Judging Systems

The Decimal System—Equality of Value in Characters—Actual Process of Scoring—Return to Comparison Judging—Analysis of the Features of the Comparison and Score-Card Methods Showing the Limitations of Each Which Prevent Its Acceptance to the Complete and Permanent Exclusion of the Other

ONE immediate result of the action of the American Poultry Association in withdrawing the illustrated edition of the 1888 Revised Standard from sale, declaring it obsolete, and substituting an edition minus the illustrations, was the publication of "The Philosophy of Judging" by I. K. Felch and H. S. Babcock, with illustrations by J. Henry Lee, the best known poultry artist of the time. This book was in its introduction a protest against the action of the Association, and in the main an exposition of the "Decimal System" of score-card judging followed by an apparently unintended demonstration of the fact that the "scale of points" as used in the Standards of the American Poultry Association, is not a factor in score-card judging as commonly practiced.

The first edition of the 1888 Standard besides giving profile outlines of a number of breeds (not nearly all) had provided in the instructions to judges, that in considering "Symmetry" the judge should be guided by these profile outlines. These outlines were not particularly good even for their time, and for the judge to follow his own interpretations, or the common interpretations, of correct standard shape, section by section, and then mark symmetry by comparison with them was absurd and could only lead to confusion. The idea of illustrations in the Standard was sound, but the method of carrying it out was defective. The illustrations in "The Philosophy of Judging" showed the color pattern in the outlines, and are rather better than the outlines in the obsolete Standard; also nearly all breeds were illustrated. However expert breeders of the time may have regarded these illustrations, they were greatly appreciated by novices and amateurs generally as the only available series of the kind, and supposed to be representative of the ideals of three eminent authorities.

H. S. BABCOCK
Author with I. K. Felch of "The Philosophy of Judging"

The "Decimal System" of Score-Card Judging

As has been pointed out, the regulation system of score-card judging was in reality a misapplication of the decimal plan of dividing and valuing sections, brought about by the difficulties of combining old and new ideas. With further consideration and experience in judging by his original plan, Mr. Felch and others had reached the conclusion that a different division of the sections to be considered could be made in such a way that the total number of sections to be considered would be 10, and that the division could not be made in such a way that each of these sections could appropriately be given a value of 10 in the scale of points. No argument is needed to show that such results are within the bounds of possibility—and I venture to say that if custom and usage had not fixed in fanciers' minds the idea of unequal valuation of characters which, as has been shown, had its origin in the partiality of judges and breeders in England for overdevelopment of peculiar or conspicuous characters, the idea that all characters are of equal value would have been accepted in America without hesitation. I base this assertion on the fact, which will be fully brought out in this discussion of judging, that, aside from the theoretical application of the idea of inequality of values in sections, and in characters, in the division of the scale of points in the Standard, all Standard breeding and judging in America is based on the principle of **Equality of Value in Characters.** Sections and characters do not necessarily correspond, but in this matter the ideas about primitive characters determined the more advanced ideas of the values of sections among a group of breeders and judges who took a different view of the value of characters.

J. HENRY LEE
Regarded as the leading American poultry artist until the arrival of F. L. Sewell

B. N. PIERCE
Prominent judge, editor and poultry artist

It was shown that the Brahmas which furnished the basis of arrangement of the official score card were given 10 detail sections, and the three general sections made 13 in all. The first problem in making a strictly decimal score-card system was to reduce these thirteen sections on the score card to 10. This was accomplished by omitting "symmetry" as superfluous; by combining "weight" and "condition" as one section (on the card); and by making two sections of the head and appendages instead of three. No good reason, other than that it arbitrarily met the requirements of the plan, could be given for considering weight and condition as one section corresponding to the detailed sections; and this created a prejudice against the plan from the outset. In all the other sections the 10 points allotted to a section were equally divided between "Form" and "Color." Thus:

Decimal Scale

Weight and Condition	Weight 5	Condition 5	Total	10
Comb and Crest	Form 5	Color 5	"	10
Head with Adjuncts	" 5	" 5	"	10
Neck	" 5	" 5	"	10
Back	" 5	" 5	"	10
Breast	" 5	" 5	"	10
Body—including fluff	" 5	" 5	"	10
Wings	" 5	" 5	"	10
Tail	" 5	" 5	"	10
Legs and Feet	" 5	" 5	"	10
Total	50	50		100

The advantages claimed for this division of sections and points were that applyng it equally to all breeds and varieties did away with confusion and error arising out of differences in the "scales of points" for different breeds and varieties; and simplified and standardized the process of cutting for faults, giving the fractional cuts always the same denominator or a series of denominators easily reduced to a common denominator at sight.

The opposition to the decimal system of scoring was based principally on the theory that by the method of percentage cuts based on the numerical valuations given sections the same results could not possibly be obtained when two different "scales of points" were used. This would have been the result if the percentage system were actually used in both cases. The truth was that it was not used in either. In demonstrating the difference in the use of the two "scales of points" when the percentage method of estimating the values of cuts is used, the authors of the "Philosophy of Judging" showed that in a certain case where a bird scored in the customary manner had scored 89 the application of the decimal scale of points of the same percentages of defect as were indicated in the cuts on the official card would have given the bird a score of 90. But in the subsequent chapters where the lists of cuts for common faults in many different breeds and varieties are given they abandon the percentage cuts idea and give lists and scales of SPECIFIC CUTS, except that occasionally a remark is made calling attention to differences that would exist if the percentage system was applied to the different "scales of points." Their lists of cuts are simply the customary specific cuts which had come into use in judging originally intended to be on a basis of percentages on the Standard "scales of points," but which—as has been pointed out—developed on a simple basis of specific cuts simply because the other plan is entirely too cumbersome to be practical.

That is the reason why, while theoretically the scores given by the two systems should be different, and those on the decimal score card sometimes higher and sometimes lower than the others, according to the difference in the "scales of points," most of the judges who used both cards scored birds the same by either, while where differences did exist they did not correspond to differences in the "scales of points," but the scores by the decimal system regularly ran a little higher than by the other. The real cause of this may be described as mechanical. There were six more spaces in which to mark cuts on the regulation card, and unless a judge was keeping this in mind, and carefully uniting his cuts in the sections which on the decimal card covered the points for a larger number of sections on the other card, he was quite apt to fail to record all the cuts that he should.

As originator and principal advocate of the Decimal Score Card Mr. Felch tried first to have it substituted for the other. This precipitated a long and sharp controversy, in which examples of poor judging under both score-card systems were so thoroughly exploited that general confidence in the value of any score-card method was greatly impaired; but the deciding argument with most of those who had to vote on the question of recognition of the decimal system both when the design was to make it the official system, and afterward when all that was asked was equal recognition, always rested on the assumption that the existing "scales of points" in the Standard were the actual basis of score-card judging, and that the cuts for defects were scientifically computed in percentages of the values of the sections.

The instruction to make cuts in percentages of the values in the "scales of points" disappeared from the Standard in the 1905 edition, simultaneously with the recognition of comparison judging as a method having the approval of the American Poultry Association, and with the issue of a substantial list of instructions for cutting for defects on the scale of specific cuts in common use in score-card judging. A very short list of these had appeared in the 1898 Standard. The extension of that list and the recognition given comparison judging in the next edition made the retention of the theory of percentage cuts so palpably absurd that it was omitted, though the idea that the values in the various "scales of points" are the fundamental factor in judging is still retained and emphasized.

The Actual Process of Score-Card Judging

In the original method of scoring poultry by giving a few credit points for excellence, comparison of different degrees of excellence in the birds competing led to the adoption of a range of values in credits in the favored features which was in most cases from 1 to 2 or 3. In the Halsted Standard of 1867, out of a total of 208 items in lists of points, there are only 28 instances where the highest credit exceeds 3; and, in most of these, giving a point a possible credit of 4, 5 or 6 plainly happened because of the small number of points considered, and the obvious inconsistency of giving certain features in one breed more credits than the same features in another. There was really no occasion to make the points total 15, unless they were used as the basis of awards of sweepstake prizes. But the usual method in scoring favored features, or features really given much consideration, was with a short series of unit credits.

When the plan of deducting from an assumed score standing for perfection in the specimen, and in every section and character, was adopted, the same principle of a series of marks was employed but inverted—giving cuts instead of credits. Consideration of the possible numbers of these cuts, and of the fact that in the system of marking performance from which the system of score-card judging was derived a mark of 85 to 90 or a little better stood for highly creditable work, will show why cuts on a scale of ½ rather than on a scale of 1 were early adopted. It is often said that the customary score for the ordinary good bird is too high—that it does not allow for the differences in quality between these and the very superior specimens which are still not perfect. I think that this view is due to putting a little too much emphasis on the idea that the perfection indicated in the Standard is unattainable, and to scoring the best birds in the strongest competition too low, and also perhaps in part to a lack of appreciation of the fact that

VETERAN BREEDERS, EXHIBITORS AND JUDGES

First row, left to right: J. Y. Bicknell, I. K. Felch, Philander Williams, Geo. O. Brown. Second row: Sharp Butterfield, John Glasgow, Henry Hales, Wm. McNeil. Third row: Henry Ball, Geo. W. Mitchell, Geo. V. Fletcher, Geo. E. Peer. Fourth row: C. A. Emry, Ben S. Myers, N. R. Nye, C. H. Rhodes

the nearer a specimen approaches the score of 100 the greater the real value of each point and fraction of a point in its score.

But the thing we have to consider in coming to an understanding of the fixing of a certain scale of specific cuts in score-card judging is that when this plan was devised and put into use it would have been a positively foolish thing to adopt a scale of cuts for defects by which if all faults were considered critically, as American breeders and judges were disposed to consider them, the best specimens of the day would have scored away below the figures which commonly represented excellent quality. It must be kept in mind also that the system of scoring, and the ordinary "scales of points" were not developed in connection with a "Standard of Perfection." Throughout the thirteen years following the appearance of the first edition the American standard for poultry was the "Standard of Excellence," and on the descriptions so considered the customary cuts established were appropriate.

Lack of familiarity with the real origin of the American system of scoring would appear to be the reason for its failure to find acceptance in England. Even so great an authority and so able a student and writer as Lewis Wright was apparently unable fully to understand the method or to grasp its principles, though by continued study of problems of judging he did eventually come to recognize the American system as better suited to common practice than the methods he worked out by applying his first ideas of it to the scales of specific cuts as commonly used by the best English judges in comparison judging where the only object is to rank the competing birds properly. In this it is immaterial whether the marks be called credits or cuts: the only difference is that when they are called credits the highest mark wins, and when they are called cuts the lowest mark wins. It is simply a question of the point of view taken in marking.

On the basis of the specific cuts customary in England, Mr. Wright found that 100 points as a total were not enough for all the cuts that could be made. So he tried to find a total that was high enough to cover all possible cuts. The details and direct results of his observations do not concern us here as they made no impression upon American ideas of judging except as acquaintance with his ideas may have intensified in some minds the idea that the total of possible cuts in any section should correspond exactly with the value given the section in the "scale of points," and if the cuts exceed that or even equal it, the judge is in the absurd position of entirely eliminating the whole section. Again and again persons calling attention to such cases, or possible cases, in discussing the subject have said "This takes away the entire section: you have cut off the head, or the wings, or the tail or the feet—as the case might be: the process by which you did this is ridiculous. If your total of 100 points represents value, and the points as allotted section by section represent values; then in a valid system of scoring for defects from the ideal bird desired by the fancier, you must always provide that when you have cut for all the faults from the fancier's point of view—when you have found the lowest possible score for Standard values—you must still have left a valuation which represents the actual commercial value of the bird for other purposes."

This idea is in harmony with the view of the relations of utility and fancy that has always predominated in America. Wright's idea that the points in the schedule were points of excellence as distinct from points of utility, and that the essential thing was to have points enough to cover the possible cuts, is in accord with the English fancier's attitude that considers fancy apart from utility. Both ideas contain principles that are applicable to a consistent scheme of judging by score card; but neither is correct as interpreted and applied. We shall find as we proceed that in the evolution of ideas of judging the right applications of the principles are being worked out.

The Return to Comparison Judging

There were from the first adoption of the score-card system of judging some judges who did not like it. When the advocates of score-card judging began to quarrel among themselves, each section pointing out the errors and absurdities in the work and ideas of the other, they greatly helped the efforts of those disposed to return to the comparison method. The situation encouraged comparison judging propaganda, and the increasing difficulties of making awards in large classes at the leading shows when all the birds in the class had to be examined in detail eventually furnished the occasion for substituting comparison for score-card judging at the Madison Square Garden in 1891.

After the comparison method came into quite general use at the large shows and at many of the smaller shows, its advocates began to represent its first adoption by the big shows as based upon the general superiority of the system. The truth is that the determining point was the time required to make the awards of prizes. This was stated in the announcement of the first Boston Show by the present association. It had also been stated by Mr. Robert Colgate, president of the New York Poultry Association, when the announcement was made in the poultry press in September, 1890, that the next show at the Garden would be judged by comparison. The exact words of his statement were: "as by this method the judging can be got through with and the prizes awarded much more quickly, thus obviating what seemes to me a very just cause of complaint." The cause of complaint which brought matters to this head was that at the preceding Garden Show some of the judges did not complete their work until Friday, the day before the close of the show.

The adoption of the comparison method of judging at the leading shows in the East, and the preference of some judges for it, the obvious faults of the score-card method, and the feeling of many leading breeders that it was immaterial which method was used, provided the awards were placed right, operated to give the comparison method a good standing and to prolong the general controversy in which every party was most diligent in showing up the mistakes of others, and the only really constructive efforts to better the situation were the "explanatory" score cards that appeared from time to time, and the painfully slow progress of the Standard toward the principles of equality of characters, and of specific cuts for faults. The explanatory score-card methods were mostly too cumbersome for general use.

Although the comparison method was easily reestablished in the large shows, the American Poultry Association never gave it any recognition, either direct or indirect, in the Standard until the 1905 edition. Its attitude since that time is most aptly described as toleration rather than approval. The instructions to judges using the comparison system consist largely of admonitions against yielding to the tendency to slackness, to personal interpretation of the Standard, to partiality in considering different characters and sections, and in gen-

PROMINENT BREEDERS AND EXHIBITORS OF ASIATICS

First row, left to right: J. W. Shaw, H. N. Rollins, John Rumbold, C. P. Nettleton—all Light Brahma. Second row: J. D. Nevins, W. W. Browning, A. Anderson, C. W. Case—Buff Cochin, except Anderson, White and Partridge. Third row: Chas. I. Balch, Hervey C. Wood, W. A. Hendrickson, C. W. Everitt—all Light Brahma. Fourth row: C. A. Ballou—Dark Brahma; D. P. Shove, J. P. Keating, O. L. Putnam—Light Brahma

eral to cautions against the errors most peculiar to that method of judging, and which the score-card method of judging was expressly devised to remedy. These instructions even go so far as to insist in a note at the conclusion that: "Under the comparison system judges must deduct the full valuations of the cuts in all sections, where a specified cut is made under the heading of "Cutting for Defects.'"

Thus in effect the American Poultry Association demands that under certain circumstances, which will occur very frequently in practically every class and competition, the judge using the comparison system shall partially score the specimens in which certain faults appear. As the list of such faults now numbers about fifty, it is perfectly obvious that if a comparison judge follows this instruction he might better use the score-card method straight and apply it to every section. Nothing but confusion could result from a mixture of the systems in judging a class and applying them in this irregular manner. It has already been intimated that the systems ought to be combined to make one efficient and consistent system, and the possibilities of that will be treated in the next chapter; but to attempt to combine them at haphazard, with each judge determining for himself the proportions of each system to be used is impractical.

Considering the score-card and comparison methods of judging AS THEY ARE, and in their historical and logical connections in this country we find:

1—That comparison judging is the primary method of judging the relative merits of different individuals of the same variety, and that it originated prior to the establishment of standards.

2—That the occasion for the making of standards arises from the inconsistency of judging with the best specimen in the competition, or the image of the best specimen he has seen in the judge's mind, as the standard of measurements.

3—That when a standard is made, describing each breed and variety in detail, the methods of marking, or scoring, which before seemed suitable, are found inadequate, and it becomes necessary to devise methods providing for a simple, consistent and practical method of recording the values of characters.

4—That to the present time the only method meeting these requirements that has been developed is the score-card method of making specific cuts, on a simple scale, from an assumed valuation of 100—representing the ideal of excellence in each breed and variety.

5—That the adoption of a Standard of ideal excellence, and of a careful and thorough system of score-card judging to be used in applying it, operates to constantly improve the quality in every point given special attention, and also to increase the number of such points, with the result that the process of judging, by any method, becomes slower and more difficult.

6—That when the improvement in quality, and the increase in size of classes reached the point where the judging at shows often extended far beyond the time limits fixed for the completion of judging, show managers returned to the comparison method as the method by which the judging could be done within the customary time limit.

7—That while this plan at first accomplished the desired object, with further improvements in the quality of standard-bred exhibition poultry and increases in the size of classes, and with the values of winnings at important shows increasing so that the responsibilities as well as the difficulties of the judge's work have increased, the comparison method has in turn become inadequate, and the requirements of good judging demand the return to the score-card method.

8—That, as in the first instance, comparison judges have made private systems of markings for their own use in making awards, these being generally of such a character that they are comprehensible only to the judge and those to whom he gives the key to his system.

9—That this method of judging gives the individual judge more latitude than is consistent with a reasonable uniformity in judging and the maintenance of the authority of the Standard.

10—That the practice of using the score-card method for minor shows, and the comparison method for the larger shows at which the best breeders compete and the best birds are shown is detrimental to the development of a better system and of better practice in score-card judging; because it fails to provide the proper basis for the application of the Standard by the score-card method,—recorded judgments of the best birds—in the strongest competitions by the most competent judges.

11—That the use of the comparison system at the major shows deprives every exhibitor of a bird that though failing to win a prize (perhaps by a single small point which the judge himself hesitated over) was substantially of the same quality as the winners of all credit for his showing, while the use of the score-card system gives all exhibitors due credit for the merit of every bird shown.

12—That the use of the comparison system tends to substitute for the Standard as the measure of excellence in poultry, the reputation of the show and of the judge, often giving the benefit of these to very ordinary specimens winning in weak competition or in no competition.

13—That the score-card system as used does not meet the requirements of judging in close competition, because the method of using it does not provide regularly for the correction of the normal errors occurring in its application.

14—That an efficient, consistent and dependable system of judging must necessarily combine the features of the comparison and score-card methods.

15—That under existing conditions of competition, the importance of efficiency in judging and the values of recorded judgments on the classes at the best shows to the exhibitors and to all breeders interested in knowing the quality of the birds and of the competition at those shows, should be considered before the advantages to the managers of the show and the convenience to visitors at the show of having the awards made quickly.

16—That the idea that the values of large shows judged by comparison to exhibitors who do not win, and to the interested poultry breeders throughout the country can be secured to them by adequate press reports supplying what the methods of judging fail to give is a fallacy, because the press has its own limitations in the handling of show reports.

17—That the one practical way to give worthy exhibitors failing to win prizes the benefits to which they are entitled is to give them the opportunities to know how closely their birds compared with the Standard and with the winners, and to make any proper use of this information in their advertising and their literature and correspondence.

PROMINENT POULTRY JUDGES

First row, left to right: Charles McClave, Adam Thompson, F. H. Shellabarger, D. T. Heimlich. Second row: J. F. Crangle, T. F. McGrew, Richard Oke, J. H. Drevenstedt. Third row: Chas. T. Cornman, W. Theo Wittman, J. Harry Wolsieffer, F. Bohrer. Fourth row: Thomas F. Rigg, W. R. Graham, J. C. Punderford, D. J. Lambert

CHAPTER XIII

Logical Developments in Judging Practice

True Character and Relations of Scales of Points and Scales of Cuts—The Use and Misuse of Disqualifications—How Condition Should Be Considered in Judging—Correct Adjustment of Scales of Cuts and Consistency in Judging Dependent Upon the General Use of a Scientific Simple System of Scoring

IN the two preceding chapters we have traced the evolution of methods of judging with due recognition of both good and bad features in all systems used, but with most emphasis on the faults and conditions which have prevented the development of a single system of universal applicability and permanent value. In this chapter the true status of several things that have not been rightly understood will be shown, the full application of various progressive movements and tendencies indicated, and the logical lines of further development made clear. It is appropriate that the author should say here that he has never approached the study of judging from the standpoint of any idea or theory of his own as to the correct method, or the best method. His object has always been to learn what he could of the methods in use, to examine their results impartially, to find out what features in ordinary practice have permanent value, and to consider how these can be united in better methods. In studying the subject with this in view he has been strongly impressed by the evident tendency of practice to develop along correct lines, and has noted that the greatest obstacle to natural progress is false theories that have been accepted without even the form of a demonstration. The greatest of these is the idea that the "scale of points" is basic in the Standard and a prime factor in judging.

Scales of Credits—"Scales of Points"—and Scales of Cuts

Some observant readers may have noted that throughout this book the phrase "scale of points" has always been used in quotation marks, indicating the existence of a reason for treating it as an expression not used in the common meanings of the words contained in it. That reason is that the word **scale** in this connection is used erroneously, and this erroneous use of it is at the root of general misunderstanding of matters relating to the judging of poultry by the Standard.

A **scale** is a graduated series of numbers, measures, tones, etc., in which no two members have the same value, and in which the difference in value of members in consecutive positions is determined on some fixed principle.

The term "scale of points" was not used in the earliest standards. In them the headings for the tabulated points were simply "Points in Games"—or whatever the breed might be. The expression "Scale of Points" first appears in the 1875 edition of the Standard of Excellence and there it is used only with the tables for Buff, Black

PROMINENT SOUTHERN AND PACIFIC COAST JUDGES
First row, southern, left to right: F. J. Marshall, H. B. Savage, D. W. Owen, Walter Burton. Second row, western: Henry Berrar, H. H. Collier, Elmer Dixon, O. G. Hinds

and White Cochins. All the tables for other breeds and varieties have the old form heading. In the 1883 Standard the term "Scale of Points" is used all through, and from that time we find the idea implied in the proper and common meaning of the word scale attached to the list of points, and the fallacy which it bred becoming firmly established as a supposed cardinal principle in judging.

The lists of points never had the characteristics of "scales" except in so far as judges might graduate their individual **scales of cuts** and by degrees brought about first general use of a common scale in which the unit of difference is a cut of ½, and then by degrees the acceptance of this scale by the American Poultry Association as the Standard SCALE OF CUTS which judges are instructed to apply. In the latest edition of the Standard the list of specific cuts for defects carries ½ as the unit or factor in the scale of cuts, but in the general instructions to judges it is stipulated that no cut of less than ¼

PROMINENT WESTERN JUDGES
First row, left to right: Geo. D. Holden, E. C. Branch, J. C. Johnston, A. B. Shaner. Second row: F. L. Platt, D. E. Hale, O. L. McCord, Reese V. Hicks. Third row: C. P. Van Winkle, Russell F. Palmer, W. M. Coats, V. O. Hobbs

decisions when giving credits on the basis of the lists totaling each 15 points. Where this was done it might be said that the judge had his scale of credits. Nor was there any feature of score-card judging as provided for in the 1875 Standard that could properly be described as a "scale." It was the development of a system of specific cuts in score-card judging that created various shall be made. That is as fine a scale as most persons can use in scoring by the book and their mental standard, but in deciding ties by comparison, and then making the cuts show the difference further division of the unit —for the particular case—would follow as a matter of course.

The logical use of a "scale of points" in a score-card

system of judging is seen when we note that in taking points—as needed elsewhere—from the general sections, Symmetry, Weight and Condition, in the matter of weight the American Poultry Association stopped when the cuts for weight that disqualified a specimen equaled the points allotted to weight in the "scale of points." This is a practical application of Wright's view that **Points** should represent the difference between the valuation placed on the ideal and the lowest valuation of a specimen or of a section that was recognized as conforming to the Standard specifications. The principle as now applied to weight is capable of application to other sections on a basis that will solve in a logical and just way

but it arose at a very early stage of the history of poultry exhibitions. It has its origin in the theory of the inequality of the value of characters, and in the ideas of those who are partial to particular characters and sections, and—therefore—naturally lean the other way when faults which seem to them particularly objectionable are concerned.

Most of the disqualifications fixed in the Standard are manifestly reasonable. Nearly all of the disqualifications in the first Standards were for unsightly things which anyone at all would notice. But it is interesting to find that in the Rouen Duck, which has the color pattern of the Wild Mallard, and in the early days of the poultry fancy was one of the few varieties of poultry that was

PROMINENT BLACK MINORCA BREEDERS AND EXHIBITORS
First row, left to right: O. J. Andruss, Geo. H. Northup, Chas. G. Pape, L. C. Mishler. Second row: S. T. Campbell, Frank Mc Grann, Rowland Story, A. Didriksen

one of the most difficult problems in judging—the best method of determining and dealing with disqualifications.

Disqualifications—Their Use and Misuse

The matter of disqualifications has been more fruitful of controversy and more productive of trouble and dissatisfaction than all other phases of judging combined. The bone of contention is the question of the justice and wisdom of disqualification for trivial superficial faults, when gross faults, and sometimes very important faults are allowed to pass with such cuts for defect as they merit in the opinion of the judge. The trouble is aggravated by the relation of small superficial faults to the practice of faking.

The impression prevails widely that disqualification for trivial faults is a comparatively recent development due to advanced ideas of quality and closeness of competition. Those things have no doubt affected the application of the practice of disqualifying for slight faults,

really good in color when judged by a critical standard. "Bills, clear yellow, dark green, blue or lead color," a disqualification read in 1867 just as it does in 1921. The greenish yellow bill desired in the male, and the brownish orange bill desired in the female, are certainly more handsome with the standard colors, but when the Standard calls for such mixed colors in bills in birds with the complex color patterns of the Rouen drake and duck these other colors are bound to appear frequently in otherwise very fine specimens, and a rigid disqualification for a normal color variation is absurd. To disqualify a dark green bill when the Standard calls for a modified shade of the same color is especially absurd. This strictness in applying the Standard for color in the case of the Rouen Duck originates in giving extraordinary attention to the bills of ducks—making a fad of color of bill. There were not many of these extreme disqualifications in the first Standards, but the precedents for the rigid use of petty disqualifications were there.

PROMINENT WOMEN EXHIBITORS

First row, left to right: Mrs. Nettie Metcalf, Buckeyes; Mrs. Louisa White, White Rocks; Mrs. Pearl Cuddeback, Turkeys; Miss Frances E. Wheeler, Pekin Ducks. Second row: Mrs. E. W. Mahood, R. I. Reds; Mrs. W. G. Robinson, Light Brahmas; Mrs. Geo. Russell, Brown Leghorns; Mrs. W. G. Curd, Barred Rocks. Third row: Mrs. A. G. F. Stice, Barred Rocks; Mrs. Donna Hanly, Partridge Wyandottes; Miss Norma Witheft, White China Geese; Mrs. Eli Fowler, Bronze Turkeys. Fourth row: Mrs. J. T. McMahan, Bronze Turkeys; Mrs. Dan C. Amos, Turkeys; Mrs. Chas. V. Keeler, White Wyandottes; Miss Eva M. Culp, Bourbon Red Turkeys

And the disqualifications that cause most controversy are generally of the same character, though mostly of later origin. To appreciate the status of the ordinary color disqualifications when they were first applied, we must consider both the fact that most of the color varieties were recently made, and had some very bad faults, and the fact that breeders were not nearly as critical as they are now. Many, indeed, were not critical at all, and would entirely overlook foreign feathers in plumage that to a trained eye were conspicuous and unsightly. We must consider too the attitude of the breeders, exhibitors and judges of the time toward the removal of disqualifications. According to the testimony of those who have been closely associated with developments along this line from very early dates, the original object of making such faults as false-colored feathers in parts of the plumage where they could be easily removed without spoiling outline or surface, and small stubs and down on the shanks of cleanlegged breeds not far removed from feather-legged ancestry, disqualifications, was as much to punish an exhibitor FOR NOT REMOVING THEM BEFORE BRINGING THE BIRD TO A SHOW as for the purpose of forcing them to stricter selection to eliminate such faults by "breeding them out."

The ethics of the question have been discussed in Chapter III. Here we are interested in the best method of dealing with ALL FAULTS on the same general principle and without taking an arbitrary attitude in any case. It is only by finding and applying such a principle that the problem of disqualifications can be solved. As it is now, a tiny speck of down between the toes of a specimen of a clean-legged breed disqualifies the entire pen—no matter how good otherwise: while every bird in the pen could be so poor in quality that no good judge would give it a second glance, and yet not a bird in it have a disqualification. The disqualification for down is an extreme one, but there are others that come very near it. The disqualification of pens for the disqualification of a single bird is the extreme example, but the principle is just the same in comparison of cases of individual specimens. Also it continues to be the same when we compare faults in different characters and different sections of a specimen.

There is an inconsistency in "capital punishment" for slight faults, while other serious faults simply reduce the grade of a specimen, that no amount of sophistical argument to justify the "law" can explain away. The common assertion that the little faults severely punished as disqualifications are peculiarly objectionable, and especially persistent in breeding is not sound, and is not borne out by the practice of breeders in the use of birds having such small faults; nor is it in accord with the principle of equality of importance of characters and of faults, which it has been shown over and over in this discussion has had a most persistent influence upon the development of judging practice and of the Standard of Perfection.

The rule that may be formulated from established principles and practice and that will apply equitably, and with the uniform tendency to all-round improvement in standard poultry is this—WHENEVER THE TOTAL OF CUTS FOR DEFECTS IN ANY SECTION REACHES THE VALUE PLACED ON THAT SECTION IN THE SCALE OF POINTS THE SPECIMEN SHALL BE DISQUALIFIED.

The application of this rule does not at all interfere with the rule requiring that a specimen must reach a certain score to be entitled to a prize. Before considering it further we must take up the matter of the division of characters in sections.

TABLE NO. 1—SHOWING THE ALLOTMENT OF POINTS TO SECTIONS IN PLYMOUTH ROCKS, LEGHORNS AND BRAHMAS IN EDITIONS OF THE STANDARD 1875-1915

Explanation.—Where the points allotted a section are divided for shape and color, the total is given in the column directly under T, and the number of shape in the column under S, and for color under C.

	Symmetry	Weight or Size	Condition	Comb	Head S C T	Beak S C T	Eyes S C T	Wattles and Ear-lobes S C T	Neck S C T	Wings S C T	Back S C T	Tail S C T	Breast S C T	Body and Fluff S C T	Legs and Toes S C T
Plymouth Rocks 1875 American Class	12	14	9	8	7				6	8	6	6	6	—(10)—	8
1890	8	6	6	8	3 3 6				6	4 6 10	4 4 8	4 4 8	5 5 10	5 3 8	8
1898	8	6	6	8	3 3 6				6	4 6 10	4 4 8	4 4 8	5 5 10	3 3 6	8
1905	8	6	4	8	3 3 6			2 4 6	8 6 9	6 4 10	6 6 12	4 5 9	5 5 10	3 3 6	3 3 6
1910	4	4	4	8	2 2 4	2 2 4	2 2 4	2 3 5	3 5 8	4 5 9	5 6 11	5 5 10	5 3 8	3 3 6
1915	4	4	4	8	2 2 4	2 2 4	2 2 4	4 6 10	4 6 10	5 5 10	5 5 10	5 5 10	5 3 8	3 3 6
Decimal Scale	5	5	10		—(T·10	8·5	C·5)—		5 5 10	5 5 10	5 5 10	5 5 10	5 5 10	5 5 10	5 5 10
Leghorns 1875 Mediteranean Class	10	10	10	15	7				15	5	5	5	5	—(5)—	5
1890	8	5	6	10	4 4 8				10*	3 4 7	4 4 8	3 4 7	4 4 8	6 4 10	3 3 6	7
1898	8	10	5	10	2 2 4				10*	3 4 7	4 4 8	3 4 7	4 4 8	6 4 10	3 3 6	7
1905	8	8	5	10	2 4 6				10*	3 4 7	4 4 8	3 4 7	4 4 8	6 4 10	3 3 6	7
1910	4	4	4	10	2 4 6	2 2 4	2 2 4	4 6 10	3 5 8	4 4 8	3 5 8	5 4 9	4 4 8	3 2 5	3 3 4
1915	4	4	4	10	2 4 6	2 2 4	2 2 4	4 6 10	3 5 8	4 6 10	5 4 9	5 4 9	4 4 8	3 2 5	3 3 4
Decimal Scale	5	5	10		—(T·10	8·5	C·5)—		5 5 10	5 5 10	5 5 10	5 5 10	5 5 10	5 5 10	5 5 10

* Wattles 4. Ear-Lobes 6.

	Symmetry	Weight or Size	Condition	Comb	Head S C T	Beak S C T	Eyes S C T	Wattles and Ear-lobes S C T	Neck S C T	Wings S C T	Back S C T	Tail S C T	Breast S C T	Body and Fluff S C T	Legs and Toes S C T
Light Brahma 1875 Asiatic Class	10	18	8	8	5				5	10	8	7	7	7*	5*	7
1890	8	6	6	8	3 3 6				6	4 6 10	4 4 8	4 4 8	5 5 10	5 3 8	8
1898	8	6	6	8	3 3 6				6	4 6 10	4 4 8	4 5 9	4 4 8	5 5 10	3 3 6	8
1905	8	6	5	8	3 3 6				6	4 5 9	4 4 8	4 4 8	4 4 8	5 5 10	3 3 6	8
1910	8	6	4	8	3 3 6			2 2 4	2 3 5	4 5 9	4 4 8	6 5 11	4 5 9	5 5 10	3 3 6	8
1915	4	4	4	8	2 2 4	2 2 4	2 2 4	4 6 10	4 6 10	6 4 10	5 5 10	6 4 10	3 3 6	3 3 6
Decimal Scale	5	5	10		—(T·10	8·5	C·5)—		5 5 10	5 5 10	5 5 10	5 5 10	5 5 10	5 5 10	5 5 10

* Breast and body 7. Fluff 5

Natural and Technical Division by Sections

In looking at any complex structure or organism, we naturally observe it as constituted of certain parts which the eye and mind note separately as well as collectively. Thus in any bird we note the head, neck, back, breast, body, wings, tail and legs, as principal parts. As far as it goes this natural division, which any normal mind would make intuitively, corresponds to the technical divisions as made in the Standard descriptions and on score cards to be used in judging. It differs from them in not so fully analyzing the head and head points where these are abnormally developed—as to some extent they are in nearly all improved breeds of fowls.

The regulation score card makes three divisions of the head parts, and the decimal score card makes two. The regulation "scale of points" gives to all head parts a total of 24 points: the decimal "scale of points" gives to all head parts a total of 20. It can be seen at a glance at Table 2, that simply as a result of arranging the number of points allotted to different sections to cover the value of common combinations of specific cuts in them, the points given to sections in the regulation "scale of points" for the American and Asiatic classes have been gradually brought to the decimal score card totals for shape and color, and further that the tendency is more and more to bring them to the decimal score card idea of an equal division of the points given a section between shape and color.

The more closely possible divisions are studied, the more plainly it will appear that the regulation card gives more than the appropriate proportion of points to head parts; that the decimal score card division which gives all head parts twice as many points as the other natural sections is a better division; and that the division of the head parts for the decimal scale of points and score card is a bad one. By this division the comb, or the comb and crest are given the same number of points as head, beak, eyes, face, ear lobes, and wattles, or wattles and beard.

A more equable division is to consider comb and wattles, or crest and beard—with which the comb and wattles are usually rudimentary or small, as one section, and the head and ear lobes as another. This separation of "Wattles and Ear Lobes" will no doubt at first appear inappropriate to many who have always considered them as grouped together in descriptions and in judging, but it is more appropriate than the present division, for comb and wattles regularly correspond in size, color and texture quite closely, or at least it is the intent of the Standard that they should do so, but the ear lobes have a peculiar structure of their own, may be, and often are different in color from comb and wattles, and the face tends to take the color and to some extent the character of the skin of the ear lobes.

Consideration of Size and Weight in Judging

Size is the most important attribute of breed character, and the only measure of size for practical use in judging poultry is **weight, with condition taken into consideration**. Size may be kept at any desired standard, approximately, without a weight provision, as long as a breed is almost entirely in the hands of fanciers who have keen preceptions of type and symmetry. But we have now nearly half a century of experience with and observation of the state of different breeds that plainly shows that whenever a breed is widely distributed and becomes popular with all classes of poultry keepers, the first essential to the maintenance of the size and type regarded as ideal is very close adherence to an established standard of weight. Laxity in the use and application of weight standards is the first cause of confusion of somewhat similar breed types.

In general, the standard weights for the different breeds, as fixed by the American Poultry Association as the ideal weights for the type and use of the breed, have been judiciously and wisely chosen. In fact, as to all **ideals** in American standards it may be said that the more severely one analyzes them, the more he is impressed with the good judgment in practical matters as well as in matters of taste that is shown in breed and variety descriptions or specifications. Some inconsistencies in Standard weights can be found, but they are not of great importance. But in the matter of departures from standard weights the Standard has allowed, and now allows, far more latitude than is consistent with clear distinctions between types and with the uniformity of size and weight which ought to exist in every standard breed.

To appreciate the extent of the latitude which the Standard partly allows and partly gives occasion for, we must note that its method of treating variations from

---

TABLE NO. 2—SHOWING THE ALLOTMENTS OF POINTS IN PRINCIPAL CLASSES IN THE 1915 STANDARD

Explanation.—Where the points allotted a section are divided for shape and color, the total is given in the column directly under T, the number for shape in the column under S, and for color under C.

Class	Symmetry	Weight or Size	Condition	Comb	Head S C T	Beak S C T	Eyes S C T	Wattles and Ear-lobes S C T	Neck S C T	Wings S C T	Back S C T	Tail S C T	Breast S C T	Body and Fluff S C T	Legs and Toes S C T	Crest S C T	Texture of Feather
Amer.	4	4	4	8	2 2 4	2 2 4	2 2 4	2 2 4	4 6 10	4 6 10	5 5 10	5 5 10	5 5 10	5 3 8	3 3 6		
Asiatic	4	4	4	8	2 2 4	2 2 4	2 2 4	2 2 4	4 6 10	4 6 10	6 4 10	6 4 10	6 4 10	5 5 10	3 3 6		
Med.	4	4	4	10	2 2 6	2 2 4	2 2 4	4 6 10	5 3 8	4 6 10	5 5 10	5 4 9	4 4 8	3 2 5	2 2 4		
Eng.	4	4	4	8	2 2 4	2 2 4	2 2 4	2 2 4	4 6 10	4 6 10	6 4 10	6 4 10	6 4 10	5 3 8	5 3 8		
Polish	4	4	4	2	2 2 4	2 2 4	2 2 4	2 2 4	3 3 6	4 6 10	4 4 8	4 5 9	4 4 8	3 3 6	2 2 4	10 5 15	
Habg.	4	4	4	10	2 2 4	2 2 4	2 2 4	5 5 10	4 4 8	4 6 10	5 5 10	4 6 10	4 4 8	3 3 6	2 2 4		
French	4	4	4	8	2 2 4	2 3 5	2 2 4	2 2 4	4 4 8	4 6 10	5 4 9	5 4 9	6 4 10	3 3 6	2 2 4	8 4 12	
Cont.	4	4	4	8	2 2 4	2 2 4	2 2 4	4 4 8	3 5 8	4 6 10	6 6 12	5 5 10	4 5 9	3 3 6	2 2 4		
Games	10*		6	2	4 1 5		4		2	5 3 8	4 6 10	4 3 7	5 3 8	4 3 7	4 3 7	10 4 14		6
Orien.	4	4	5	6	(T-5 8-3 C-2)	2 2 4		2 2 4	4 4 8	4 4 8	6 4 10	4 5 9	4 4 8	3 3 6		4
Malays	10*12*5			8	—(T-11 8-5 C-6)—	2 3	3 3 6	4 4 8	3 3 6	3 3 6	3 3 6	2 2 4	3 3 6*		
Seb. Bant.	4	2	4	8	2 2 4	2 2 4	2 2 4	4 4 8	4 4 8	4 6 10	4 6 10	6 6 12	5 5 10	4 4 8	2 4 4		
Miscel. Bant	4	2	4	8	2 2 4	2 2 4	2 2 4	4 6 10	4 6 10	6 4 10	5 5 10	5 5 10	5 3 8	5 3 8		
Pol. Bant.	4	2	4	2	2 3 4	2 2 4	2 2 4	2 2 4	3 3 6	4 6 10	4 4 8	4 5 9	4 4 8	3 3 6	3 3 6	10 5 15	
Silkies	4	4	4	5	(T-6 8-3 C-3)	1 2 3	2 4 6	4 3 7	3 3 6	4 3 7	4 3 7	3 3 6	3 3 6	3 3 6	10	9
Sult.	4	4	5	4	3 3 6		3	4 3 7	3 3 6	4 3 7	5 3 8	3 3 6	3 3 6	9 3 12	8 4 12	
Ducks 1	4	6	10		2 2 4	2 2 4		4 4 8	4 6 10	8 4 12	2 2 4	10 4 14	12 4 16	2 2 4		
Ducks 2	4	6	10		2 2 4	2 2 4		4 3 7	3 3 6	4 3 7	6 4 10	6 4 10	6 2 8	2 2 4	15
Ducks 3	4*	6	4		4 3 7	5 2 7	2	8 4 12	3 3 6	6 3 9	6 3 9	6 3 9	2 2 4		
Geese	4	6	10		(T-10 8-4 C-6)	2 2 4		6 3 9	6 6 12	8 4 12	2 2 4	8 4 12	10 4 14	2 1 3		
Turk.	4	18	4		2 2 4		2 2 4	4*	3 2 5	4 6 10	4 6 10	4 8 12	5 5 10	3 2 5		

standard weights is one that inevitably works to put the general average weight of the breed considerably below the standard, while the standard weight and the general average weight should be the same, and would be if variations from weight were discounted uniformly by one principle. The object of the Standard is to establish the weights it fixes as the most common weights and the general average weights. But when its rules for judging allow such heavy reductions of the standard weights before disqualifying for low weight, and give no discounts on overweight, the natural effect is to widen the range of weights. This comes about because those who maintain the standard weights as average weights in their flocks must necessarily breed so that about half their stock will tend over standard weight and the other half in the opposite direction. The weight of the stock as a whole is balanced at the point fixed by the Standard just as most other qualities are.

The error of the method of dealing with weight is that it tends to make the standard weights the maximum and the disqualifying weights the minimum weights for a breed, when the standard weight should be the medium between the two extremes. The practical way to correct the error is to cut in the same proportion for overweight as well as for underweight, also increasing the values of specific cuts for weight double. Thus instead of cutting two points per pound for underweight, the cut would be at the rate of four points per pound for a variation from weight either way. Then on the present valuation given "weight" in the "scale of points" a specimen that went a full pound either under or over weight would be disqualified. By the decimal system with valuations of 5 points per section the application of this rule would be more satisfactory because it would be more lenient both at the maximum and minimum disqualifying points.

As it is now, the Standard places no limit on size except as the awards are affected by the provision that when birds are tied for honors the one nearest standard weight shall win. The real check on excessive size is the general dislike in America of anything approaching coarseness in standard poultry. Under the rule suggested the range of weights in standard Plymouth Rock cocks would be from 8¼ to 10¾ pounds. Now it is from 7½ to 12 pounds.

The rule, as it would relate to weight, is not ideal, because it does not apply equally to different breeds and weights, but it reduces the degrees of inconsistency by 50 per cent, and would increase the uniformity of size and weight in all varieties at least 50 per cent.

Consideration of Condition in Score-Card Judging

Under all present systems of judging by score card "Condition" is given a general cut made on the general appearance of the bird, before detailed sections are considered. Consequently, it often happens in practice that the judge examining a bird section by section finds one or more occasions for further cuts for condition. What the judge does in that case depends upon the marks already made, and whether he is in a hurry or working leisurely. If he has a cut of 1 for general condition, it is easy to put ½ after it; if a cut of ½, is it easy to put a 1, 2 or any figure appropriate before it. But if an erasure is necessary, and the judge is pressed for time, or if he is working with a clerk to mark cards and correction might easily lead to confusion and further error, he is very apt to put the extra cuts for condition on the sections showing faults of condition, which is not only wrong in principle, but exaggerates his estimates of faults properly chargeable to the section, and is confusing and misleading to those who try to make the scoring square with either general practice or the judge's known general tendency in cutting for faults. A very considerable proportion of common inconsistencies in judging is due to this sort of "juggling" the cuts in the interest of speed, and to avoid the necessity of stopping to make erasures.

The difficulties of the situation can be avoided, and conditions to which the rule suggested for disqualifications will apply made, by simply providing a special column for cuts for condition, section by section, and making the final cut for condition the sum of all the sectional cuts.

Correct Adjustment of the "Scale of Cuts"

A number of times recently judges have called attention to the fact that common practice in applying the scale of specific cuts given in the Standard gave in many cases total scores that were manifestly too low—that is, the score as obtained by the score card was much lower than a competent judge's impression of the quality of the bird would place it at sight. In considering details of practice in score-card judging, we have to recognize that the final test of the value and general accuracy of any system of scoring is the closeness with which the results of judging by the card approach expert instantaneous appraisal of the quality of a specimen—provided closer inspection shows no bad faults not visible on the surface. The judges point out that where no such hidden defects are discovered and the bird still falls short of an the final cut for condition the sum of all sectional cuts.

The root of the difficulty in such cases appears to lie in taking ½ as the minimum specified cut, and so virtually making it the unit of measurement in scoring—scaling down—for faults. It seems to apply very well to ordinary stock and to competitions where the quality is uneven and the winners are easily placed. But in the superior grades of stock, and where the closeness of competition compels close scrutiny of every point in which specimens differ, the general use of the minimum cut of ½, and of that as the factor in the short series of cuts which are commonly made will lead to systematically heavier cuts on good birds in strong classes. The only way to avoid that difficulty is to make the minimum cut ¼ and make that fraction the factor in the scale of cuts. Doing this would simply be to apply generally the minimum cut introduced in two cases of specific cuts in the Standard list in the last revision, and recognized as necessary in the instructions to judges in that edition.

Theoretically it would be supposed that to make the minimum cut ¼ would mean that such cuts as ¾, 1¼, 1¾, etc., would become common, and that where now the judge in the ordinary course of cutting for faults makes three common grades of faults, which he cuts ½, 1 and 1½, then he would have twice as many grades within the same range of values of defects. But on what we know of common judging practice there is no reason to suppose that it would work out that way. What experience and observation indicates as likely to happen is: First, that judges will continue to make three common grades of the faults they find, and will mark them ¼, ½, and 1, with heavier cuts for larger defects than would commonly be met in ordinary good-quality exhibition stock. The result would be lighter cuts and higher scores on this class of stock without materially affecting the scores of specimens which should be cut more severely. The practical adjustment of the whole "scale of cuts" to fairly represent values in all grades of standard stock depends—as has been pointed out—on beginning with correct scores for the best birds in strong competition.

Another effect of making the ¼ cut the minimum cut and the factor in the scales of cuts would appear when scores as first made for a class in which competition was close were tested and corrected upon comparison of the birds plainly in the running for the prizes whose relative positions could be determined only by careful consideration and estimate of values. In this stage of judging the judge would divide the minimum ¼ cut as often and as far as necessary to establish the differences between competing birds. In this connection it should be said that with competent and careful scoring in which all sections and characters are given proper consideration, and the minimum cut is ¼, the wholesale tying of scores which was once a common feature even in comparatively small shows under some judges would not often occur, and in a large proportion of cases of ties the tie could be broken by the rules now obtaining for such cases, or where this failed by a similar simple rule, so often that rescoring would be necessary only for the leading birds in strong competition when observation of the birds as placed by the score card showed the need of a review of them to make certain that the awards were right.

ing five "sections" of them as now. This change in the division of the bird into sections is in the interest of more rational judging of exhibition specimens and also of placing relatively more emphasis on other parts of the bird and less on head points.

SYMMETRY as a special section is omitted from the card, because symmetry is simply the result of correct shape and proportions in all sections.

The division of the columns under "Condition," etc., provides for recording the specific cuts, with the value of each indicated by a sign, in the spaces under "Cuts," and in the spaces under "Sum" placing the total value of all cuts made.

The difference between this card and the regulation card is in FORM, not in anything essential. Its object is to indicate every cut and the value of it, so that anyone taking a bird and its score card can locate the cuts. By the regulation card it does not appear whether the numerical cut recorded in any division of any section represents one fault or several faults.

The recording of specific cuts in a small space is made practical by using a simple code of signs for the fractions ¼ and ½—a dot (.) for ¼; and a dash (—) for ½. Thus if in the space for "Cuts" under "Shape" and in the section "Comb and Wattles," the record was a dot, a dash, and a figure 1, it would signify that the comb and wattles had three faults, one of which was cut ¼, one ½, and one 1, and the total to be written in the next space under "Sum" would be 1¾.

The minimum cut prescribed in the Standard is ¼ of a point —or simply ¼. By making it the minimum cut the Standard establishes ¼ as the specific cut for the character that is nearly perfect, but that should have a very slight cut.

This cut being established, it follows that the next distinct degree of fault must be cut ½; the next ¾; the next 1, and so on. And the SCALE OF CUTS is really on a ¼ point basis, except that in ordinary practice it is not necessary to make fine distinctions except where the faults are least. So we may say that the ordinary working SCALE OF POINTS in score-card judging is as follows:

For the smallest degree of fault demanding a cut............1/4
For a more pronounced yet moderate fault...................1/2
For conspicuous but not extreme fault........................1
For grosser faults...2/3

Simple automatic general rules for disqualifying and debarring birds for serious fault may be made by making a total of five cuts in any one section disqualify, and applying rules for debarring for a given number of cuts, in any division—as condition, shape, color,

IMPROVED PRACTICAL
STANDARD POULTRY SCORE CARD
FOR SINGLE BIRDS

Name of Association _____

Place of Show __Syracuse, N. Y.__

Date of Show __Sept. 13-18, 1920__

Exhibitor _____

Variety __Barred Plymouth Rock__

Entry No. __24__ Band No. _____ Weight __8__ Cat _____

SEX AND AGE __Ckl.__

| | CONDITION | | SHAPE | | COLOR | | | | TOTALS |
| | | | | | PATTERN | | SHADES | | |
	CUTS	SUM	CUTS	SUM	CUTS	SUM	CUTS	SUM	
Comb and Wattles or Crest and Beard			.	¾					¾
Head, Beak, Face, Eyes and Ear Lobes			.	¼			.	¼	½
Neck	.	¼			—	½	.	¼	1
Back			.	¼	—	½			¾
Tail			—	½	.	½	.	¾	1¾
Wings					.	¼	—	½	¾
Breast			.	¼	.	¼	.	¼	¾
Body and Fluff			.	¼	.	¼	—	½	1
Legs and Toes									
TOTALS		¼		2		2		3	7¼

A TOTAL OF 5 POINTS IN ANY SECTION DISQUALIFIES

5 Debars from 1st Prize | 5 Debars from 1st Prize | 5 Debars from 1st Prize | 5 Debars from 1st Prizes

10 DEBARS FROM 2nd PRIZE

SCORE __92¾__

_____ JUDGE

_____ SEC'Y.

A NEWLY ARRANGED SCORE CARD

In a proper judgment of birds as compared with ideal standards COMPLETE ANALYSIS of the bird is of as much importance as appropriate cuts for the faults observed. Such an analysis can be made only by systematic examination and the only way one can be sure he has made such an examination where so many characters are involved as in a standard fowl, is by recording his cuts in as much detail as is necessary. The score card above shows how this can be done with a card making only a few changes in form in the regulation score card.

The special features of this card are—the separation of "Condition" from "Shape" and "Color," thus distinguishing plainly on the record between constitutional faults, or permanent faults, and superficial and temporary faults; the division of "Color" to more precisely indicate the character of the faults of color observed; and the combination of all head points in two sections instead of mak-

IMPROVED PRACTICAL
STANDARD POULTRY SCORE CARD
FOR EXHIBITION PENS

Name of Association _____

Place of Show _____

Date of Show _____

Exhibitor _____

Variety _____

Entry No. _____ Age _____

To compete for 1st prize every bird in a pen must be eligible for a 1st prize.
To compete for 2nd prize every bird in a pen must be eligible for a 2nd prize.
The highest scoring female shall be taken as the standard of comparison in matching.

	MALE	FEMALES				Total Individual Scores
Band Nos.	1	3	2	5	4	
Individual Scores	92	94	93½	93	92½	465
Uniformity in Following	CUTS SUM		CUTS SUM	CUTS SUM	CUTS SUM	
Size and Condition	— ½				— ½	
Type and Conformation	/ 1	PEN STANDARD NO CUTS			— ½	
Character of Head			/ 1		/ 1	
Character of Color Markings				. ¼		
Even Shade of Color			— ½	— ½		
TOTALS	1½		1½	¾	2	5¾

SCORE OF PEN—Total individual score less pen cuts. __459¼__

_____ Judge

_____ Secretary

NEW SCORE CARD FOR PEN

Applying the same ideas to pen scoring we have the card as shown herewith. Because the birds in a pen, and the females in particular must match, one hen—the best— should be taken as the standard. The birds' total individual scores count as shown on the card. Then deductions are made for each bird (except the female used as the pen standard) for faults in matching in the particulars listed in the wide column at the left. The sums of all these cuts, deducted from the total of individual scores gives the score of the pen. The assumed scores in this case represent a very well-matched pen.

INDEX

Angle of Vision, Effect of 110

Beaks, Crooked 33
Beaks and Bills, Treatment of 119
Bleaching, Detecting 134
Bleaching White Birds 133
Body, Wrong Carriage of 33
Brahmas, 1875 Description of Light 155
Brahmas, 1867 Description of 155
Brahmas, Points in 154
Brahmas, Points in Light 157
Breasts, Crooked 32
Breeder and His "Handiwork," The 27

Catch a Fowl, The Right Way to 50
Catch, Penning Birds in Small Space to 52
Catching Poultry, Gentleness in 50
Changes in Color with Age, Examples of Common 38
  In Blacks and Whites 40
  In Buff and Red Varieties 39
  In Parti-Colors with Buff, Bay or Red Ground 39
  In the Black and White Mottled Varieties 38
  In the Ermine Color Pattern 38
  Where Undesirable White Appears First 40
Chicks, Faults Noticeable in 32
Chicks Not to Crowd, Teaching 31
Chicks: Which to Cull and When 30
Color Faults in Fowls 78
  Barred Varieties 94
    Dominiques 97
    Hamburgs (Penciled) 98
    Rocks 94
  Black Varieties 80
  Blue or Blue-Laced Pattern 91
  Buff Varieties 81
  Campines 98
  Cornish, The Dark 87
  Ermine Color Pattern 98
  Games, Black and Red Exhibition 89
  Laced Varieties 89
    Bantam, Sebright 91
    Buff Laced 91
    Laced Polish 91
    Laced Wyandottes 90
    Red Laced White 91
  Miscellaneous Color Patterns 100
    Bantams, Game 101
    Bantams, Gray Japanese 101
    Buttercup 101
    Dorking, Colored 101
    Faverolles, Salmon 101
    Games, Birchen 101
    Games, Exhibition 101
    Leghorns, Red Pyle 101
    Polish, White-Crested Black 101
  Mottled, Spangled and Speckled Varieties 92
    Spangled 93
    Speckled 93
    Sussex, Speckled 93
  Partridge Varieties 86
  Penciled Varieties 85
    Black-Red 85
    Black-White 86
  Red Varieties 83
  Silver and Silver Gray 89
  Silver-Penciled Varieties 86
  Stippled Varieties 87
  Games and Game Bantams—Black and Red Exhibition 89
  White Varieties 78
"Color Feeding," Special 41
Color More Difficult to Hold Than Shape 37
Color, Some Effects of Common Feeds on 44
Combs, Faults in 33
Combs, Making Over Bad 24
Combs, Surgical Operations on 118
Condition in Score-Card Judging, Consideration of 174
Cooping the Birds at the Show 143
Coops, Conditioning and Training 57
Coops, Returning Birds to 52
Coops, Shipping 141
Coops, Taking Birds from 52
Culls, Good Breeders have Few 29
Cutting for Defects, The Per Cent Method of 158

Decimal Scale 162
"Decimal System" of Score-Card Judging, The 161
Deformities, How Wrong Conditions Cause 30
Differences in Young and Old Birds, Normal 37
Disqualifications and Defects: 59
  Back 62
  Beaks and Bills 61
  Beards and Muffs 61
  Bills, Beaks and 61
  Breast, Body and Fluff 64
  Combs 61
  Crests 61
  Ear Lobes 61
  Eyes, Heads and 62
  Face 61
  Heads and Eyes 62
  Legs and Toes 65
  Legs, Smooth 66
  Legs, Feathered 67
  Neck 62
  Tail 63
  Toes, Legs and 65
  Wattles 61
  Weight 67
  Wings 63
Disqualifications and Defects, Breed 67
  Anconas 75
  Andalusians, Blue 74
  Bantams, Booted 78
    Brahma 78
    Cochin 78
    Game 77
    Japanese 78
    Malay 78
    Mille Fleur Booted 78
    Polish 78
    Rose Comb 78
    Sebright 78
  Brahmas 71
  Buckeyes 70
  Buttercups 77
  Campines 77
  Cochins 71
  Cornish 76
  Crevecoeurs and La Fleche 75
  Dorkings 75
  Dominiques, American 70
  Faverolles 75
  Frizzles 78
  Games, Exhibition 77
  Hamburgs 77
  Houdans 75
  Javas 70
  La Fleche, Crevecoeurs and 75
  Langshans 72
  Leghorns 72
  Malays 78
  Minorcas 73
  Orpingtons 75
  Plymouth Rocks 67
  Polish 77
  Redcaps 77
  Rhode Island Reds 70
  Silkies 78
  Spanish, Black 74
  Sultans 78
  Sumatras, Black 77
  Sussex 75
  Wyandottes 69
Disqualifications—Their Use and Misuse 170
Dorkings, Points in Colored 154
  Points in Silver Gray 154
  Points in White 154
Down, Removal of Stubs and 24
Down, Treatment for Stubs and 122
Dryer, Construction of Rawnsley 133
Ducks: Color and Color Pattern Faults 103
  Black and Black-and-White Varieties 104
  Blue Varieties 104
  Buff Varieties 104
  Fawn and White 105
  Rouen Penciled 106
  White Varieties 103
Disqualifications and Defects: 102
  Aylesbury 102
  Black East India 103
  Blue Swedish 103
  Buff 103
  Call 103
  Cayuga 103
  Crested White 103
  Indian Runner 102
  Muscovy 103
  Pekin 102
  Rouen 102

Ear Lobes and Faces, Treatment of 118
Eyes Should Match in Color 84
Examination, Order in Systematic 58
Exhibition Form, Character of Soil and 48
Exhibition Pens, Matching 111
Exhibition Stock Needs More Room 28
Exhibitor, The Novice's Chances as an 16
Exhibitor's Kit, The 143
Exhibits of Poultry, The Earliest 7
Exhibits in Single-Bird Classes, Uniformity of 112

Fading, Prevention of 25
Feathers, Effect of Heat on Growth of 46
  Fluffing and Bending 136
  Giving Luster to 137
  Humane Way to Pull, The 56
  In Black Birds, Dyeing White 138
  Removing of "Foul," and Coloring 24
  Removing Tips or Edges of Different Color 138
  Splicing 137
  Splicing Tail and Wing 24
  When to Remove Unmolted and Damaged 56
Feeding for Rapid Increase in Weight and Growth of Plumage 113
Feet, Treatment of Shanks and 119
Geese: Disqualifications and Defects
  African 106
  Canadian or Wild 107
  Chinese 107
  Egyptian 107
  Emden 106
  Toulouse 106

Hatching, Timely 28
Head Parts, Damaged 62
Head Points, Fitting, Fixing, and Faking 117

Individual Attention, Importance of 29
Inspection, How to Hold Birds for 52

Judge, Thoroughness of the Poultry 19
Judges, Suggestions to 157
Judging by the Standard, Methods of 18
  Consideration of Condition in Score-Card 174
  Consideration of Size and Weight in 173
  Improvement in Score-Card 160
  In Progress, Matters Relating to 145
  Merits and Faults of Score-Card 20
  Standard Descriptions as the Basis of 17
  Status of Comparison 22
  The Actual Process of Score-Card 62
  "Decimal System" of Score-Card 161
  The Return to Comparison 164
  Why Different Methods Give Slightly Different Scores in 20

Leghorns, 1867 Description of White 156
Leghorns, 1875 Description of White 156
Leghorns, Points in White 156
Legs, Coloring and Staining 121
Legs, Ordinary Cleaning and Polishing of 120
Lice on Development of Exhibition Stock, Effect of 46
Lights, Effect of Different 109

Molt, Supervising the 53
Molting Is an Advantage, When Premature 56
Molting, Partial 54

Plumage Color, Uncertainty of Development of 34
  Dyeing Buff and Red 138
  Effects of Sun and Shade on 47
  Purple Barring in Black 138
Posing, Training and 114

Quality, The Heritage of 28
Quills, Removing Black from White 138

Sanitary Conditions, General 49
Scale of Cuts, Correct Adjustment of the 174
Scales of Credits—"Scales of Points"—and Scales of Cuts 168
Scoring, How Standard Descriptions Affected 154
Selection of Adult Birds, Early 35
Self-Consciousness in Beautiful Birds 27
Selling Faked Specimens, About 26
Shanks and Feet, Treatment of 119
Show, At the Close of the 147
  Care of Birds After Return from 152
  Cooping the Birds at the 143
  General Care of Birds at the 145
  Grooming the Birds at the 144
Shows at Agricultural Fairs, Great Poultry 14
  Benefit Near-by Exhibitors, How Permanent Large 13
  Character of Competition of Different Classes of 15
  Great Exposition Poultry 14
  Methods of Shipping to 139
  The Beginning of Permanent Big 9
  The Risks in Promoting Poultry 8
  Timing Shipments to 139
Side Sprigs, Regarding the Removal of 23
Single Birds and Matching Pens, Trying Out 109
Size and Weight in Judging, Consideration of 173
Soil and Exhibition Form, Character of 48
Spanish Fowls, Points in 154
Spurs, Trimming 121
Standard Poultry, "Points" in 154
Standard, The "Best Bird" As a 160
Stubs and Down, Removal of 24
Stubs and Down, Treatment for 122
"Symmetry" Historically and Technically 159

Teaching Chicks Not to Crowd 31
Toenails, Trimming 122
Training and Posing 114
Turkeys: Disqualifications and Defects 107
  Black 108
  Bourbon Red 109
  Bronze 108
  Narragansett 108
  Slate 108
  White Holland 108
Turkeys and Waterfowl, How to Catch 51
Type, How to Hold Correct 36
  The Adult Type is the Standard 36
  With Age, Usual Changes in 36

Unfair Practices, Review of Efforts to Prevent 25
  Further Observations on the Difficulties of the Question 25

Washing and Bleaching, Relations of 23
  Colored Fowls 135
  Poultry for Exhibition 122
  White Orpingtons at Morris Farm 131
Waterfowl, How to Catch Turkeys and 51
Wings, Slipped and Twisted 33
Wings, Wrong Carriage of 33